KU-410-473

Editorial Manager Chester Fisher
Senior Editor Judith Maxwell
Editors Bridget Daly
Brenda Clarke
Series Designers QED (Alaistair Campbell and Edward Kinsey)
Designers Jim Marks and Nigel Osborne
Series Consultant Keith Lye
Consultant Tony Osman
Production Penny Kitchenham
Picture Research Leonora Elford and Janice Croot

© Macdonald Educational Ltd, 1979
First Published 1979
Reprinted 1980, 1981
Macdonald Educational
Holywell House
Worship Street
London EC2A 2EN

2081/3200
ISBN 0 356 07005 0

Designed and created in Great Britain

Printed and bound by
New Interlitho, Italy

The Mechanical World

Ron Taylor

Mark Lambert

Macdonald

Contents

Man and Technology

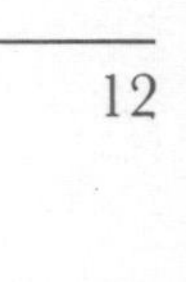

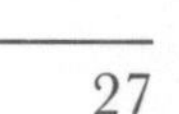

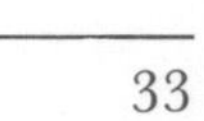

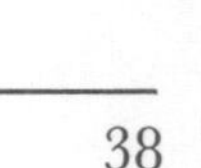

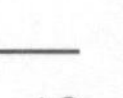

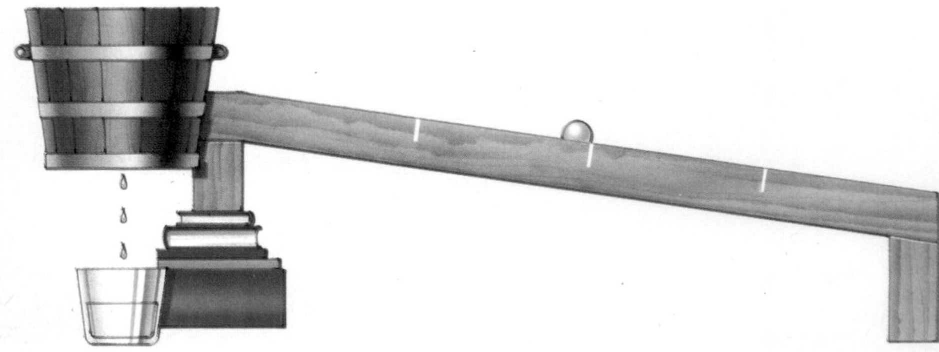

Transport through the Ages

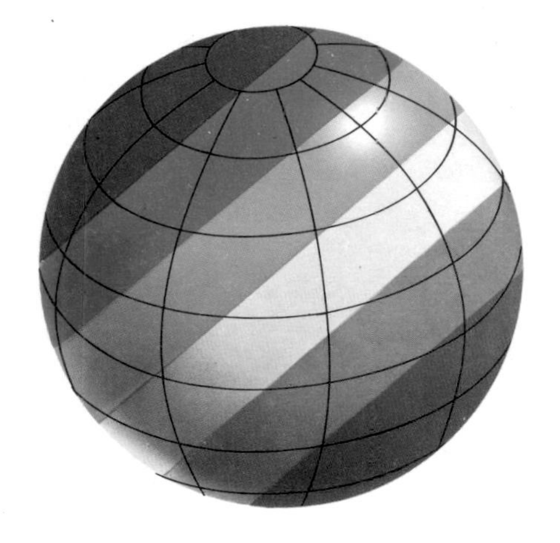

World of Knowledge

This book breaks new ground in the method it uses to present information to the reader. The unique page design combines narrative with an alphabetical reference section and it uses colourful photographs, diagrams and illustrations to provide an instant and detailed understanding of the book's theme. The main body of information is presented in a series of chapters that cover, in depth, the subject of this book. At the bottom of each page is a reference section which gives, in alphabetical order, concise articles which define, or enlarge on, the topics discussed in the chapter. Throughout the book, the use of SMALL CAPITALS in the text directs the reader to further information that is printed in the reference section. The same method is used to cross-reference entries within each reference section. Finally, there is a comprehensive index at the end of the book that will help the reader find information in the text, illustrations and reference sections. The quality of the text, and the originality of its presentation, ensure that this book can be read both for enjoyment and for the most up-to-date information on the subject.

Man and Technology

Ron Taylor

Introduction

We are now living in an age of technology as scientific discoveries and principles are applied to raising our standards of living, increasing our wealth, and extending our knowledge. **Man and Technology** tells the story of man's ingenuity, which arises from his innate intelligence and imagination – qualities which distinguish him from all other animals. His great catalogue of achievement begins in the mists of time with the emergence of man the tool-maker. It continues through the Stone, Bronze and Iron ages and the rise of civilization, the Scientific Revolution of the 1500s and 1600s, and the Industrial Revolution of the late 1700s and 1800s. **Man and Technology** brings the story up to date, into the age of space exploration and automation. Technology has its darker side. For example, it has led to pollution, the squandering of natural resources, and a lack of job satisfaction experienced by many workers in automated factories. In the future, man's ingenuity must also be applied to moderating and controlling the abuses that technology has caused.

The history of human technology begins long before the start of recorded history, when man the tool-maker, faced with a hostile environment, used his intelligence and imagination not only to survive but also to prosper.

The Dawn of Technology

Above: Long before human beings appeared on the scene, tools of a kind were used by animals in pursuit of their food. Here, a sea otter cracks the shell of a tasty clam, using a stone. Although the sea otter selects its stone tool, it does not shape it as we do our tools. Perhaps the only animal besides man that can do this is the chimpanzee. In captivity, chimps will fit sticks together to reach food that is otherwise beyond their grasp.

Man the toolmaker

The history of man as a technologist is a far longer one than was suspected a century ago, when the first HOMINID fossils were unearthed. In 1925 the Taung skull was discovered in South Africa. This was human, if only just, and at about two million years old, it predated human remains previously found by at least a million years. Taung man, in fact, was the first of the AUSTRALOPITHECINES to be discovered, and the Australopithecines, as Dr Louis LEAKEY later found in Tanzania, used tools.

Admittedly, these tools were of the simplest kind: pebbles and other small stone artefacts which had been shaped roughly by hand. They were turned up by Dr Leakey and his wife, together with fossil fragments of the bones of many more Australopithecines, in the Olduvai Gorge of Tanzania. Many of these, and also more recent finds in East Africa, go back in time even further than Taung man, to the dawn of human prehistory, about three and a half million years ago.

How did the anthropologists decide that the small stones were, in fact, Australopithecine tools or weapons and not just chance finds? Their evidence is threefold. First, the stones were not only found closely associated with the human remains but also showed clear signs of having been shaped intelligently. Small flakes of stone chipped off in the tool-making process lay nearby. Second, *Australopithecus,* although possessing a brain little larger than a chimpanzee's, walked and ran upright – as was clear from his anatomy – and could also grasp and wield stones in a typically human manner. Thirdly, near some of the Australopithecine fossils lay animal skulls bearing a depressed fracture attributable to blows from the stones. *Australopithecus* was a hunter.

Fire and hand axes

Australopithecines hunted the African plains for some millions of years but eventually gave rise to still more man-like hominids, among whom were the people who first discovered the use of fire. Firemaking must have been a very early invention of man, predating his use of all but the simplest tools. Yet it is still a specifically human activity: although some other animals can use tools, none is capable of employing fire deliberately for warmth.

Almost certainly, firemaking was first practised in the manner still employed by many modern aboriginal peoples, that is by rubbing together pieces of wood called firesticks among dry grasses or other tinder. In the great stretch of time called the lower Palaeolithic, or early OLD STONE AGE, man would undoubtedly have learned and lost the technique of firemaking time and again before the habit became widespread.

Reference

A

Abbevillian industry is a primitive stone industry belonging to the early OLD STONE AGE (about 200,000 years ago) and named after findings in France. It extended for a very long period of time after the second glacial period of the ICE AGE. The typical Abbevillian artefact is a crude stone hand axe with 1 cutting edge. This industry, also known as the Chellean, was followed by the ACHEULIAN INDUSTRY.

Acheulian industry is a stone industry of the early to middle OLD STONE AGE, also named after findings in France. It lasted from the second inter-glacial period to the third glacial period of the ICE AGE. Acheulian artefacts consist mainly of stone hand axes of more refined manufacture than the earlier Abbevillian type, having 2 cutting edges.

Aurignacian culture belongs to the later OLD STONE AGE and is named after many findings in southern France. It displaced the earlier MOUSTERIAN CULTURE and was practised by a people of modern appearance, often known as Cro-Magnon men. Aurignacian artefacts include finely-made tools and weapons of bone and flint, ornaments including bone necklaces, small sculptures in stone and bone and small cave paintings including finger tracings.

Australopithecines are man's ancestors who were first known to make tools. They lived as long ago as 3 million years and were, in general, chimpanzee-like in facial features and only about 1 metre high. Unlike apes, however, they walked upright and used simple weapons for hunting.

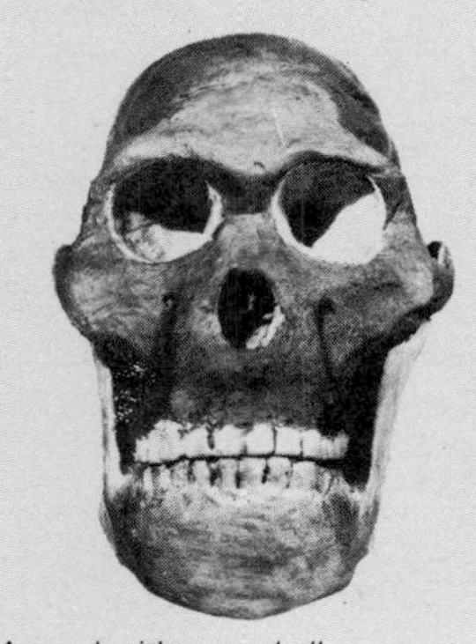

Australopithecus *skull*

H

Hand axes are among the earliest products of man's industry, his first true artefacts other than the even earlier pebble tools of the AUSTRALOPITHECINES. Hand axes range from heavy, crudely-made stone implements of ABBEVILLIAN type dating from 200,000 years ago, to more finely-made stone hand axes of succeeding cultures of the middle and later Stone Ages.

Hominids are man-like primates. In zoology, the *Hominidae* is the family of man and his extinct relatives. The oldest of these

About 600,000 years ago human beings lived at a very much more advanced stage than the Australopithecines. They not only used fire for warmth but probably also for cooking and possibly for hardening simple wooden weapons. Such people are known to anthropologists as HOMO ERECTUS, the first true member of our own genus, *Homo*.

The fossil remains of *Homo erectus* have been found in East Africa but the first important human fossil was discovered, in 1891, in central Java. At the time, he was named *Pithecanthropus erectus*, the Upright Ape man, or the Ape man of Java. Later, in recognition of his human characteristics, he was renamed.

Homo erectus also made stone weapon-tools known as HAND AXES. Though crudely fashioned, these are much more recognizably shaped for chopping than are Australopithecine pebble tools. Obviously, *Homo erectus* was a widespread species because these hand-held, sharp-edged tools have been found in many areas of the world other than East Africa and Java. They correspond to Stone Age finds originally made in France and known to archaeologists as the ABBEVILLIAN INDUSTRY and the ACHEULIAN INDUSTRY respectively.

Modern man

Four hundred thousand years ago – the date of early Acheulian finds – *Homo erectus* lived in Europe. The prehistoric stage now seemed to be set for the evolution of *Homo sapiens* or modern man. In fact, little positive trace of modern man appears for another 250,000 years, human fossil remains over this long period being sparsely scattered and often ambiguous.

When *Homo sapiens* finally did appear, in the latter half of the ICE AGE, it was not simply as modern man but as a range of types or varieties. Most of these have left only fragmentary fossils. Of these extinct varieties of *Homo sapiens*, only

Below: Until the arrival of Europeans, Australian Aborigines lived in many ways at an Old Stone Age level. However, this way of life was largely dictated by their harsh surroundings. They were hunters and food gatherers who built no permanent dwellings and made only a small range of tools and weapons. Some of their weapons, such as the spearthrower shown, were quite sophisticated. They had also domesticated a dog, the dingo, but this later ran wild and became a pest.

fossils, *Ramapithecus*, goes back in time about 14 million years. In their turn, the first hominids arose from a general stock of higher primates that also gave rise to the pongids, or great apes.

Homo erectus is the scientific name of a species of man preceding *Homo sapiens*. He has also been known, from early archaeological finds, as the Ape man of Java and Peking man. *Homo erectus* made hand axes and other primitive stone tools and weapons, and used fire for warmth and cooking.

I

Ice Age refers commonly to the glaciation of the Northern Hemisphere that occurred between 600,000 and 20,000 years ago. Late in the Ice Age modern man appeared as a cave dweller and much of his early technology, such as the making of bone fishing weapons and the use of animal skins and furs as clothing, was stimulated by the colder climate of that time.

L

Leakey, Louis (1903-72) was a British anthropologist. His great discoveries took place in East Africa. In 1959 he uncovered *Zinjanthropus*, some 1.75 million years old; in 1964 *Homo habilis*, some 2 million years old and in 1967 the fossilized remains of 'pre-man', some 14 million years old.

Leakey on site in E. Africa

Levalloisian industry was practised by man in Europe from the third inter-glacial to the fourth glacial period of the ICE AGE. It is named after a suburb of Paris where remains were found. The characteristic artefact of this industry is a stone tool of oval shape which has been carefully detached by pressure-flaking from a larger stone core.

M

Magdalenian culture was that of the great cave painters of Europe, 10-15,000 years ago in the very late OLD STONE AGE. It is named after a cave in the Dordogne, south-west France, which, besides paintings and carvings, contained bone and stone tools of advanced workmanship.

Maglemosian culture is typical of man's activities in the warmer times immediately following the ICE AGE in Europe. It is named after a bog in Denmark, which was a lake in the MIDDLE STONE AGE, where the industry was prac-

Above: Australian Aborigines prepare a meal of roasted kangaroo. The meat is placed on a fire in a hole in the ground and covered with earth.

one is well known from frequent fossil remains, namely *Homo sapiens neanderthalensis* or Neanderthal man.

Neanderthals are named after a valley of the River Rhine in north Germany, but their remains have been found also in many other countries including North Africa. Though distinct in physical appearance from our surviving species, they resembled us in a number of technological and cultural ways. For example, they buried their dead ceremoniously and methodically, and like many primitive varieties of *Homo sapiens* they were cannibals. Archaeologists know Neanderthal man by the LEVALLOISIAN INDUSTRY and the MOUSTERIAN CULTURE, which yielded stone tools of greater refinement and variety than the few types of hand axe made by *Homo erectus.*

Neanderthal man disappeared, superseded by more modern men or in part absorbed by them through interbreeding. These first truly modern men were tall, well-built and possessed high foreheads. They were typified by the AURIGNACIAN CULTURE of southern France. Not only did they make a whole range of stone tools including hand axes, points and scrapers, but also they carved personal ornaments and figurines from bone, horn and ivory, and made cave paintings. The Aurignacians could be said to have invented art.

They were followed in rapid succession (according to the time scale adopted so far) by other cave dwellers belonging, respectively, to the SOLUTREAN CULTURE and the MAGDALENIAN CULTURE. The Magdalenians represent the final flowering of human culture in the Old Stone Age: they were the great cave painters of Altamira in Spain and Lascaux in France, who lived between 10,000 and 13,000 years ago.

During the Mesolithic or MIDDLE STONE AGE which followed the ending of the Ice Age in Europe, human beings began to settle into a village way of life. They domesticated the dog, cultivated an early type of wheat and built houses of wood and reeds. As exemplified by the MAGLEMOSIAN CULTURE in Denmark, their hunting weapons now included the bow and arrow and elegantly carved bone harpoons. But it was in the Near East, in areas made suitable for human development by climatic changes following the retreat of the ice, that the first great civilizations arose.

tised. Fish spears with barbed bone points and carpenters' tools of bone are characteristic finds and traces of primitive huts have also been discovered.

Microliths are small flint tools of various shapes, characteristic of the Mesolithic or MIDDLE STONE AGE. They were made by a precise method of flaking flints in 2 directions and were often mounted on wooden handles to serve as knives, borers and scrapers. Microliths have been found in many places including Europe, China and India.

Middle Stone Age, or Mesolithic, is the name given by archaeologists to human cultures from about 8,000 to 6,000 years ago, in which man began to settle into village communities. During this time he domesticated the dog, herded cattle and cultivated cereals and other crops for the first time.

Mousterian culture is the principal culture of Neanderthal man and is named after a cave in the Dordogne, south-west France. Tools characteristic of this OLD STONE AGE culture include bifacial hand axes of the ACHEULIAN type, smaller stone points and scrapers, and a few primitive bone tools.

Old Stone Age, or Palaeolithic, is the name given by archaeologists to human cultures which cover more than 90% of human prehistory, from the AUSTRALOPITHECINES 3 million years ago to the MAGDALENIANS about 13,000 years ago. The lower, middle and upper Palaeolithic correspond to the early, middle and later Old Stone Ages.

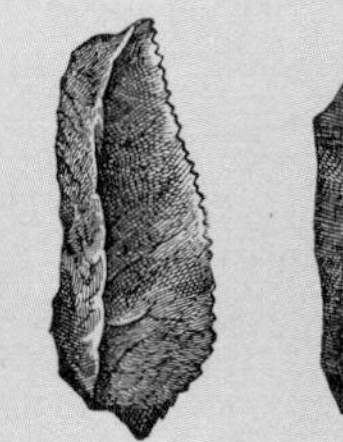

Palaeolithic flint tools

Oldowan industry refers to stone artefacts found together with the fossil remains of AUSTRALOPITHECINES. These consist mainly of crude pebble tools which show some signs of having been worked by flaking. The first major site of these finds was the Olduvai Gorge in Tanzania, explored by Dr Louis LEAKEY and his wife.

Solutrean culture belongs to Europe of the later Old Stone Age and is named after a village in France. It is characterized by finds of javelin flint points delicately flaked into the shape of a laurel leaf. The Solutreans brought the important technology of pressure flaking to its highest perfection. Their culture followed that of the AURIGNACIANS and they were succeeded by the MAGDALENIANS.

The development of farming in the Middle East freed people from the drudgery of hunting and gathering food. Trade became important and various industries and crafts began to develop in the growing villages and towns.

Man the Farmer

Hunters and fishermen

The MIDDLE STONE AGE *(see page 5)* lasted from about 10,000 to 7000 BC in the Near and Middle East, and later in Europe. It was a time of small scattered villages and encampments but also of great migrations of people and animals. Warmer climates enabled game herds to wander. As they did so, they split up and formed smaller herds, so that man was compelled to follow the migrating animals as a nomadic hunter.

Forests covered northern Europe, reducing areas of grassland and the numbers of grazing animals. Forest hunters needed better weapons to be sure of killing the well-camouflaged forest game. They improved their bows by increasing their size and power, and tipped their arrows with barbed flint MICROLITHS *(see page 5)*.

Coastal areas of the more northerly parts of Europe benefited most from the retreat of the ice, as large areas of land became available for settlement. Some of the earliest villages must have been built in these regions, though few traces of dwellings occur other than some slabs of bark which were possibly hut foundations. Peoples of this MAGLEMOSIAN CULTURE *(see page 4)* included coastal fishermen who used barbed bone harpoons and who operated from small boats, probably skin-covered coracles.

Below: Mammoths were a favourite prey of the Old Stone Age hunters of southern Russia. These men built large tents of hides, weighing them down with bones and tusks of their giant prey. Their tents often housed several families and were heated against the arctic climate.

Reference

A **Amerindians** are peoples of Mongoloid race who migrated from Asia to America during the OLD STONE AGE *(see page 5)* and became the Eskimos, the Red Indians and the Central and South American forest Indians that we know today. When discovered by western explorers in the 1500s the Red Indians in general had a Neolithic, or NEW STONE AGE, culture whereas the Eskimos and the forest Indians had a Palaeolithic, or Old Stone Age, culture.

B **Barrows** are raised graves, frequently of Neolithic, or NEW STONE AGE, origin. They were made by heaping earth and stones over the bodies and belongings of 1 or more persons. They are commonly seen in Europe as raised grassy mounds called tumuli. They are classified according to their shapes into long barrows or round barrows.

Amerindian pottery, Peru

Beaker culture is associated with the first peoples of the BRONZE AGE (or COPPER AGE) to reach Europe. It is named after the beaker-shaped pots found in their graves. Beaker folk made spears, daggers and ornaments of copper.

Bricks made of sun-dried clay were first used for buildings during early Neolithic times, perhaps as early as 8000 BC. By 6000 BC a technique of making bricks of regular shape using moulds had been perfected in western Turkey.

Bronze is an alloy of copper with tin or, less usually, another metal such as aluminium, together with small amounts of other elements. Ancient bronze was made usually by smelting together a COPPER ORE and TINSTONE. This produced an alloy which was very much harder and stronger than copper. Moreover, bronze melts at a lower temperature than copper and so can be cast into moulds more easily. The earliest date for the smelting of bronze is about 3100 BC.

Bronze Age is the name of a period of human cultural development extending over the period 3500–1000 BC. It is dated usually to include an early period before the invention of bronze alloy, when copper was the only metal employed for implements and ornaments. The Bronze Age occurred in

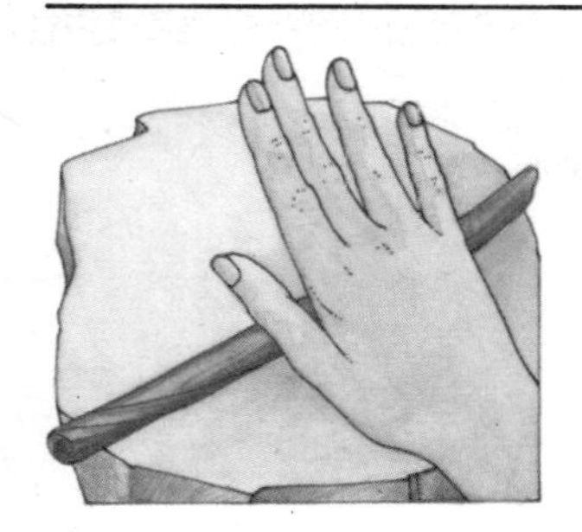

Left: The first potters found their clay as surface deposits. They removed stones and dirt, then pounded and kneaded the clay on a stone, adding water if necessary to make it more plastic. The clay could then be rolled evenly, as shown.

Left: The clay rolls so formed were worked into pots by coiling – winding the rolls round and round until they built up into the shape required. This method is still employed by craft potters to make vessels of shapes which cannot be thrown on a wheel.

The first farmers

Changes in climate and ecology also led to a diversification in food gathering as more and more kinds of wild vegetables were found to be edible. It was no longer necessary to lead a nomadic existence in places where food was readily available in the vicinity of the camp. In Iraq and Iran the gathering of wild cereal plants, growing abundantly in these drier, warmer regions, led to the domestication of Einkhorn and Emmer wheat, the forerunners of bread wheat, and another early cereal, two-rowed barley. These crops were harvested with flint sickles set in wooden or bone handles. Early agricultural communities living in these regions also herded goats and, at a rather later date, sheep, cattle and pigs.

Stone Age towns

As very early settlements became established the cultural and technological achievements of their inhabitants grew very rapidly in comparison with those of earlier or contemporary nomadic hunters and herders. Between 8000 to 6000 BC a few villages within the area known as the FERTILE CRESCENT increased in size until they became sizeable towns. These included Catal Hüyük in Anatolia and Jericho in Palestine. By the end of this period people lived in houses built of sun-dried mud BRICK, often semi-circular in shape and sometimes of two storeys. Sunken granaries and clay-brick ovens for drying corn were other architectural features, and the towns were surrounded by defensive walls of mud brick.

In these and other settlements of the time a development of major importance was a pottery industry — although this happened long before the invention of the POTTER'S WHEEL. Pottery was invented in Iran, about 6500 BC. Early pots were hand-turned or moulded of clay, often tempered (strengthened) with grit or straw. The pots often bore gaily-painted geometrical designs using metallic pigments as the colour base.

Cups, bowls, boxes and other smaller domestic containers were carved from wood. Other household articles included mats and baskets of woven straw and reeds. Although no woven garments survive from this period, it cannot be doubted that the spinning and weaving of textiles from various types of wool was already a long-established craft. Terracotta spindles, loom-boards and loom-weights have been found as direct evidence of a weaving trade. LOOMS *(see page 22)* of this time would probably have included both the horizontal pegged-out type and the more compact, vertical type that was developed from it.

Another progressive feature of early civilized life was increased trade, both between the settled communities and between them and the surrounding tribal nomads. Jericho, sited by an oasis in Palestine, was by 6000 BC a prosperous town of some 2,000 people. It is known to have imported obsidian stone tools from Anatolia and cowrie shells and turquoise from elsewhere, and to have exported minerals such as sulphur, bitumen and common salt.

Tilling the soil for crop growing was still primitive. At first, digging sticks were used to break up soil clods and to dig holes for seed-planting. These were gradually supplanted by HOES which were either forked pieces of timber, sharpened at one end, or flint-bladed implements. The latter were wielded in a similar way to the short-handled, but iron-bladed, hoes still used in rural West Africa and elsewhere.

Water storage and IRRIGATION were practised very early on, perhaps by 6000 BC in centres of population such as Jericho and Catal Hüyük, to judge by the remains of water tanks built there. The most ancient of all irrigation devices, the SHADUF, was probably already being used in

Above: A Saudi-Arabian potter at work. He spins his wheel by thrusts of his foot on a lower wheel. This was a development from the earliest type of potter's wheel, which was turned by hand. Adding a kickwheel freed both the potter's hands for shaping, or throwing, his pots. More recent types of throwing wheel are turned by a foot-treadle or by an electric motor.

Mesopotamia, Egypt, Europe and China, but not at the same time.

Bronze founding was practised in Mesopotamia from about 3000 BC onwards when the bronze alloy was cast into clay moulds. In ancient Europe, at a much later date, foundries were fed with bronze scrap that came from abroad, perhaps through trade with the East.

C

China clay, or kaolin, takes its name from geological deposits of the mineral in China which were first exploited by the ancient Chinese of the BRONZE AGE. It was used as early as 1500 BC, mainly for the production of pure white ceramics.

Bronze Age oven, Germany

Copper Age is a name sometimes given to a phase of ancient technology in which the metal copper, but not the alloy bronze, was used for tools, weapons and ornaments. Copper, occasionally found as the pure metal, is easily beaten out and shaped.

Copper ores were used by the ancient Egyptians for eye-shadow. The ore was ground down into a powder and applied with the moistened tip of a small stick. At about the same time in Mesopotamia, they were used for alloying bronze.

Curing methods were used early in the village life of the NEW STONE AGE to preserve hides and foodstuffs. Hides were smoked over a fire to kill parasites and to form a protective surface. Meats were either smoked or salted.

D

Dams and dykes were built in the Middle East from very ancient times for IRRIGATION purposes and were among the most impressive of man's early engineering feats. Mostly, however, they have not survived in good repair. One which remains standing is a stone dyke in Syria, 2 km long, built about 1300 BC.

Dyes extracted from plants and animals have been used since ancient times to dye textiles or parts of the body. They include yellow saffron from the crocus, dark blue indigo from the plant *Indigo tinctoria,* red madder from *Rubia tinctorum* and red henna from *Lawsonia inermis,* Tyrian purple from a snail and red cochineal from a beetle.

E

Embalming is a technique for preserving

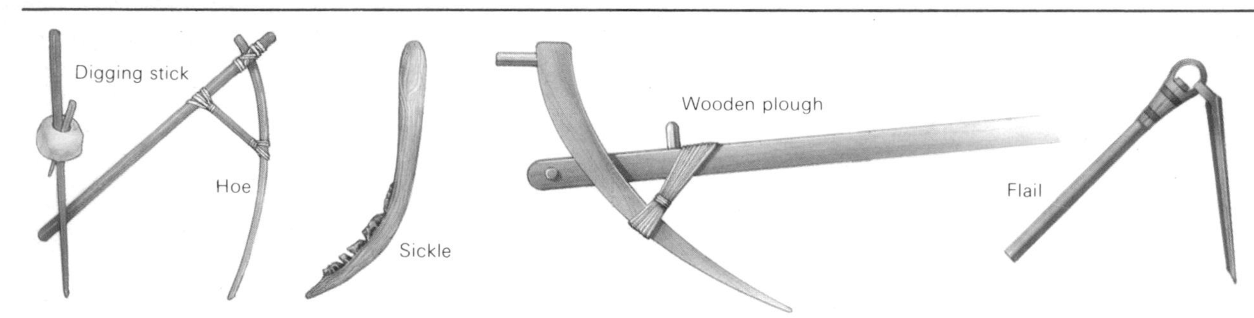

Left: First among farming tools was the digging stick, used by agriculturalists of the Stone Age to break up hard soil and to make holes for planting seeds. This gave rise to the wooden hoe, which in turn was the ancestor of the first plough. A much older tool than any of these, however, was the sickle, employed first to reap wild grasses, later to reap cultivated cereals. The very early type shown is an animal bone inset with sharp flints. The flail is another farming tool with a very long history, having been used to thresh grain from the time of early civilizations.

these towns to raise water from tanks or shallow wells. Later it was to be seen, as it still is today, over much of rural Asia, lifting and distributing water from rivers and pools into irrigation ditches.

Rivers and cities

For the first 5,000 years of the NEW STONE AGE, people generally lived in scattered villages or continued a nomadic way of life. Prosperous towns were few and far between and any greater development of civilization unknown.

When cities first arose, they did so along the banks of great rivers. This happened not only in MESOPOTAMIA, where the rivers were the Tigris and Euphrates, and Egypt, in which civilization first developed near the Nile delta. It also happened repeatedly during the following 2,000 years: along the valley of the River Indus in north-west India, along the banks of the rivers Oxus and Jaxartes (now Amu and Syr Darya) in southern Turkestan, and on the plains watered by the Yellow River and the River Yangtze in northern China.

These rivers provided the economic basis for the extension and development of civilization. They offered a means of transport for people and materials, and a reservoir of food for fishermen. More important still, they provided unlimited water for irrigation.

This was probably first learned because of periodic, sometimes catastrophic, floodings of the great rivers. These floods actually increased the fertility of the land by the addition of rich alluvial mud. Later, river-bank populations would have learned to anticipate the floods and then to control them by building embankments, DAMS AND DYKES and irrigation channels.

Unlimited irrigation meant the rapid growth of agriculture, until food production exceeded consumption. The food surplus so obtained promoted the growth of trade and attracted other industries, crafts and trades into the area. With this variety of production came increased work opportunities for the inhabitants of towns. Their numbers grew, and cities evolved.

A number of other important reasons have been given for the growth of the first cities, particularly those of Mesopotamia and Egypt. Religion undoubtedly played a great part, since the Mesopotamian cities grew up around temples. So also did the development of slave labour for work in the fields. Prisoners of war proved very useful in this capacity. Previously, they would have just been killed.

Metallurgists and metalsmiths

Of the new technologies practised in the cities, none was more important, especially for future history, than the melting and shaping of metals. Not that metals were at all new to the inhabitants

Left: Ethiopian tribesmen pounding grain to make flour. Man began this kind of activity perhaps 9,000 years ago, when he first cultivated cereals. Originally, these were wild grasses. By crossing species and selecting those with large, tasty grains, early peoples produced bread wheat and other cereals we eat today.

corpses from decay. Embalmers first removed the soft inner parts of the body through the anus, then repacked the body with preservative herbs, overlaid the body with cloths soaked in wine and finally wrapped it round tightly with tarred bandages.

F **Fertile Crescent** is a name given by archaeologists to the curve linking the valleys of the rivers Nile and Euphrates, along which the first great city civilizations developed.

Flint mining is known to have been practised by many peoples from the Old Stone Age onwards, but from the northern Europe of the NEW STONE AGE it is often the only remaining sign of inhabitation.

Food crops of the NEW STONE AGE and the BRONZE AGE included wheat, barley, rye, root and leaf vegetables, dates, grapes and olives. All of these were under cultivation by 3000 BC. The first crop to be cultivated was Emmer wheat, which was later crossed with other wild grasses to produce bread wheat.

H **Hoes** evolved from the digging stick, a wooden stake rammed through a stone with a hole in it. They were wielded first by 1, then by 2, labourers and from the 2-person hoe arose the PLOUGH.

Corn being harvested in Egypt,

I **Iron** is only rarely found, in meteorites, as the pure metal. In this form it was first used in pre-dynastic Egypt and in Stone Age Mesopotamia for ornaments and toys. As a more useful metal it remained unrecognized during the first 3,000 years or so of man's civilized life. When iron was first made from its ores, around 1500 BC, it was again employed for decorative work. The soft, grey metal was less suitable for weapons than bronze.

Iron ores are varied and widespread in many coun-

Above: A reconstruction of a north European Iron Age village. The Hittites invented ways of making and using iron for tools and weapons about 1400 BC. Only about 1000 years later had it reached settlements such as this at Woodbury, England. Relics of the Iron Age are scattered widely throughout Europe. They include forts and burial mounds in which much pottery and many metal implements and ornaments have been found. However, the buildings of this village, being of timber, left relatively few remains.

of town or city. For untold centuries they had known copper and IRON as unusual products of nature that could be fashioned into trinkets or even tools. But with the development of towns and cities, metallurgy as an industry was practised on a scale never before realized.

To make more use of metals, ancient metallurgists first needed to extract them from their ores. This they did by heating the ores, which were generally oxides or sulphides of the metal, in a wood fire. Pure metal appeared in the hot ashes when carbon (present as the charcoal of burned wood) reacted with the ore so as to reduce it. This process must first have occurred with an ore of copper, and most probably as the result of sheer accident.

The COPPER AGE that followed this discovery lasted for a 1,000 years in Egypt. In Mesopotamia, copper as a structural metal was soon superseded by BRONZE, an alloy (mixture of metals) which is harder and more durable. Bronze was first made by reducing together COPPER ORES and TINSTONE in a wood fire. Both types of ore for making bronze were readily available to the Mesopotamian cities. From precisely where remains a mystery, except that it was somewhere within the area of the mineral crescent which followed the path of the FERTILE CRESCENT.

The BRONZE AGE began in the city states of Sumeria, then already several hundred years old, about 3100 BC. Metalsmiths of this time made copper and bronze vessels, weapons and armour from sheet metal, beating this out using a hammer and anvil. For finer work, such as ornaments, they would also make use of metal wire, drawn out from hot metal with the aid of a die, a device invented at a very early time.

The casting of bronze into MOULDS came later and was developed independently in several parts of the world, most notably in Mesopotamia and in China. The products of these industries, which are often of great beauty, can be seen in museums today. But the same bronze-casting process is still extensively used today for producing art bronzes. Known as the lost-wax or waste-wax process, it was employed to produce the Benin bronzes of West Africa.

In this process a model of the object to be cast in bronze is first carved from wax, and then clay is plastered around it to make the mould. Pipes of wax are attached to the model so that they project through the clay to the surface of the mould. When the mould is heated and dried, the wax runs away leaving a hollow space inside the clay. Molten bronze is poured into the mould through the vents and it is then left to harden. Finally the clay is chipped away to reveal the finished bronze article.

Such a process can be seen illustrated on the walls of Egyptian tombs, together with artisans practising other trades of the time such as bricklaying and cabinet making. The range of metal tools then available to a craftsman was considerable. Besides the bronzesmith's tools already mentioned, he could call on rasps, bow-drills, chisels, knives and saws as necessary.

Metals and agriculture

To agriculture, the invention of bronze meant improvements such as sharper and longer-lasting blades for axes, sickles and hoes. Rather surprisingly, however, PLOUGHS failed to benefit for a very long time from the incorporation of metal parts. Although the plough, by increasing crop production, had a more profound effect on man's life than almost any other invention, it remained for several thousand years a rather primitive wooden instrument. This is further illustrated by the fact that the mould board (that part of a plough which overturns the soil, so

---

tries. It is uncertain which of these ores was first exploited by man at the beginning of the Iron Age. The ores most used in modern times include hematite, magnetite, siderite and iron pyrites.

Irrigation was of vital importance to the development of the first civilizations, sited along the banks of the rivers Nile, Tigris, Euphrates and Indus. DAMS AND DYKES were built to hold back water, which was then released in a controlled fashion through sluices into irrigation channels. From these it could be further distributed to crops by hand-operated water lifts such as the SHADUF.

J

Jade was worked as a structural material in BRONZE AGE China. As early as 2000 BC this hard, translucent green stone was mined and worked into plaques or tiles, perhaps for wall or floor decoration. Together with another hard, coloured stone (obsidian), jade has also been found in Middle Eastern archaeological sites.

M

Masonry is a craft which was more common to ancient Egypt, rich in building stone, than in Mesopotamia where mud brick was for thousands of years almost the sole building material. Masons' tools included hammers, chisels and wedges, squares for measuring correct angles and bone rods for levelling stone blocks.

Maya were one of the many peoples of Central and South America who had elaborate civilizations that lasted many centuries without really emerging from the Stone Age. Living in what is now Guatemala, between AD 300 and 900 they built religious cities covering many hectares, with palaces, temples and mausoleums of stone. At the end of this

Stone Age masonry, Orkneys

period the Maya civilization abruptly disappeared and the cities were abandoned by their inhabitants.

Megaliths are large stones erected by ancient peoples, possibly for a religious purpose. Stonehenge and Avebury Ring in Britain and the still more numerous megaliths of Brittany are examples. These megaliths are associated with peoples of the BEAKER CULTURE who migrated to northern Europe in the BRONZE AGE.

Mesopotamia is a region of what is now central Iraq, lying between the great

improving aeration and mixing) was not in general use in the Western Hemisphere until the Middle Ages (although Chinese agriculturalists had first used it 2,000 years earlier!). In ancient Egypt, Mesopotamia and even Greece, ploughs were generally either all wood or were fitted with ploughshares of flint or stone. The first people to use ploughs with metal (iron) shares at all widely were the Romans, although a few iron ploughshares have been found dating from as early as 1000 BC. A possible reason for the relative absence of bronze ploughshares is that this alloy, even when hardened by the addition of antimony or arsenic, as many early bronzes were, is much less hard-wearing than all but the softest iron alloys.

A variety of trades

A walk through the streets of a town or city of Mesopotamia of 4,000 years ago would reveal many of the activities still to be seen in the rural communities of modern Africa and Asia.

There would be no blacksmiths, but besides workers in copper, bronze, silver and gold there would be potters, weavers, carpenters and sellers of food, medicines and cosmetics.

In the courtyards of houses, women would be seen MILLING flour and the BREWING (*see page 18*) of beer using malted barley might be another visible female activity. The houses themselves would be built from sun-dried BRICKS, whereas, in an Egyptian community of the same period, MASONRY would be more evident. This difference would be very obvious when a temple or public utility building was being erected. The explanation is simple: Mesopotamian cities were situated far from sources of building stone whereas the Egyptians had ample supplies.

Pyramids and ziggurats

This fact explains why the Egyptians' greatest buildings, the PYRAMIDS, are constructed from blocks of stone, whereas ZIGGURATS, their Mesopotamian equivalents in scale if not in function, were built of mud brick.

The Egyptians quarried their stone by cutting into the rock face with copper chisels and stone picks, then splitting away the giant blocks by driving in wooden wedges. The problems to be overcome in quarrying large stone blocks by these primitive means were formidable, but an even greater problem was the transport of blocks from the quarries to the building sites. Stone was quarried close to the River Nile, and floated downstream towards the sites. Overland, between quarry and river and river and building site, the multi-tonne blocks were hauled on wooden sledges and over wooden rollers using thick papyrus ropes.

By comparison, the erection of the great temples known as ziggurats must have been a much easier affair, more like the putting-up of a large brick building today. Mesopotamian bricks were laid very differently from ours, end-to-end and angled in courses to produce a characteristic herringbone pattern like a parquet floor.

Other Bronze Ages

While the leisured classes of Egypt and Mesopotamia enjoyed their Bronze Age plenty of commodities and slaves, Europe and other parts of the West remained populated largely by barbarians subsisting by hunting, herding and food gathering. Europe's Bronze Age began only 1,000 years later, mainly imported by successive waves of migrants from the East but also, in some regions, arising with the independent discovery of bronze. Bronze Age Europeans have left behind many clues to their ways of life, including their BEAKER CULTURE, their numerous raised graves or BARROWS and, most impressive of all, the great monuments of their sun-worship, the MEGALITHS.

Further still to the west, man either never enjoyed a Bronze Age at all, or remained at this level of culture until discovered by Europeans or others, as the lives of the many AMERINDIAN tribes evidence.

But in ancient China, a fully-developed Bronze Age sprang up in about 1500 BC, with the casting of bronze urns, pots and beakers of elaborate design and intricate decoration. These beautiful objects demonstrate an advanced level of technology, as also does the Chinese pottery of this early period. Both for bronze-casting moulds and for ceramics the Chinese used CHINA CLAY, or kaolin, which conferred a pure whiteness to their pottery matched nowhere else for many centuries. Chinese craftsmen worked also in another material rarely used by others, JADE, small plaques of which are among the earliest artefacts of this civilization.

Right: Grimes Graves in Norfolk were a main site of New Stone Age industry in Britain. The flints seen being mined were used to make implements of various kinds. Flint arrowheads, scrapers and other weapons and tools found in many parts of Britain today are common reminders of peoples who inhabited the country long before the arrival of the Celts or 'Ancient Britons'.

---

rivers Tigris and Euphrates. Along the banks of these rivers arose the earliest city civilizations. In chronological order the inhabitants of these cities were: the Sumerians, Akkadians, Amorites, Assyrians and Hittites, and collectively they span the period from the New Stone Age to the IRON AGE (*see page 21*).

Milling was shown as a working activity by many early Egyptian statuettes and tomb paintings. Stones for milling cereal grains date from before 2500 BC, although they followed an even earlier milling instrument, the pestle and mortar. A later development was the rotary quern or hand-mill, in which 2 stones were used, one above the other. The corn was fed through a hole in the upper stone which was rotated with a handle against the lower stone.

Quern (hand mill)

Moulds were first employed in the NEW STONE AGE of MESOPOTAMIA for making clay building BRICKS of regular size and shape, and decorative clay seals used as owners' marks. Later on, molten bronze was cast into moulds to make axe and spear heads and decorative objects.

N

New Stone Age, or Neolithic, is a word used by archaeologists for the earliest settled human cultures. In the Near East this age extended from the foundation of the first towns, such as Jericho, about 8000 BC, up until the use of copper and bronze in Mesopotamian cities, about 3100 BC. Neolithic periods in other parts of the world, such as that in China, differed in duration, but in all cases Neolithic peoples had a settled agriculture and practised animal husbandry. Their technology was also more advanced than their Mesolithic, or MIDDLE STONE AGE (*see page 5*), forbears.

P

Papyrus was used by the ancient Egyptians as writing paper. Papyrus reed was collected from the River Nile and laid in strips to form 2 sheets, one on top of the other, the strips of the 2 sheets being at right angles. The sheets were then beaten with mallets until they combined to provide a smooth writing surface.

Ploughs evolved from HOES, the earliest being a simple forked branch. A ploughman held the forked handles and soil was ploughed by the

sharpened end of the branch drawn by a second ploughman. The plough was later drawn by oxen as shown in early Egyptian picture writing (hieroglyphics).

Potters' wheels began to be used about 3000 BC when the wheel was simply a flat, pivoted stone spun by the potter's hand. About 1,000 years later a second, heavier wheel was added below the first, which added momentum and so made spinning the throwing wheel easier. Still later this lower wheel was arranged as a kickwheel for the potter, thus leaving his hands completely free to work the clay.

Pyramids are built from stone blocks, each block dressed with great precision by a mason to its required dimensions.

R **Reaping** was at first carried out with flint-bladed sickles. These were used to reap grasses and corn throughout the Middle and New Stone Ages but gave way eventually to bronze-bladed, then iron-bladed sickles. Long-handled scythes did not appear until Roman times.

S **Shaduf** is an IRRIGATION device. It is a long pole pivoted near the middle with a bucket at one end and a counterweight at the other.

Shaduf in Egypt,

Stockraising began in the Middle Stone Age when nomads drove herds of cattle, sheep and goats, or reindeer as they moved from camp to camp.

T **Tinstone** is an ore of the metal tin, known also as cassiterite. In Mesopotamia at the start of the Bronze Age, it was most often employed together with copper ores for alloying.

W **Writing tablets** provide the earliest of all written records, those of the Sumerian scribes in ancient MESOPOTAMIA. These tablets were made of sun-dried clay and were engraved with a reed stylus.

Z **Ziggurats** are great temples in the shape of a stepped, or terraced, pyramid, built by the ancient Mesopotamians. They were made mainly of mud brick. The great ziggurat of Ur in modern Iraq, dating from 2000 BC, has been excavated almost completely.

The ability to count was an essential element in the development of civilization. But the ancient Greeks, Chinese and others soon began to study mathematics for its own sake and this led to many scientific discoveries.

Signs and Numerals

Counting

The ability to count must be as old as man himself, although in the beginning his counting would have been of the 'one, two, many . . .' kind still used by a few food gathering peoples living simple lives in remote places. As man came to settle down and to own things, his need to keep a record of his possessions led him first to make tallies, such as notched sticks and knotted cords, and later to invent COUNTING SYSTEMS which allowed him to handle large numbers more easily.

All counting systems use groups of numbers which are repeated during arithmetical calculations. To early man, the numbers of his fingers and toes were the most obvious groups to select and for this reason most counting systems came to be based on numbers that were multiples of five. The DECIMAL system, now used almost universally, is an obvious example, having the counting base of 10.

The FINGER COUNTING system did not end with primitive man but was still used as a method of ready reckoning among tradesmen of the Middle Ages and survives to this day in parts of the East. Some ancient counting systems, however, seem very long-winded to us. When adding by the decimal system we count from one to 10, then 'carry over', but the ancient Mesopotamians counted from one to 60 before carrying over. That is, the number base of their arithmetic was not 10, but 60. However, by reckoning time in seconds and minutes we continue to cling to the old Mesopotamian system.

The shopkeepers and bookkeepers of ancient civilizations also profited from the use of aids to ready reckoning, other than their fingers. First came counting stones, or CALCULI, which were used widely in the Middle East and in Europe, and counting sticks, which were similarly employed by the ancient Chinese. In both cases there arose from these devices the counting frame known as the abacus, which was invented independently in several countries and which is still in general use in eastern market places – that is, in places where it has not been supplanted by cheap electronic calculators.

Above: Chinese children work out a problem in arithmetic, using an abacus. This computing device appeared more or less independently in several countries. The Chinese abacus has 7 beads on each wire, in groups of 5 and 2 separated by a bar. Counting in decimals is, however, easier on the Russian abacus, which has 10 beads on each wire. To set up the number 715 on the Russian abacus, you simply separate 5 beads on the first wire, 1 bead on the next, and 7 beads on the third wire. Additions and subtractions can then be made by adding or taking away beads, in the usual decimal fashion.

Uninventive arithmetic

In the first civilizations the art of reckoning by arithmetic was put to uses that we now find unimaginative or uninventive.

To ourselves, living in a scientific age, an invention begins with an original idea, and is first expressed by its inventor as a scheme, design or plan, together with any necessary data and mathematical calculations. Nothing could be less like the first uses of arithmetic that were recorded on clay tablets in CUNEIFORM WRITING by the Mesopotamians, more than 5,000 years ago. Far from using their arithmetic as a tool to explore

Reference

A **Alexandrian School** is the name of a group of Greek artists, philosophers and scientists who worked in the Egyptian city of Alexandria from its founding in 332 BC until the early centuries of the Christian era. The scientists contributed much knowledge to the fields of mechanics, ASTRONOMY, geography, chemistry and medicine.

Algebra is a branch of mathematics in which operations in arithmetic are generalized by the use of letters to stand for numbers. An important use is the finding of unknown quantities, which is done using algebra much more effectively than with arithmetic. The name 'algebra' is Arabic but the first uses of algebra date from much earlier than the Arab empire. The Rhind papyrus of Egypt, dating from about 1700 BC, contains an algebraic solution to a problem.

Angular measurement was first recorded by architects of ancient Egyptian temples. A frequent example was the angle between a plumb line and a near-vertical wall.

Archimedes (287—212 BC) was a Greek scientist and a member of the ALEXANDRIAN SCHOOL. He created the sciences of mechanics and HYDROSTATICS, advanced knowledge in several branches of mathematics, and invented war machines. He was killed by a Roman soldier during an invasion of his native city, Syracuse.

Archimedes, represented in The Mechanic's Magazine, *London, 1824.*

Architecture stimulated many early feats of mathematics. These included the calculation of the volumes of the Egyptian pyramids, performed with almost incredible accuracy. GEOMETRY first arose as a means of solving problems in architectural design and division of land.

Astrology is a system of beliefs in which human af-

Right: Numbers began with finger-counting. The earliest written numerals look like notches on a stick, knots on a string, or other kinds of tallies. A sign for zero, which allowed decimal calculations and numbering, did not appear before the Hindus, and was introduced to the West by the Arabs.

Modern Arabic/ European	1	2	3	4	5	6	7	8	9	10
Babylonian	[illegible]	[illegible]	[illegible]	[illegible]	[illegible]	[illegible]	[illegible]	[illegible]	[illegible]	[illegible]
Egyptian	[illegible]	[illegible]	[illegible]	[illegible]	[illegible]	[illegible]	[illegible]	[illegible]	[illegible]	[illegible]
Greek	Α	Β	Γ	Δ	Ε	F	Ζ	Η	Θ	Ι
Roman	I	II	III	IV	V	VI	VII	VIII	IX	X
Ancient Chinese	一	二	三	四	五	六	七	八	九	十
Mayan	•	••	•••	••••	—	[illegible]	[illegible]	[illegible]	[illegible]	[illegible]
Hindu	[illegible]	[illegible]	[illegible]	[illegible]	[illegible]	[illegible]	[illegible]	[illegible]	[illegible]	[illegible]
Arabic/European of the 1400s	[illegible]	[illegible]	[illegible]	[illegible]	[illegible]	[illegible]	[illegible]	[illegible]	[illegible]	[illegible]

their world, the Mesopotamians used it only to draw up lists of kings and to itemize property – which included slaves.

In Egypt, too, the advance of mathematical knowledge and ability was at first linked more with everyday, practical affairs than with exploration of ideas for their own sake. The ancient Egyptians performed multiplications by a system of repeated addition, and division similarly by repeated halvings. This long-winded but effective arithmetic was applied typically to calculating areas of land in the division of fields for agriculture and to working out volumes and other dimensions during building operations.

Both in Egypt and in Mesopotamia many of the methods of ALGEBRA and GEOMETRY arose from these and other practical considerations. A famous example is the PYTHAGORAS theorem of geometry, named after the Greek mathematician and mystic who lived about 550 BC. The theorem was certainly invented long before him by the pragmatic Egyptians and Mesopotamians, and also, probably independently, by the ancient Chinese.

Egyptian mathematics reached its most astonishing accuracy in calculations for the design of PYRAMIDS *(see page 11)*. The oldest pyramid, that of King Zoser, built between 2750 and 2500 BC, is more than 146 metres high, covers an area of more than five hectares and yet is built to an overall accuracy of about 1 centimetre.

The religious ideas of ancient man and his notions of the universe – his COSMOLOGY *(see page 28)* – could have stimulated his efforts in mathematics, for example, in the preparation of CALENDARS. These recorded very precisely such events as the phases of the Moon and the positions, at various times, of certain bright stars. This enabled the length of time for the Earth's year to be calculated. Nearly 5,000 years ago Egyptian astronomers and astrologers, serving both the civil and the religious authorities of the state, drew up a detailed calendar for a year of 365 days, divided into 12 months of 30 days each plus a five-day addition which was made at the beginning of each year. This calendar was the basis of later Egyptian revisions, which attempted to remove that five-day discrepancy. It was also the basis of the Etruscan, Roman (Julian), Christian, Moslem and Jewish calendars.

Measurement and units

To make any measurement, we need to use standard units. The first measurements of ancient man were those of length and his earliest units were based on the lengths of parts of his body. Best known of these are the cubit (the length of the forearm from elbow to fingertips) and the foot. Since forearms and feet vary in length so widely, these were not very good standards.

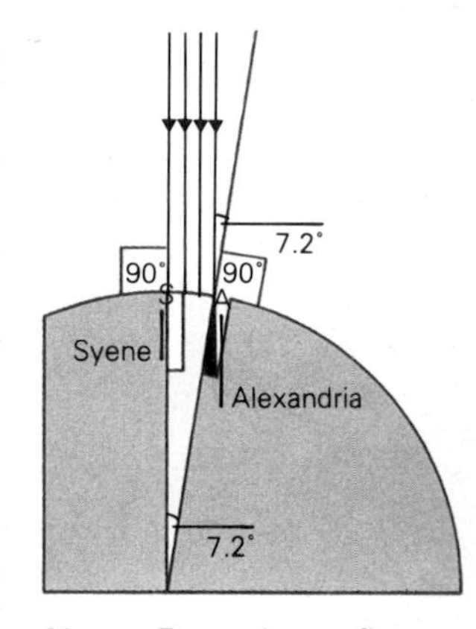

Above: Eratosthenes first measured the Earth's circumference, in the 100s BC. He noted that when the Sun is overhead at Syene, it is 7.2° from the vertical at Alexandria. 7.2° is one-fiftieth of a circle (360°), so the circumference is 50 × the distance between the towns.

fairs are influenced by the relative positions of the Sun, Moon, stars and other heavenly bodies.

Astronomy is the study of the universe or, as it once was, the heavens. It had a profound influence on the development of the other physical sciences and on mathematics, but until late medieval times it was usually mixed up with ASTROLOGY.

C

Calculi were stones used in ancient times for making arithmetical calculations. They were ancestral to the abacus, which is essentially formed from calculi strung on wires.

Calculus is a branch of mathematics which is used to solve problems of rates of change. In ancient times it was partly anticipated by ARCHIMEDES, but the methods of calculus used today were invented much later, and independently, by Isaac Newton in England and Gottfried von Leibniz in Germany.

Calendars record the division of the year into its days, months and seasons, together with the Earthly and heavenly events that take place in them. The first calendars of ancient Egypt and Mesopotamia, because they were so connected with

The Shepherds Kalender, *1600s*

the cosmologies of the time, were among the most important concerns of mathematicians, astronomers and astrologers.

Conic sections are geometrical curves formed when a right circular cone is sliced at various angles. A *parabola* is formed when the cone is sliced parallel to its side. A steeper slice produces a *hyperbola,* and a shallower slice an *ellipse.* If a slice is made parallel to the base of the cone, a *circle* is formed.

Counting systems rely upon a base number. The best-known is the base 10 employed for the DECIMAL system. Other number bases include: 2, used for the binary system in computers; 60, a base first employed by the ancient Mesopotamians and still present in our reckoning of hours, minutes and seconds; and 20, familiar to us as 'a score', but also once used widely by the Maya of Central America.

Cuneiform writing was the early Mesopotamian practice of writing with a sharp stylus on a clay tablet so as to produce wedge-shaped letters and numerals.

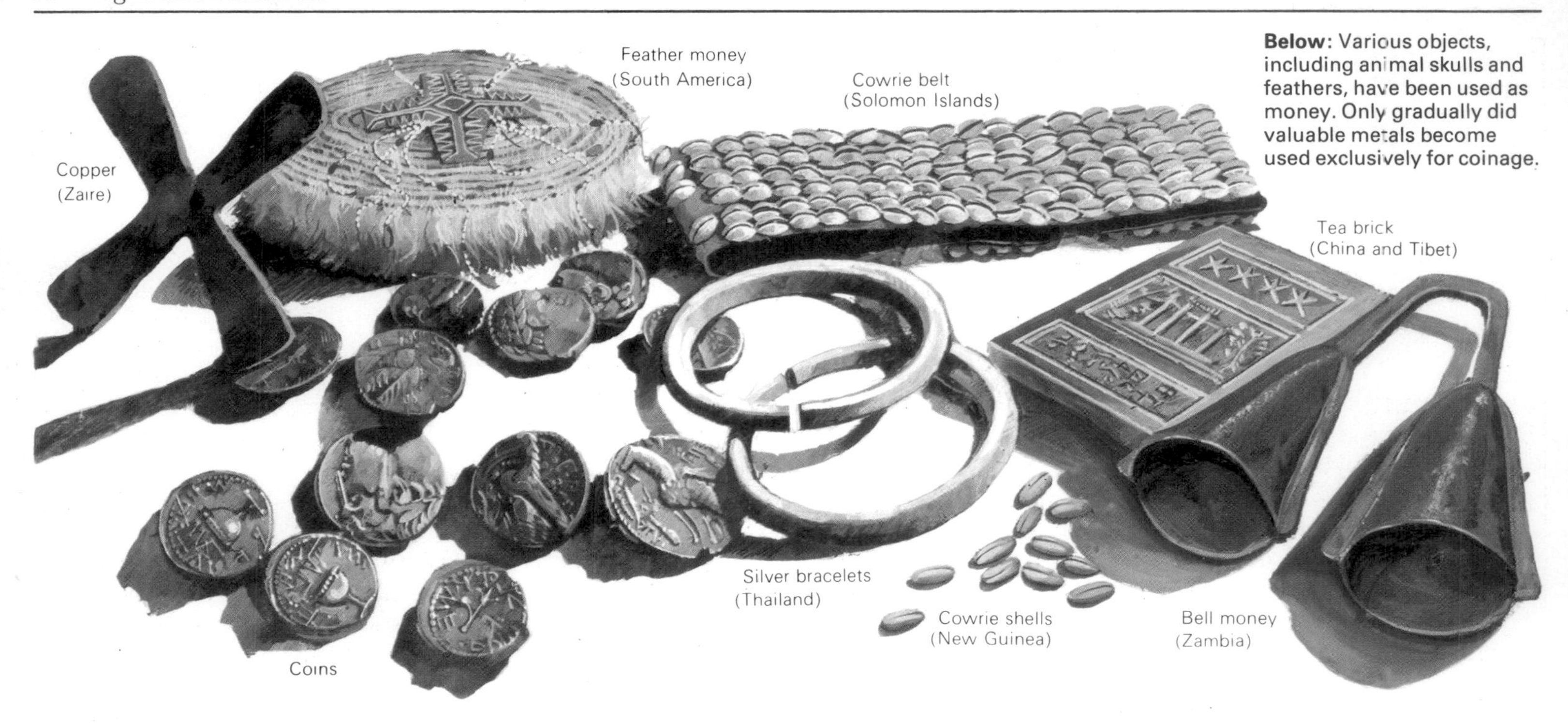

Below: Various objects, including animal skulls and feathers, have been used as money. Only gradually did valuable metals become used exclusively for coinage.

Measurements became more accurate with the invention of official, independent standards; for example, the Egyptian royal cubit – which did not vary as much as the length of each individual Egyptian arm! In time, such standards were also applied to area and volume, encouraged by the need for calculations in trade, agriculture, ARCHITECTURE and SURVEYING. With the invention of the weighing balance, or scales, standardized weights were also used in commerce.

From such practical considerations arose the first systems of mathematical knowledge. The ancient Egyptians and Mesopotamians left behind no mathematical textbooks but they had attained a thorough knowledge of some branches of the subject, such as MENSURATION.

Mathematics and mystics

The next phase of the development of mathematical knowledge is the most brilliant one in man's history. The Hellenic Greeks, unlike the Egyptians, Mesopotamians, Phoenicians and others before them, established mathematics as a separate science by studying it for its own sake.

Even so, at an early date, in the important work of Pythagoras and his school in southern Italy, the idea of number was a partly mystical one. These Greeks believed that the processes of nature somehow depended on numbers. For example, they linked together the ideas of musical harmony, numbers and geometry. They observed that a vibrating string or drumskin emits musical notes which lie at various harmonic intervals from one another. These intervals can be represented by simple arithmetical ratios such as 1/2, 2/3, and 3/4. They further demonstrated that such ratios also occurred in geometry as the relationship between parts of triangles and other figures.

The work of the Pythagoreans profoundly influenced future thinking. In philosophy and cosmology it led, for example, to the idea of the harmony of the spheres – a universe of concentric spheres with Earth at its centre, the distance between each sphere and its neighbour being a natural, or harmonic, interval. In mathematics, their work stimulated the pursuit of knowledge of geometrical figures and so foreshadowed the work of EUCLID and the foundations of modern geometry.

The Pythagorean school split up after it had discovered, but failed to explain, irrational numbers. These are numbers such as PI and the square root of two, which cannot be shown as ratios of whole numbers. (Pi, for example, is expressed as the repeating decimal

Below: Early explorers could sail vast oceans aided by navigational instruments. These arose after developments in mathematics, particularly trigonometry, which deals with angles and distances. Here a navigator of 1508 uses a cross-staff to measure the Sun's altitude by the angle it makes with the horizon.

D **Decimals** were invented by Hindu mathematicians about AD 400, when they replaced a COUNTING SYSTEM based on the number 60 (sexagesimal system) with one based on 10. This involved a very early use of a notation for ZERO.

E **Equations** originated in the work of Diophantus, a Greek mathematician of the ALEXANDRIAN SCHOOL. Algebraic equations were first written out by the Italian, Leonardo Fibonacci and his contemporaries about AD 1220. The symbols we now use were formulated by Descartes in 1637.

Euclid (*c.* 300 BC) was a Greek who founded a mathematical school at Alexandria. His 13-volume work on GEOMETRY, *Elements*, is still a major treatise.

Euclid, drawn 1661

F **Figurate numbers** are numbers shown as a flat shape such as a triangle or a square, or as a solid figure such as a cube or a pyramid. A simple example is the number 6 which can be shown as a triangular number of 1+2+3 dots. They were much studied in antiquity.

Finger counting must have been among the very first methods of reckoning in arithmetic. The Greeks and Romans had systems of finger counting in which fingers of the right hand represented large numbers and those of the left hand small numbers.

G **Geometry** is the branch of mathematics concerned with points, lines, planes and solid figures, and their properties and mutual relations. Geometrical calculations were made, for example, by the ancient Egyptians for land measurements. But geometry as a mathematical science was almost wholly the work of the Hellenic Greeks, particularly Euclid.

Graphs are diagrams that show how one variable quantity changes as another, or others, change. The various values of these quantities are measured along scales, or co-ordinates, usually at right angles to one another. Graphical relationships can take the form of straight lines, curves or bar graphs.

Groma was an instrument for land SURVEYING used by the Romans. Its working

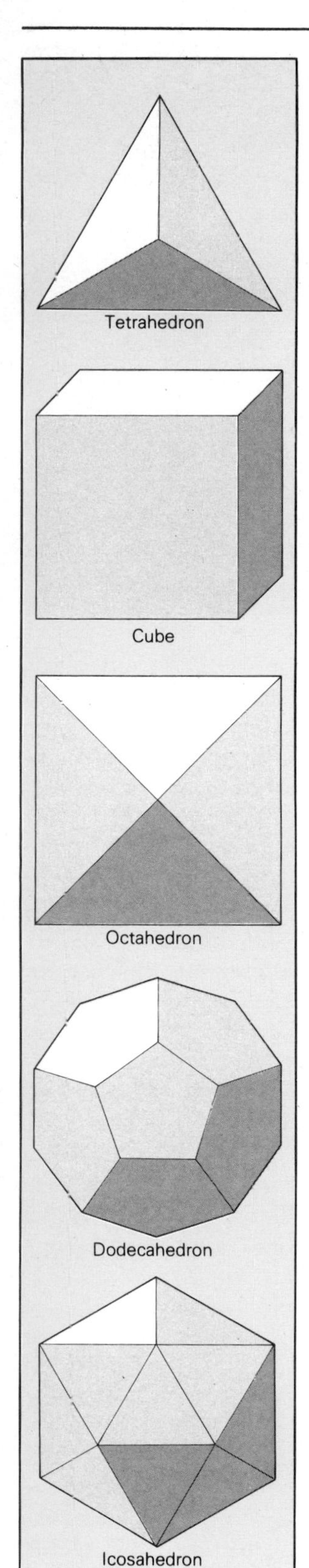

Left: Greek geometers were fascinated by the five regular solids, each of which has a number of surfaces, or faces, of identical shape and area.

Right: Zeno (c. 495–c.430 BC) is best known for his paradox, of Achilles and the tortoise. If Achilles races the tortoise, giving it a short start, he can never overtake it, because whenever he reaches the point where the tortoise once was, the tortoise has always moved on by a certain amount. Such paradoxes stimulated Greek mathematics, in this case to prove that Achilles *will* overtake the tortoise.

3.141592 . . . .) But this discovery of irrational numbers was itself the first step in the development of a new and important branch of mathematics known as NUMBER THEORY.

In ancient China, numbers also held a mystical fascination. Like the ancient Greeks, Chinese mathematicians studied FIGURATE NUMBERS (numbers that can be shown as dotted geometrical shapes). They believed these shapes to represent universal principles such as maleness and femaleness. They also delighted in magic squares, in which rows and columns of different numbers all add up to the same total. We know of these, and other, more practical, examples of Chinese mathematics from ancient books such as the *I Ching.* Quite possibly, Chinese and Greek mathematicians may have influenced each other through information carried along the great trade routes.

Practical influences

Greek mathematics was never wholly given up to mystical ideas and from a very early date concerned itself also with practical applications. Even before Pythagoras, the Ionian Greek philosopher Thales (lived *c.* 600 BC) applied his geometrical knowledge to navigation, by inventing a method of triangulation by which the distances of ships from each other, and from the shore, could be calculated. His contemporary, Anaximander, applied his geometrical knowledge to the making of the first MAP of the known world.

After the Pythagoreans, Greek mathematicians concerned themselves less with mystical ideas and more with pure and applied mathematics. In pure maths, Euclid (lived *c.* 300 BC), an early member of the ALEXANDRIAN SCHOOL, drew up his *Elements* of geometry. This laid the groundwork for the modern-day teaching of the subject. Apollonius of Perga (lived *c.* 230 BC) was another great pure mathematician. Until his time, geometrical constructions and their applications had been limited to what can be done with a rule and compasses. Apollonius's study of CONIC SECTIONS dealt with curves such as the ellipse and the hyperbola which cannot be drawn in this way. This geometrical study is still taught in schools but more important still, it vitally influenced developments in physics, astronomy and engineering at the time of the Renaissance.

ARCHIMEDES, a contemporary of Apollonius, was the greatest of all Greek scientists and the one who applied mathematics most widely. Among his many accomplishments is the invention of the science of mechanics. He applied his mechanical knowledge to the design of war machinery but his theories on the actions of pulleys and levers have had far greater consequences for the scientists and engineers who followed him. He also invented the science of HYDROSTATICS, discovering the principle of SPECIFIC GRAVITY and the laws of floating bodies which form the basis of ship design. As a pure mathematician, Archimedes made major advances in geometry, calculated an accurate value for PI, and came near to discovering CALCULUS and to inventing LOGARITHMS.

---

principle was similar to a theodolite. Horizontal sightings were made along a top piece, towards distant upright poles. Verticals were provided by plumb lines hanging from the top piece.

H

Hydrostatics is the science dealing with the physical properties of water and other liquids at rest, including the laws of floating bodies. It was essentially the invention of ARCHIMEDES, about 250 BC.

L

Logarithms (logs) are mathematical functions used to shorten complicated problems in arithmetic. When numbers are to be multiplied together, their logs are added, and when they are to be divided, their logs are subtracted. Higher powers of numbers are found by multiplying their logs, and roots of numbers by dividing their logs.

M

Maps were made by ancient geographers, most notably by PTOLEMY of Alexandria in his work *Geographia.* The first accurate maps of Europe were compass charts of Genoese sailors, which appeared about AD 1250.

Mensuration is the branch of mathematics dealing with length, area and volume. The ancient Egyptians employed it widely for measuring buildings and land. The Greeks later developed it as part of their GEOMETRY and arithmetic.

N

Number theory is a branch of mathematics which deals with the types and properties of numbers.

P

Pi is a Greek letter, also written π. In mathematics it stands for the ratio of the circumference of a circle

Ptolemy, drawn 1493

to its diameter. Calculation of the value of π has been a concern of mathematicians at least since the ancient Greeks. ARCHIMEDES calculated its value very accurately, as lying between $3\frac{10}{70}$ and $3\frac{10}{71}$.

Pneumatics is the branch of physics that deals with the properties of air and other gases. A number of machines working on pneumatic principles were invented by engineers of the ALEXANDRIAN SCHOOL.

Ptolemy (lived AD 100s) was an Alexandrian astronomer, mathematician and geog-

For all mathematicians and astronomers of this time, the strongest practical influence was that of navigation at sea. This required exact information for the computation of courses and the drawing-up of maps and charts. It demanded not only close observation of stars and other heavenly bodies but also the solution of geometrical problems that arose from these observations.

The geometrical problems to be solved were mostly those of triangles. This involved ANGULAR MEASUREMENT, which had been known for a very long time, together with the lengths of sides of triangles. From these measurements, the branch of mathematics known as TRIGONOMETRY was born. This is still the basis of all navigation and surveying. The astronomers Hipparchus (lived *c.* 130 BC) and PTOLEMY (lived *c.* AD 130) and the geometer Menelaos (lived *c.* AD 100) can be credited with these advances in knowledge. Menelaos invented spherical trigonometry, a branch of the subject particularly applicable to navigation, in which the sides of triangles are not straight but the curves of the Earth's surface.

As already mentioned, the ancient Egyptians employed methods of geometry and algebra without ever discovering the mathematical sciences with these names. Just as geometry was the invention of the Greeks, so, at a later date, was ALGEBRA. More than anyone else, Diophantus of Alexandria (lived *c.* AD 250) can be credited with the invention of algebraic EQUATIONS, by which 'unknown' quantities can be determined from the relationships connecting them.

This invention had its greatest historical effect more than a thousand years later, when it was used by GALILEO *(see page 29)* and NEWTON *(see page 30)* to express the foundations of modern physical science in wholly mathematical terms.

The invention of zero

Meanwhile, long before the days of Galileo, the main progress in mathematics shifted, with the decline of the largely Greek Alexandrian School, eastwards to India.

The Greeks had advanced geometry to a complete mathematical science, yet in arithmetic there remained no easy and efficient method of counting. Both the Roman system and that used in the Middle East had their origins in Mesopotamian arithmetic and were cumbersome to use, particularly for large numbers.

Right: Astronomers of the 1600s view heavenly constellations with the newly invented telescope. In the hands of Galileo Galilei, this instrument helped to revolutionize man's idea of the universe.

The mathematicians of India, particularly Aryabhata (lived *c.* AD 500) and Brahmagupta (AD 588–660), invented modern arithmetic when they discovered an efficient way to show the value of a digit, or whole number, by its position in a larger number. Thus, in the number 10, the digit 1 has the value 10, and in the number 100, it has the value 100. This discovery required the invention of a sign for ZERO and it formed the basis of DECIMALS.

Hindu numerals are the ones we still use for arithmetic today. They reached the West indirectly, together with other familiar mathematical techniques such as GRAPHS, via the mathematicians of the great Arab empire that flourished from AD 700 to 1300. The Arabs were the custodians both of Greek science and Hindu mathematics. Arab mathematicians, most notably al-Khwarizmi (AD 780–850), who wrote an important treatise on algebra, influenced Europe through translations of their works.

Most famous of these translations was that of Adelard of Bath (*c.* AD 1100) who studied the works of al-Khwarizmi at the University of Cordova, while disguised as an Arab. Leonardo of Pisa, or Leonardo Fibonacci (AD 1180–1250), also lived in the Arab world but as a factory manager in North Africa, where he learned the practical value of Arabic numerals and was chiefly responsible for transmitting these to the businessmen of Europe.

---

rapher. His *Almagest* described a system whereby the Sun, planets and stars revolved around the Earth. His *Geographia* was an early treatise on MAP-making, cataloguing places and their co-ordinates.

Pythagoras (lived 500s BC) was a Greek philosopher, mathematician and traveller who set up a moral and religious school in southern Italy. This school is famous for its investigations into the relations of numbers. His theorem proves the relationship between the sides of any right-angled triangle.

Pythagoras, drawn 1833

S

Specific gravity is the ratio of the weight of any volume of a substance to that of the same volume of water. It is thus a standard measure of density. ARCHIMEDES used this principle to show that a king's crown was not pure gold because it displaced too much water. Thus, it was alloyed with lighter metals.

Statics is the branch of physics, and of mechanics, dealing with the action of forces on bodies at rest.

Surveying was an activity of the ancient Egyptians, who measured land for purposes of agriculture and building. The Romans surveyed land with an instrument called the GROMA. Later surveyors made accurate MAPS with the use of the magnetic compass and other instruments.

T

Trigonometry is a branch of mathematics dealing with the properties of triangles. It is used to discover the lengths of the 3 sides and the sizes of the 3 angles. It was invented by mathematicians of the ALEXANDRIAN SCHOOL and has since been used in surveying, navigation and engineering.

Z

Zero was the invention of Hindu, or possibly of Chinese, mathematicians, about AD 400. The Hindu use of a symbol for zero in their written arithmetic allowed them to 'carry over' when adding and to 'borrow' when subtracting. This made their calculations much less cumbersome.

The advent of the Iron Age marked the start of great technological advances in the ancient world. In Asia, China led the world in technological invention during Europe's Dark Ages.

Civilization and Technology

Ironmaking

Iron as first used by man fell to him as a gift from the skies in the form of meteorites. From a very early time he would have recognized these as metallic objects and exploited them as such. This was, at first, a rather unimportant part of his technology, as explained by the fact that meteorites of noticeable size are rare objects. By comparison, IRON ORES *(see page 9)* are plentiful and widespread: nevertheless, for thousands of years they remained unrecognized as sources of the metal.

Even when, by accident or inspiration, an early metallurgist first made metallic iron by reducing an iron ore in a wood fire, he could not have proceeded to melt and cast the metal in any quantity in the way that, no doubt, he could have melted and cast bronze.

Bronze alloys melt at temperatures of about 1,000°C, which an open wood fire, if fanned by air from BELLOWS, can provide. Iron, however, melts at the much higher temperature of 1,539°C and so is much more difficult to liquefy in this way.

The first ironsmiths worked in the late Hittite Empire, about 1400 BC. Iron ores, when reduced in their fires, yielded a spongy, slaggy metal that the ironsmiths hammered while it was still hot until they had removed most of the slag and had formed the metal to their requirements.

Iron made in this way, by repeated hammering and hot forging, is known as WROUGHT IRON. For the following 2,500 years in most parts of the ironmaking world, 'wrought iron' was synonymous with 'iron'. It was used for the armour and weapons of the Assyrian armies that, in biblical times, 'came down like a wolf on the fold'. More than 1,000 years later it was still in use by the Franks for the manufacture of their elegant swords, among the few interesting products of technology of the European DARK AGES. Between and after these periods, wrought iron also provided the material for such varied products as tools for agriculture and craftwork, chariot

Below: In their manufacture of stone bowls and other examples of fine stoneware, ancient Egyptian masons employed this stone-hollowing drill. Its cutting bit was a sharp-edged flint crescent, which was mounted on a shaft weighed down with stones. This tool had evolved from the even earlier bowdrill.

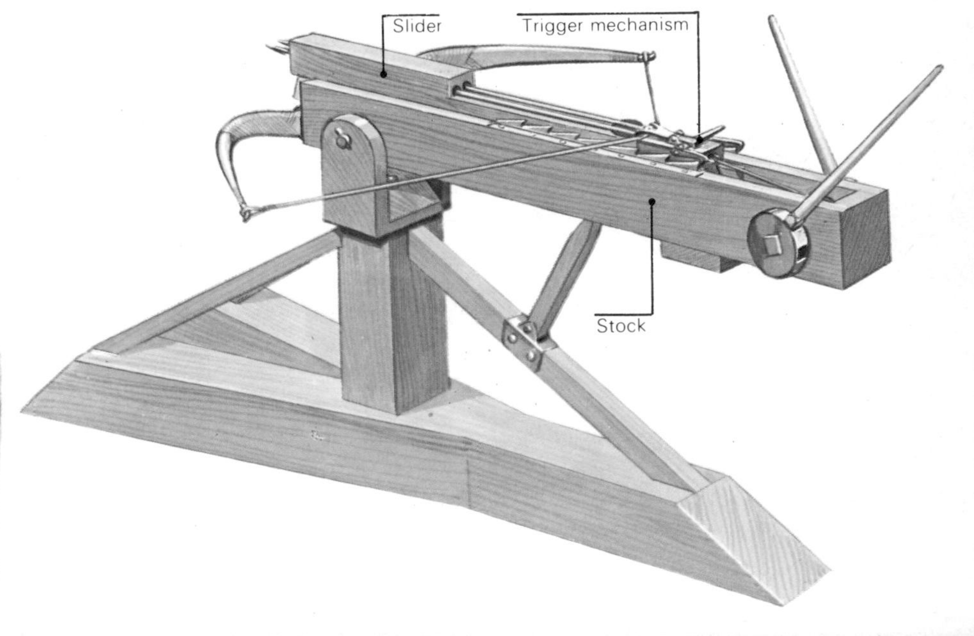

Right: One of the smaller Greek siege weapons was this crossbow, which fired two arrows at the same time. A man's strength was not enough to bend its powerful bow, which was pulled back by a winch and ratchet mechanism.

Reference

A **Alloys** were first made using copper and tin, which together make bronze. Other types of early bronze contain antimony or zinc. ELECTRUM is a natural gold-silver alloy valued by the Greeks and other ancient peoples. Brass, an alloy of copper and zinc, was first made in Roman times. CEMENTATION STEEL was a very early alloy of iron with carbon.

Aqueducts were a major feature of Roman technology. The principle of the Roman arch lent lightness to these great structures. At one time aqueducts into Rome carried 1,000,000 cubic metres of water each day. Siphons made from lead pipes carried water over raised obstacles: the siphon, however, was a Greek invention.

The Pont du Gard is a Roman aqueduct

Archimedean screw is a water-lifting device used for irrigation from the time of the later dynasties of Egypt to the present day in Asia. Thus, it was probably not actually invented by Archimedes *(see page 12)* although he and the Alexandrians helped to bring it into common use by explaining its working principle.

B **Bellfounding** has been known as a craft since AD 800 in Europe, when bells and organ pipes were first cast from bronze. This technology gave rise in medieval times to the founding of bronze cannon.

Bellows were first employed to raise the temperatures of hearths and furnaces by the ancient Mesopotamians and then by the Egyptians, Greeks, Chinese and Romans. Foundrymen in developing countries today often use bellows similar to the earliest types, having drums of animal skin compressed with the hand or foot or by use of a stick or lever. Chinese bellows of the 200s BC were fitted with flap

Right: This tomb painting shows Egyptian artisans using such tools as bellows and tongs. The ancient Egyptians had an impressive array of tools for use in agriculture, building and manufacture. However, Egypt also had the longest of all Copper Ages. For over 2,000 years copper was the only metal used widely for tools and weapons, as Egypt's lack of tin ores prevented the development of bronze alloys.

wheels, machines and scientific instruments.

Hittite ironsmiths, or a subject people called Chalybes, first made at this time a type of steel known as CEMENTATION STEEL, using similar methods to those they employed in their manufacture of wrought iron. The principal difference between the two processes was that steel required the addition of more carbon, in the form of charcoal which was hammered together with the iron.

With the fall of the Hittite Empire in 1200 BC, ironmaking technology began to spread east and west. In some hundreds of years it had reached Europe, whose IRON AGE thus began. However, this was at first very patchily developed. For example, the ancient Greeks made surprisingly little use of iron. But by Roman times wrought iron was the accepted material for all applications of metals that involved hard and prolonged wear, such as the PLOUGHSHARE, or received repeated physical shocks, such as the sword blade.

In the meantime, more profound developments in iron technology were taking place in a part of the world remote from and almost unknown to Europe. The Chinese, in about 400 BC, succeeded in melting and casting iron.

At first, this CAST IRON was a brittle material because it contained impurities such as phosphorus and silicon. However, the high phosphorus content, by lowering the melting point, actually facilitated the melting of the iron. Not long afterwards the Chinese succeeded in producing a relatively tough, malleable cast iron that gave good service as iron vessels, tools and weapons. Because they could mould it to almost any desired shape, the Chinese had much more freedom to use their iron. One example is a cocking and firing mechanism for their CROSSBOWS similar to that used on many relatively modern hand guns.

Chinese iron technology, arising quite independently of that of the Hittites, relied principally on the efficiency of the air bellows employed to provide forced draught to the iron-melting FURNACES. These bellows were sometimes astonishingly sophisticated pieces of equipment.

While the Chinese worked in cast iron, countries to the west continued to make wrought iron, a state of affairs which continued with few exceptions for the next 1,500 years until the invention in the West of the late-medieval blast furnace. The few exceptions seem to have occurred mainly in southern India, which provided Imperial Rome with its much-valued seric iron. This reached the Romans in the form

valves and operated by double-acting pistons. Such sophisticated technology allowed the Chinese to make CAST IRON more than 1,000 years before this became possible in Europe.

Brewing was known to the ancient Mesopotamians, who made many kinds of beer from fermented, malted grains. However, it was not until late medieval times that a process using hops to flavour and preserve beer was invented in Germany.

C

Cast iron is made by melting iron then pouring, or casting, it into moulds. This did not happen in Europe until the later Middle Ages because furnace temperatures were not

Turning barley to malt

high enough to melt iron, which liquefies at 1,539°C. Until this time the only forms of the metal in use were wrought iron and steel. However, the Chinese, by employing more efficient air BELLOWS to raise furnace temperatures, were able to melt and cast iron as early as the 300s BC.

Cement was used from ancient Mesopotamian times to line and seal water tanks and channels. Later it was used in the form of mortar for bricklaying. The Romans built walls from a hard cement made from volcanic earth mixed with lime, which they mixed together with stony rubble.

Cementation steel is a very hard iron alloy invented by the earliest ironsmiths, usually identified as the Chalybes, in about 1400 BC. They hammered red-hot wrought iron together with carbon in the form of charcoal, until the carbon penetrated the iron and combined with it to form cementite, a very hard component of steels. Some centuries later the cementation process was improved by quenching the hot steel in water, so hardening it further. This had not occurred to earlier metalsmiths probably because quenching copper and bronze actually softens them.

Cire perdue is the name for the lost-wax process of casting bronze or precious metals.

Cloisonné is the name of an enamelling process in which precious stones or other decorative materials are held in hollows by metal wires that are fixed by WELDING to a metal plate. It was first practised by the Sumerians and later by the Egyptians.

of small cast ingots or cakes which they believed to have come from farther east, perhaps from China. Seric iron is, most probably, the same thing as WOOTZ STEEL, the cast ingots of which were named from the word for iron in Canarese, a southern Indian language.

Iron in peace and war

The first machine of any size – if one discounts the sledges and rollers used by the Egyptians to transport their blocks of building stone – was the SIEGE TOWER of the Assyrian army, with its iron battering-ram. The defenders of a besieged town or city might respond to this attack by attempting to seize and deflect the battering ram with a grapple also made of wrought iron. These uses of iron on a moderately large scale were more than matched by the Greeks some centuries later in the 400s BC, when they built the Parthenon using cantilever beams of wrought iron. It is strange that, apart from this, the Greeks failed to benefit from a wider and more imaginative use of this metal.

Even if their spears were tipped with iron, Greek soldiers continued to wear bronze armour, and in several other ways bronze remained the favoured metal of the Greeks and peoples to their east for uses to which iron was possibly better suited. This 'bronze habit' persisted even into early medieval Europe, where one of the few examples of large-scale metallurgy was BELLFOUNDING and the making of organ pipes, processes which used bronze but not iron. One reason for this was the inability of the West to melt and cast iron.

Smaller-scale applications of iron were, however, quite common at this time. Iron tools, notably the hammer, anvil and hinged tongs used by the ironsmiths themselves, were in fairly general use in progressive parts of Europe as early as 500 BC. Roman soldiers of the later Empire wore iron cuirasses and wielded iron swords, and the Romans, like the Greeks, learned somehow the art of making large beams of wrought iron, as in the T-shaped girders which helped to support the dome of the Baths of Caracalla in Rome.

During the Dark Ages much technology, including iron founding, lapsed. This affected even the habits of warfare, so that the manufacture of iron armour had to be reinvented in the time of Charlemagne (AD 742–814), when the pattern-welded sword was the highest expression of metallurgy (*see* WELDING).

Above: A craftsman of Victorian times blows a glass vessel and then works it into the required shape by rolling, or marvering, it against a metal surface. This type of glassblowing technique has a very long history, dating back to the Near East of the AD 200s.

Fine metalwork

Egyptians of early dynasties and their contemporaries in Ur and other Mesopotamian cities were already adept at working the precious metals GOLD and SILVER into intricate shapes, principally those of decorative ornaments and religious statuettes. ELECTRUM, a natural ALLOY of gold and silver, also provided these and other ancient peoples, such as the Myceneans, with a material which is easily formed into intricate shapes without breaking and is resistant to tarnishing.

From these metals, wire and sheet were made for decorative ENAMELLING work, including the very popular type known as CLOISONNÉ.

In these techniques of fine metalwork, small pieces of beaten gold or precious stones were held in a vitreous, or glassy, base with metal wire. To hold the wire and other materials together, jewellers invented processes of SOLDERING, which were also used in the fabrication of lace-like, filigree ornaments often found in Egyptian tombs.

Techniques of fine metalwork originating in ancient Egypt persisted for 3,000 years or more, for longer even than the Egyptian civilization itself, to be used for cloisonné and filigree metalwork by the Etruscans, Greeks, Romans and Celts.

Right: The beautiful work in precious metals made by the ancient metalsmiths of Peru has lately become famous. Earliest of these metalsmiths were the Chavin workers in gold and silver. They started working at the time of the fall of Babylon in Mesopotamia, and were still active in early Christian times.

Cranes and hoists were much used by builders in Rome but probably originated elsewhere. A primitive type of crane is depicted on a Babylonian tomb painting of about 1900 BC. Roman cranes employed compound pulleys, an invention of the Greeks, to lift heavy weights. The motive power of some cranes was provided by a TREADMILL.

Crossbows ranged in size from hand-held weapons (including an early one called the 'scorpion') and larger siege crossbows mounted on a frame, to the siege catapult and the ballista. The last was a large catapult designed to throw heavy stones and other missiles. Large crossbows were drawn back for firing by one or more strings wound on a winch; smaller weapons had a lever for this purpose.

Ctesibius was an Alexandrian engineer (*c.* 100s BC). His most famous invention was the *hydraulis,* a musical organ with pipes supplied with air from bellows placed inside a large vessel and compressed by water flowing around them. His other inventions included a crossbow having pistons operated by compressed air, a water pump for fire-fighting, and a water clock.

Ctesibius's water clock

Dark Ages refers to a period of European history when progress, particularly that of science and technology, virtually ceased because of widespread political and social instability. It lasted from the fall of Rome, AD late 400s, to the rise of the Christian empires, about AD 1000.

Electrum is a naturally occurring alloy of gold and silver, light yellow in colour. Together with copper, bronze, silver and gold it was used from early BRONZE AGE *(see page 6)* times for making statues and jewellery.

Enamelling is a craft in which glassy, often brightly coloured surfaces are applied to materials such as ceramics and metals. It has a very long history. Egyptian tablets and beads of stone, surfaced with a turquoise or lapis lazuli glaze, date from before 2000 BC. Greek, Etruscan, Roman and Celtic CLOISONNÉ enamels are from later times.

Fermented foods have been made by nomadic peoples since the Middle

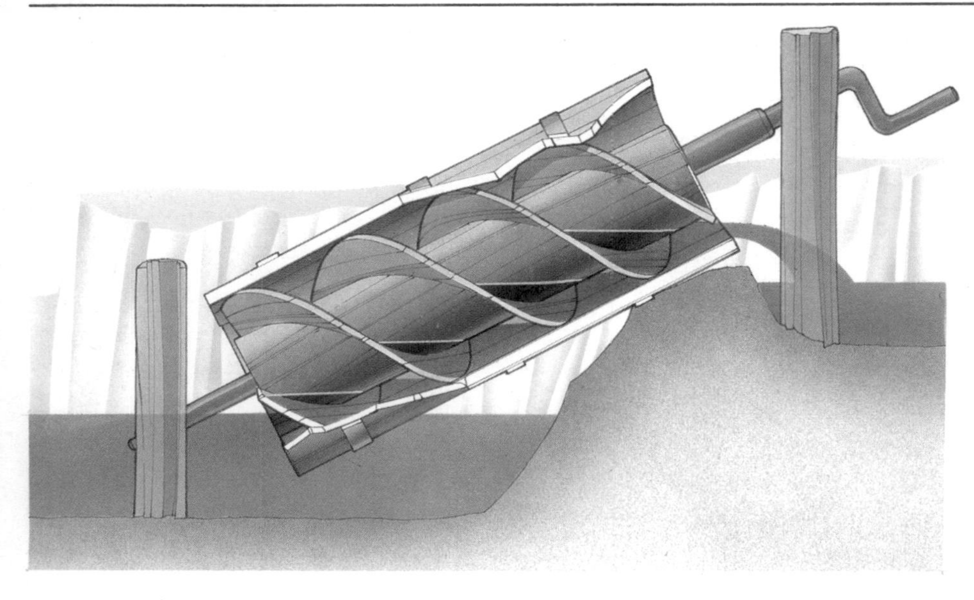

Above: The Archimedean screw is a water-lifting device, or pump, which works by trapping and raising water as the screw is turned. Despite its name, it certainly originated much earlier than Archimedes, having been used for irrigation by the ancient Egyptians and Mesopotamians. It is still to be seen at work in many Asian paddy fields.

Contemporary with the Etruscan and Greek civilizations, but a world away in their origins and general way of life, were the people of the Chavin culture of Peru, whose fine metalwork in gold and silver matched in quality that of any in Europe and the Near East. Among the Chavin people metallurgy was encouraged by the rich deposits of metals available from the nearby Andes, just as in the Europe of Greek and Roman times it was encouraged by greatly increased mining activity.

Early miners

The mining of metals began when men first recognized their presence in rocks and alluvial deposits. For this reason, gold, which is found in nature in its pure form, was the first metal to be mined, in Egypt in about 3000 BC. At a later date the Egyptians are known to have mined copper and turquoise in Sinai.

As already remarked, not much is known of the mines that supplied the Mesopotamians with the ores of tin and copper that they needed for their bronze industry. Etruscan iron mines, dating from about 900 BC, are better known, chiefly because they were later taken over and greatly expanded by the Romans.

GREEK MINING was also extensive, having been started by the Myceneans and expanded by the Hellenes. Hellenic Greeks needed silver in large quantities with which to make their MONEY. To get it they mined chiefly a lead ore, GALENA, which contains a little of the precious metal. Silver was extracted from the lead by a process in which the ore was first concentrated, or enriched. Then it was heated in a porous clay crucible, together with bone ash and other substances. The lead combined chemically with these, freeing metallic silver.

This was only one variant of a process known generally as cupellation, after the cupel, or crucible. A form of cupellation had been used earlier by the Egyptians to extract silver, with the aid of salt, from electrum, to yield pure gold.

ROMAN MINING operations dwarfed those of any previous industry. As in the Greek mines, but to a far greater extent, slave labour was employed. Just as much as the metal deposits, this forced labour was the basis of an Empire-wide industry. Miners sometimes worked 100 or more metres down, lighting their way with oil LAMPS and attacking the face of the ore-bearing seam with iron picks. The air they breathed was renewed through ventilation shafts, and water was removed from mine workings with the aid of a type of water-wheel called a scoopwheel, and also with the ARCHIMEDEAN SCREW better known for its use in agricultural irrigation.

Little is known for certain of mining in the Dark Ages of Europe that followed the fall of the Roman Empire. However, it is hardly likely that with abandonment of the Roman mines, mining stopped altogether. It probably persisted in central Europe just as it had before the coming of the Romans, when the HALLSTAT INDUSTRY was concerned, to some extent, with metalliferous mining. It is known that long after the Romans had gone, SAXON MINING flourished in the Harz Mountains of central Europe.

Machines large and small

For 750 years, the Assyrian Empire waxed and waned until its collapse in 609 BC. During their long reign the Assyrians had thought of machines chiefly as engines of war, a trend followed, more or less, by the conquerors who succeeded them. Besides the siege towers already mentioned these war machines included a number of types of enlarged CROSSBOW such as the catapult, ballista and onager.

HERO (or Heron) and CTESIBIUS, two Greek engineers, invented quite sophisticated war machines. The latter is said to have designed a

Stone Age: these include such fermented milk foods as yoghurt and kumiss. During fermentation, chemical products such as alcohol (in wine), acetic acid (in vinegar) or lactic acid (in yoghurt), are produced by the action of certain microorganisms. Because of these products, fermented foods often keep longer, which may explain why they were first made.
Furnaces were employed by ancient metalsmiths to heat and ALLOY metals and to reduce them from their ores. They were fuelled with wood or charcoal and temperatures were raised by air blasts from BELLOWS. An early example from Europe was the Catalan furnace which had 2 pairs of bellows and in which iron was heated for forging. Only in China were furnaces made hot enough to melt iron for casting, by the use of more efficient bellows.

G **Galena** (lead sulphide) is the principal ore of the metal lead. This glittering grey, crystalline mineral was mined from ancient Mesopotamian times (about 2000 BC), for lead and for the silver associated with it.
Gears were described by the Alexandrian engineer HERO in his engineering writings as a means of transmitting power. They remained little, if at all, used at the time because crude workmanship made them too inefficient. However, large wooden gears are known to have been used in Roman times to transmit power to millstones in early water mills.

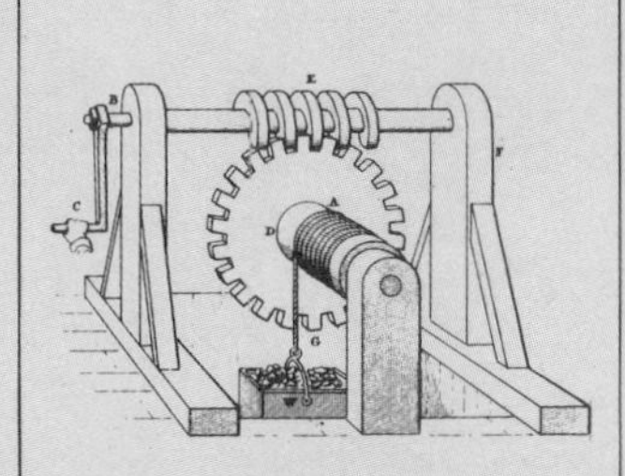

Gear wheel on a windlass

Glass is made, traditionally, by heating together sand, lime and soda until they fuse to form the familiar hard, transparent material. Glass was first made by the ancient Egyptians in this way; their charcoal-burning furnaces could just provide the necessary temperature of about 1,000°C. The colour of their glass was at first caused by impurities but later the Egyptians deliberately added small amounts of other substances to lend the glass particular colours. Copper compounds, for instance, gave red or green colours.
Gold is a precious metal always found naturally in its pure or native state. It was highly valued by many BRONZE AGE *(see page 6)* peoples, including the ancient Egyptians and the Chavin

sort of large air-gun. Philo of Byzantium is believed to have drawn up plans for a catapult loaded with a bronze spring, although these, like so much of the work of ancient Greek engineers, have not survived. The onager was a type of catapult where the 'spring' was a skein of twisted cords which fired a missile when the cords were released suddenly. Torsion catapults of an earlier type had been used by Alexander the Great (356–323 BC) and by Trajan (*c.* AD 53–117) in their sieges.

The application of machines for peaceful purposes also has a long and varied history. Like the war machines, these depended for their invention on man's intellectual grasp of fundamental working principles.

These principles are embodied in a few SIMPLE MACHINES which, taken together, provide the basis of operation of all more complicated mechanical devices. A very ancient example is the irrigation device, the SHADUF *(see page 11)* which is a straightforward application of the lever, one type of simple machine. Other early examples of levers are the hoe and the chisel, tools which at first had blades of copper or bronze. The spade, another simple lever, had to wait for its invention until the improvement of iron enabled a blade of sufficient strength to be made.

Right: Archimedes in his bath – a medieval artist's idea of a famous legend in the history of science. Lying on the floor is a crown, that of Hiero II, King of Syracuse. Archimedes' problem was to discover whether this was pure gold, or gold diluted with silver. While in his bath, he noticed that his body displaced a certain volume of water. This inspired him to place weights of gold and silver, each equal to that of the crown, in water. As gold is denser than silver, the gold weight had a smaller volume. The crown itself displaced more water than the gold, less than the silver. Thus, it was diluted or alloyed.

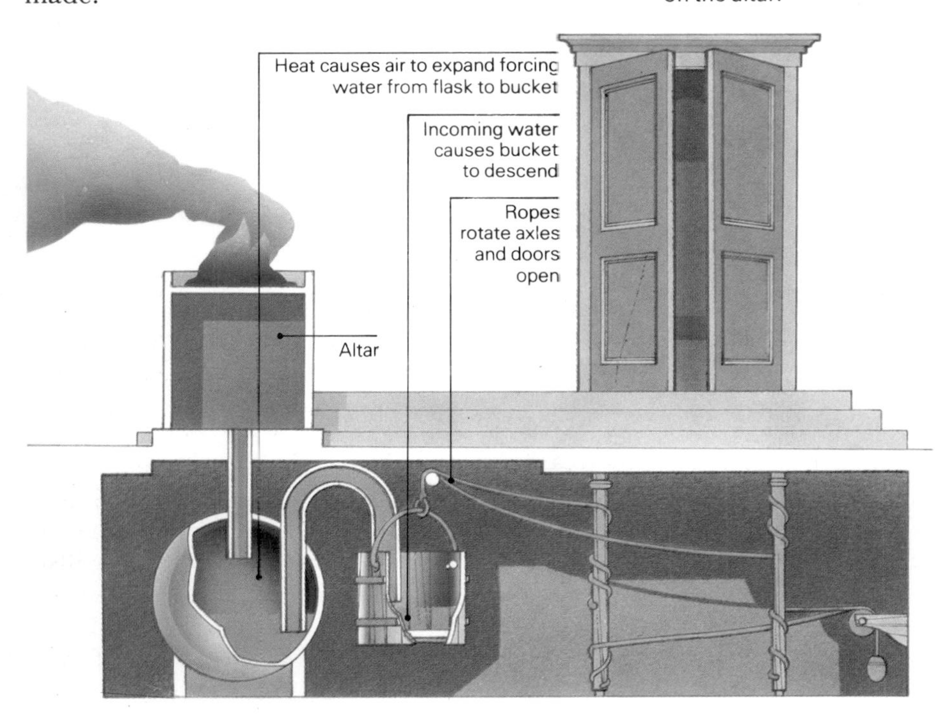

Below: Greek engineers of Alexandria invented many ingenious devices, but these were rarely put to important uses. This one, invented by Hero, opened temple doors after a fire had been kindled on the altar.

Egyptians and Greeks extracted the juices of fruits such as grapes for wine and olives for oil using filter presses, an early example of which was the beam press, which was a lever device. At a later date they developed more efficient presses which used the principles of the SCREW and the pulley, two other simple machines.

Levers and pulleys provided the working basis of large CRANES used by the Romans for lifting heavy stones during building operations. Very similar cranes were still in use in the Middle Ages.

The wheel-and-axle, a fourth type of simple machine, is most obviously applied in wheeled vehicles, but the history of technology has many other important examples. Wood-turning with LATHES is a subject depicted by artists of the later Egyptian dynasties. After a few centuries these wheel-and-axle machines had been improved to cut stone and glass.

TREADMILLS were a feature of Roman mining, where they provided the motive power for drainage scoopwheels, and of Roman building, where they powered the pulleys of the large cranes. More significant to history were the large wheels fitted to WATERMILLS in which, from ancient to modern times, corn was ground to make bread flour. The millstone itself, and also the POTTER'S WHEEL *(see page 11)* are further important examples of the wheel-and-axle.

The Romans, although they employed more machinery than their predecessors, invented very little of it themselves. When they did do so the result was likely to be rather crude and clumsy, as in the case of their early version of a reaping machine.

---

people of Peru, for its untarnishable yellow lustre and because it is so easily worked. Then, as now, gold was found by prospecting deposits in the beds of rivers and by mining rocks.

Greek mining dates from earlier than 1000 BC, in Mycenean times. By 600 BC the lead ore GALENA was mined by slaves belonging to Athens, mainly for the extraction of SILVER, which is present as 1 part in 600 in the lead ore.

H

Hallstat industry was the first evidence of the IRON AGE in central Europe. Dating from about 900 BC, it is characterized by finds of leaf-shaped iron and bronze swords, axes with winged heads, and smaller objects, including safety pins, of iron and bronze, all found in Austria and Moravia.

Hero (*c.* AD 200s) was a Greek engineer of great inventiveness who studied and worked in Alexandria. His inventions include a steam turbine, a pneumatic machine for opening doors and a screw cutter, none of which was used for any serious purpose at the time. He also wrote copiously on engineering.

Hypocaust is the name of the underfloor space in Roman buildings into which hot air was directed for central heating. The hot air rose and escaped into rooms through flue-tiles.

Hypocaust, Cheshire, England

I

Iron Age is the name of a period of human cultural development extending over the period 1000 BC–AD 100 in Europe. It began with the smelting of iron from its ores by the Hittites and with the destruction of their empire spread slowly eastwards to India and China and westwards into Europe. Many human cultures of the past, including those of America before the arrival of Europeans, and Africa before the Arabs, never experienced an Iron Age.

K

Kilns are ovens for making bread, drying corn or strengthening glass, or furnaces for melting glass or firing pottery or bricks. The first pottery kilns were built by the Mesopotamians before 3000 BC but kilns were not used for glassware before the Middle Ages, when Venice was a centre for the glass industry.

L

Lamps carved from stone, burning oil by

Above: Chinese wooden printing blocks. The Chinese invented printing, like so much else, their first use of movable type going back perhaps to the AD 700s. However, these blocks date from the 1800s, showing that once printing had reached the West, it progressed there much more quickly than in China.

Left: The Chinese were the first papermakers. About AD 100, they began to use the fibrous bast, or phloem, from beneath the bark of trees, to make paper in the manner shown. **1** The fibres were first mashed evenly with water. **2** The mashed pulp was spread on a mesh or screen, and then compressed to remove most of the water. **3** Matted fibrous paper sheets were shaken from the screen to dry. This method was still in use in the Far East as late as the 1700s.

The inventions of the Greek engineers of Alexandria also came to very little. A famous example is the steam turbine of Hero which, if it was used at all, was used as a toy. The scientific principles underlying several other Alexandrian inventions, if widely applied, could have led to major progress in water engineering, such as pressure pumps for land drainage and sewage disposal.

In one of his written works, Hero of Alexandria refers briefly to what might have been a WINDMILL. However, nothing more certain is known about this machine until it appears, about AD 650, as a wind-driven irrigation pump in Iran. The windmill was not put to the task of grinding corn until four to five centuries later. Windmills probably evolved from much earlier watermills, which had similar driving shafts.

The influence of China

For about 1,500 years China led the world in technological invention. In the West, this period extended from the decay of Greek civilization almost until the Renaissance.

The influence of Chinese technological progress on countries farther west was very patchy,

means of a wick, were used by cavemen 17,000 years ago. But even by Roman times the design of such lamps had not much improved. Resinous wood was sometimes used as fuel for lamps, as in early lighthouses.

Lathes are cutting and shaping tools. The earliest type was a pole lathe operated by treadles that used the tension of a bent wooden bow to spin a piece of wood against a chisel.

Leather is prepared from animal skins using various methods for preserving, hardening and softening the skins. Ancient peoples cured leather by smoking or salting it and softened it by rubbing in fat. From ancient

Soaking skins in lime

Egyptian times alum (sulphate of aluminium and another metal, such as potassium) was used for preserving leather. A more successful process, also of great antiquity, is tanning, in which skins are soaked in solutions of oak bark or other vegetable or mineral substances.

Locksmiths worked at an established craft in ancient Rome, making metal latches and locks using iron tools such as a hammer and anvil and pincers. The Romans were the first to develop a lock that required a particular shaped key to open it. Iron lock keys of a much earlier date have been discovered in Egypt and in that country sarcophagi as old as 2000 BC were fitted with wooden tumbler locks.

Looms began in the MIDDLE STONE AGE *(see page 5)* as horizontal, pegged-out ground looms. These were largely replaced in the first civilizations by vertical looms. In one type the warp threads hung from a beam, being weighted at their lower ends. In another type, more like that still used in rural Africa and Asia today, the lower ends of the vertical warp threads were pulled tightly around a lower beam.

Magnetic compass was invented by the Chinese in about AD 500, when it took the form of a spoon compass. The spoon, of lodestone or magnetized iron, balanced on a bowl and swung to point north–south. From about AD 850 the Chinese used compasses with a magnetized iron needle, which gave rise to the first European compass of the 1100s.

Money in ancient times took

Above and **below**: The ancient Chinese were skilful instrument makers. Shown is a Chinese seismograph, an instrument for detecting and measuring Earth tremors. In each dragon's mouth a ball is poised. An Earth tremor will cause one or more balls to fall into the open mouths of the frogs below. The severity of the tremor is measured by the number of balls so displaced.

mainly because trade over such long distances, and through so many intermediaries, did not permit a thorough exchange of ideas. It was not until AD 1275 that an acute observer from the West, Marco Polo, was permitted to visit China and stay there long enough to make a thorough report. By that time Chinese technology was in decline.

This failure of ideas to travel probably explains why the Chinese method for melting and casting iron remained exclusively theirs. On the other hand, their technology of PAPERMAKING, invented not later than AD 100, eventually filtered through to Europe. This also happened with SILKMAKING, another Chinese invention, except that this took even longer than papermaking to reach the West: silkworms were smuggled from China to Byzantium (now Istanbul) about AD 550 but silk was not made there in any quantity for another six centuries.

Chinese inventions which involved a less complex technology had a better chance of reaching the West. Fairly certainly we owe the wheelbarrow and the harness and stirrup to the Chinese and probably also the ship's rudder and the lateen, or triangular, sail.

However, the Chinese invention of PRINTING with wooden blocks, which predates Gutenberg's Bible (AD 1400s) by about 700 years, does not seem to have been the ancestor of Western printing with movable type. Block printing was used in the AD 700s in India and in Egypt. It died out in Egypt and was reintroduced to the West via the Islamic countries. By AD 1294 paper money was being printed in Iran and the first instance of a picture print in Europe was a St Christopher painted in AD 1423. Chinese PORCELAIN undoubtedly inspired many Western imitations. These were for some centuries mainly unsuccessful because they lacked the Chinese expertise of 2,000 years work in fine ceramics.

During the long period of its technological ascendancy, China also encouraged many mathematicians and astronomers, whose discoveries, if not far-reaching by modern standards, were at least usefully thorough. Their work also influenced the development of a number of scientific instruments including navigational aids such as the MAGNETIC COMPASS. OPTICAL LENSES, although first familiar to us from descriptions by Arab scientists, were very possibly first made in China. Just as influential was the Chinese invention of the ESCAPEMENT MECHANISM of clocks *(see page 42)* and no list of this sort would be complete without mention of GUNPOWDER *(see page 35)*. Since the full extent of our indebtedness to ancient China has yet to be explored, we can expect this list to grow longer with time.

Bricks and stone

The oldest buildings of any size are tombs, temples or fortified places, but by Roman times a large building could also be the villa of a prosperous citizen, a place of civic assembly or a public baths.

The Romans quarried stone for their buildings in a manner similar to that used by the ancient Egyptians, namely by splitting the rock face with wedges inserted into chiselled slots. Many Roman buildings were made to look much less like fortifications by the use of arches built of brick or stone. For their building design generally, the Romans owed much to the earlier Greeks, but the arch they took from another earlier people, the Etruscans.

Fired bricks for arches and other building

---

various forms, from cowrie shells and beads to belts made of feathers. From about 2000 BC copper ingots were widely used and after about 700 BC long, narrow iron bars were currency. The first coins were produced in Asia Minor in about 700 BC. They were made from gold or silver, heated until soft and stamped with an iron die bearing the symbol of the state or king.

O **Optical lenses** were first used by Arab scientists of AD 1000 or earlier, probably as the result of anatomical studies of the eye. Spectacles in which lenses were mounted were a later development depending on the availability of manufactured GLASS of good quality: this occurred in Venice in the early 1300s.

Spectacles, c.1500s

P **Papermaking** was an invention of the ancient Chinese, who were making the material for writing on at the start of our Christian era. Twelve hundred years later the Arabs brought paper to the West. Early European paper was made from rags of linen, a new material, and was of very good quality.

Parchment is a material prepared from animal skins as a surface on which to write. Supple, untanned skins of young animals were used, washed and stretched on a frame. Parchment takes its name from Pergamum in western Turkey, where, about 300 BC, it was made as a cheaper substitute for Egyptian papyrus.

Pharos lighthouse was the most famous of ancient lighthouses and one of the Seven Wonders of the World. It stood on a peninsula near Alexandria in Egypt, having been built there by the Greeks in 280 BC. About 80 metres high, it threw a beam visible to sailors more than 50 km away. Its lamp was a fire of resinous wood, the light from which was reflected by polished metal mirrors.

Ploughshare is the cutting blade of a plough. Tillage was made very much more efficient by this refinement. Flint and hardened iron were the materials most favoured for early ploughshares, copper being too soft and bronze, apparently, too easily worn. The first ploughs to be fitted with iron shares were used in Palestine, about 1100 BC.

Porcelain is a hard ceramic material made with china

Above: A waterwheel in Syria, situated by a ruined aqueduct, which it once fed with water from buckets on its rim. The ancestors of this wheel go back to the Roman waterwheels built by Vitruvius, an architect and engineer of the time of Julius Caesar. That such ancient waterwheels were also the ancestors of windmills is indicated by the similar gearing arrangements of ancient waterwheel mills and the first windmills built in the 600s in Persia.

structures were invented long before Roman times. Glazed bricks (with a glassy surface) were used extensively for the building of Babylon, in about 700 BC. The manufacture of fired brick on a large scale had been made possible by the development of more efficient KILNS.

Such improvements in building techniques were reflected in the quality and elaboration of domestic architecture. The home life of a well-to-do Roman citizen could be more luxurious than that of a rich man living either before or after his time. This can be illustrated by the heating arrangements of his house. Whenever he felt so inclined, the citizen could take a hot bath, water being supplied from a boiler. In chilly weather he would remain warm because his rooms were centrally heated with warm air rising from the HYPOCAUST below.

SANITATION and drainage systems might also be found in his villa but these and other progressive features of domestic civil engineering originated much earlier than the Romans. Two thousand years before their empire was founded, the cities of the Indus Valley civilization were built to a systematic plan that incorporated public drainage systems. The wide streets of the city of Mohenjo daro, which divide up the city neatly into blocks, are bordered by deep, brick-built drains which are better than many of those to be seen today in some Eastern capitals.

The Myceneans of Greece and Minoans of Crete also had sanitation systems and bequeathed to the later Greeks and Romans knowledge of such admirable additions to the home as bathrooms, urinals and sloping rain-water gutters, together with all necessary earthenware piping. Storm-water drains leading to main sewers were features of Etruscan towns and the city of Babylon before being adopted by the planners of the city of Rome.

Perhaps the most impressive of all Roman buildings are the AQUEDUCTS constructed to carry the very large flows of water demanded by Rome and other large cities. Many of these survive today, two impressive examples being that at Nimes, southern France, and that at Segovia, central Spain, which still brings water into the town.

In certain ways Roman town planning reached a high level of enlightenment. The public buildings of Rome included both hospitals and libraries. Rome had no fewer than 28 libraries in the year AD 350, many of which stocked tens of thousands of books and PARCHMENT scrolls.

The largest building works of the earlier Phoenicians were harbours and docks, few of

---

clay or kaolin, feldspar and quartz or flint. It was invented for making fine ware by the Chinese, about AD 700, and reached the West, via the Arabs, a few centuries later.

Postmill was an early type of windmill in which both the sails and the superstructure containing milling machinery were pivoted on an upright post driven into the ground, and could turn into the wind.

Printing was undoubtedly the invention of the Chinese, who before AD 700 used wooden printing blocks on which the characters were carved in relief and inked.

Postmill in Germany, 1620

Earlier the Romans had printed textiles with wooden patterns but they never printed words. Printing appears to have developed independently in the West, in Italy in the late 1200s.

Q

Quarrying was first carried out on a large scale to provide stone blocks for the pyramids. These stones were removed from the rock face with the aid of chisels and wedges and, although weighing as much as 100 tonnes, were transported away on log rollers by manpower, without the aid of pulleys. The Romans developed a saw driven by water power which they used to cut up quarried stone.

R

Roman mining was extensive and large-scale. Their metalliferous mines were scattered all over Europe from north Britain to the Balkans and exploited gold, silver, iron, lead, tin, copper, mercury and zinc deposits and ores. The deeper mines were drained by the use of the ARCHIMEDEAN SCREW and a wheel, with buckets attached to it.

S

Sanitation was a feature of cities of the Indus Valley, Mycenean buildings and many Roman estates, which had baths, piped drainage and even water closets. However, it was never a widespread feature: the ancient and medieval worlds were largely without sanitation.

Saxon mining became a major industry in the Harz Mountains of central Europe earlier than AD 1000. In the later Middle Ages metal mining spread throughout Europe and coal also began to be mined.

which have left visible traces. Also long disappeared is the most monumental building of the ancient Greeks, not a temple nor an amphitheatre but the PHAROS LIGHTHOUSE, constructed near Alexandria in 280 BC and finally destroyed in AD 1375, when it was levelled by an earthquake. This was only one of many lighthouses built around the Mediterranean shores.

Large buildings of brick and stone are rare survivors from the DARK AGES of Europe. Those that have left traces are mostly churches of central and southern Europe. In rural northerly regions builders reverted to using wood, as their remote ancestors had done. The revival of architecture in Europe had to wait many centuries until the building of the first great Gothic cathedrals.

Trades and luxuries

Daily life in an ancient city called upon the products of many crafts and trades, some of which could be described as luxuries. These might include toilet articles made from gold or ivory, woven rugs and hangings and numerous delicate items of furniture provided by a thriving woodworking craft.

The ancient Egyptians, as their tomb paintings show, made wide use of cosmetics. They stored these and other household substances in small containers of wood, ceramics, ivory and GLASS. Glassmaking was probably an Egyptian invention of about 1500 BC, when its first products included small vessels made by a dipping process. Glass was melted, then a core of sand wrapped in cloth was dipped into the glass, withdrawn, and the glass allowed to cool, when the core was removed.

Glassblowing was a much later invention made in the Near East about 100 BC, when glass vessels were blown into clay moulds. A mechanized version of such a process is used nowadays for mass-producing glass bottles and light bulbs. A century or so later, glass vases, goblets and bottles were blown by a method very

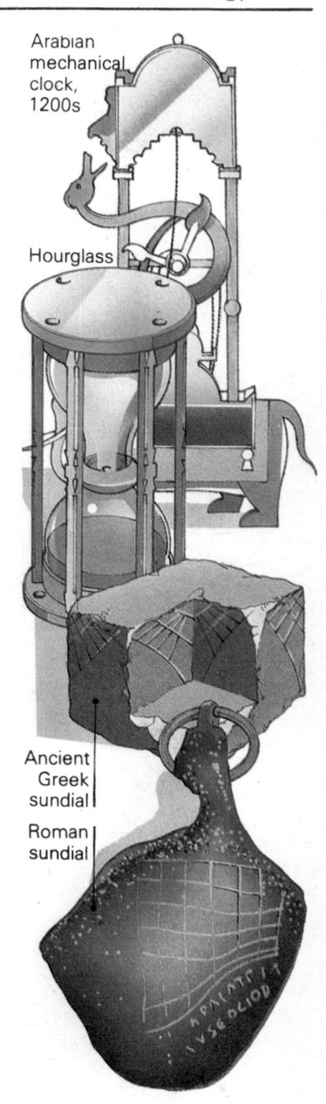

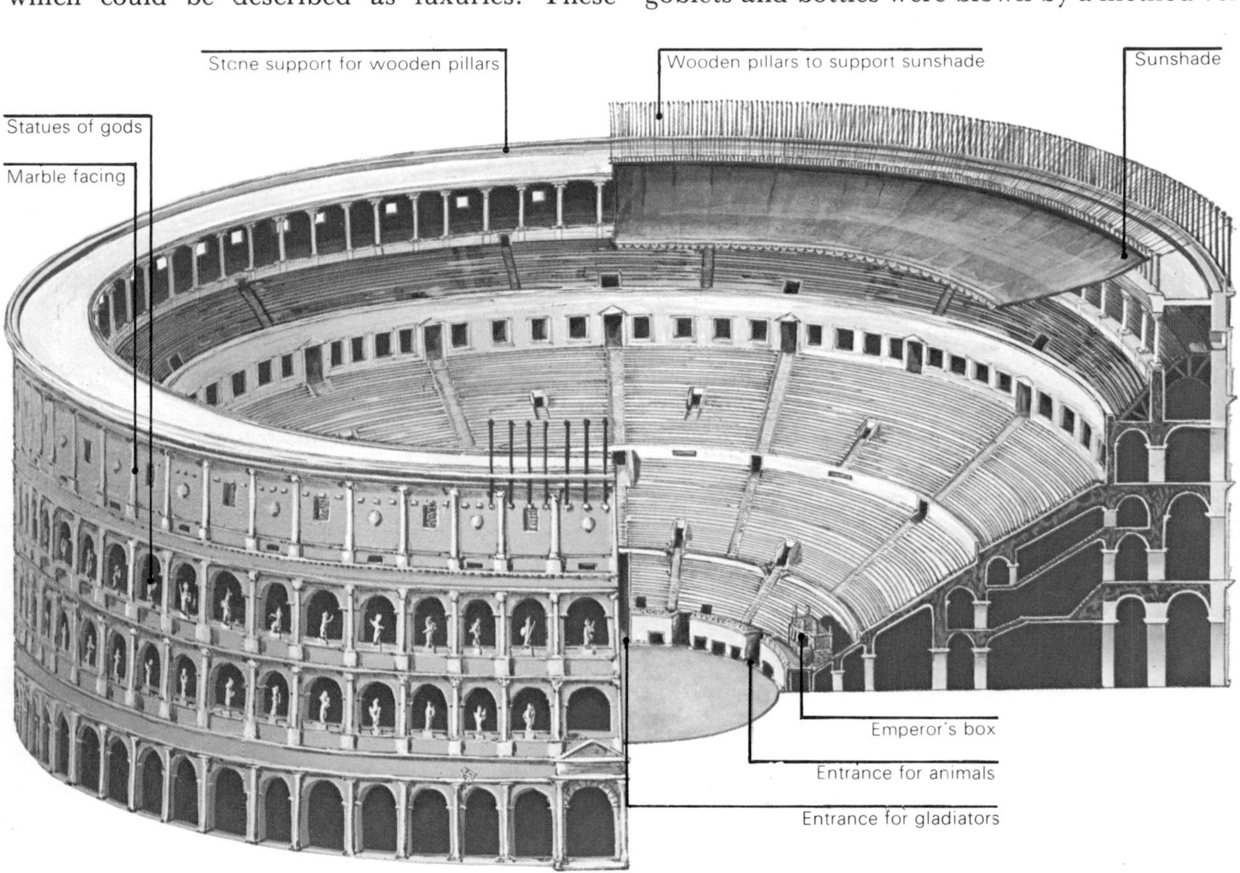

Above: History has seen a wide assortment of devices for measuring time. Sundials derive from the even earlier Egyptian shadow clock, perhaps the first of all timing devices. Hour-glasses are also still used today, if only as egg-timers. The mechanical clock was invented by the Chinese, about AD 700, and later taken up, first by the Arabs, then by Europeans.

Left: The use of the arch by the Romans is shown by this cutaway view of the Colosseum. It explains why their largest buildings were much more lightweight than those of the Egyptians.

Screws were used from early Egyptian times in the form of wine presses. Metal screws for securing wood were widely used by Greeks and others and HERO of Alexandria invented a screw cutter although this was probably never used. The screw pump or ARCHIMEDEAN SCREW has been used from very ancient times to lift water.

Shoemaking was already an established craft in the first city civilizations. Leather was prepared, cut and shaped in a way similar to that now used to make heelless slippers or sandals. Heels and welts appeared in the Middle Ages.

Shoemaker, 1600s

Siege tower was a military development of the Assyrians. It usually took the form of an iron battering-ram on wheels, manned by many soldiers who were shielded from missiles as the tower was moved up to the attack.

Silkmaking was invented by the Chinese at the latest by 1000 BC, when their first silk fabrics appeared. It took another 2,000 years for this technology to become established in the West.

Silver was mined by the Mesopotamians, Greeks, Latins and Celts and used mainly for ornaments. GALENA, a lead ore, was also a major source of silver.

Simple machines are devices by which power or force can be magnified and transmitted over a distance. They are the basis of all more complicated machinery. The lever, screw, wedge, pulley and wheel-and-axle are simple machines used widely in ancient times, the principles of which were first explained largely by the Greek scientists.

Soldering is the use of alloys which melt at low temperatures, such as tin and lead, to join other metals.

Threshing is the separation of cereal grains from the husk. In Roman and Saxon times this was done by beating the grain with a wooden board or, later, a jointed flail, or using farm animals to tread the grain. Grain was collected free of chaff by sieving it through a basket.

Timber was always in short supply in ancient Egypt, a fact which influenced the use of papyrus reed as a

Above: The great castle of Krak des Chevaliers, near Jerusalem, built at the start of the 1200s. This and other crusaders' castles are monuments to constructive effort, perhaps surprising in view of the destructiveness otherwise shown by crusaders.

Below: The Madaba mosaic shows a town plan of Jerusalem. This is the earliest such map to be discovered. It dates from the early AD 500s and was originally part of a larger map of the Holy Land, which covered the floor of a church in Jordan.

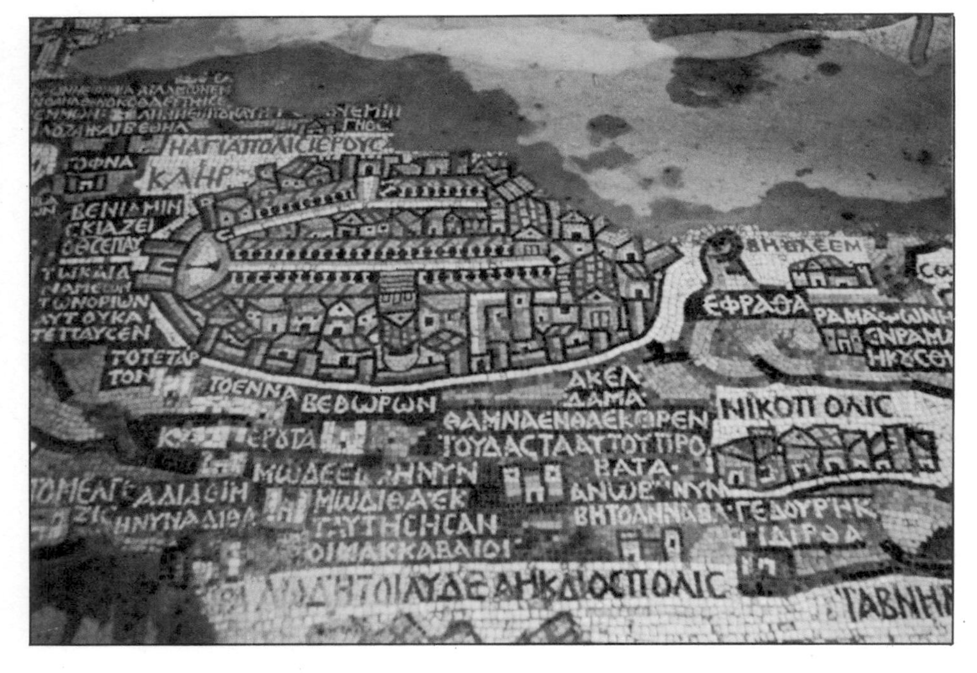

similar to that now used by a glassblowing craftsman. Glass was 'gathered' from a furnace at the end of a blowpipe, blown into a bubble and shaped with metal tools while still hot and pliable. To this main shape could be added smaller pieces of glass as decoration, and a glass foot with or without a stem.

The fermentation of grain to make beer and fruits to make wine were other luxury activities of the ancient civilizations. Winemaking provided the basis for the distillation of alcoholic spirits but it is doubtful whether spirit distillation was practised before medieval times. FERMENTED FOODS, on the other hand, have their very ancient origin among nomadic peoples and reached their richest variety in Eastern countries, particularly Japan.

The making of LEATHER also has an extremely long history. Armies of the first city civilizations wore leather armour and SHOEMAKING with leather is a craft which dates from these early times. Clothes, when woven, were at first made from wool in Mesopotamia and ancient Europe. Cotton arrived later, brought to Mesopotamian cities from India by the Syrians. Clothes woven from cotton or wool would be printed in various patterns with a selection of vegetable DYES *(see page 7)*.

Time and money

In the cities of the Roman and Byzantine empires, as in those further east in Arabia and China, the many crafts and trades reviewed above were not practised in isolation but were linked in a complex web of trade, and sometimes of finance. A MONEY economy had largely replaced the earlier system of barter and this permitted luxury by enabling a small proportion of the population to grow rich.

With this increase in the business of civilization came new concepts of work and the value of time. The measurement of daily time had its origins in the early Egyptian civilization with such devices as the shadow clock and the water clock. Later, but not much better, methods of measuring time involved the use of sundials, sandglasses (or hourglasses) and candle clocks. Accurate timekeeping had to await the development of the mechanical clock by the Chinese, and it was several centuries before Europe had its own mechanical timekeepers.

---

material for boat-building. The Mesopotamians relied on a supply of timber from Syrian forests. Greeks, Latins, Celts and later peoples of Europe did not lack timber and used it in houses and for tools.

Treadmills were used from ancient times up until the 1800s wherever there was a surplus of slave labour. Men walked on steps attached to a moving cylinder which powered machines such as CRANES.

W

Watermills were widely used to grind corn in ancient and medieval Europe. At first they had horizontal millwheels or paddles, the vertical shaft being fixed to the millstone.

Water mill

The Vitruvian mill, developed about 150 BC, had a larger, vertical millwheel, its horizontal shaft being geared down to rotate the millstone faster.

Welding is a method of joining 2 metals whereby enough heat is applied to the point of contact to melt the metals into each other. In forge-welding, the most ancient method, the 2 objects are heated and then hammered together.

Windmills appeared first in Iran, about AD 650. These early, two-storey mills had sails at a lower level than the millstone: in this they resembled even earlier watermills, from which they were probably developed. European windmills arose independently, an early type being the POSTMILL. In medieval Holland windmills were used principally for pumping water.

Wootz steel was an iron-carbon ALLOY made by heating iron with charcoal in small crucibles. Made in southern India, it was coveted by trading Romans for swordmaking, and gave rise in the Near East to patterned, or damascened, steel blades. A similar crucible process was used to make hard steel in Europe but not until the 1600s.

Wrought iron was made by the earliest of ironsmiths by hammering spongy, unmelted iron that had been reduced from its ores by charcoal, a process which continued to be used, with little change, for millennia. Wrought iron is typically tough and malleable but softer than steels containing more carbon.

The brilliance of Greek science dominated European thinking for over 1,000 years. But a new spirit of enquiry, begun in the 1400s, challenged many old ideas. It reached its peak with Galileo, who established the modern scientific method.

The Scientific Revolution

The Greek heritage

True scientific thought began with the ancient Greeks. The story of how Greek science inspired modern science is a fascinating one, but it is also a complicated story because in most cases, Greek thought was handed on only patchily and indirectly throughout the Dark Ages.

By the start of the Christian era, the greatest days of Greek civilization were long over. Greek literature and science continued to flourish at Alexandria, although after the great library there was damaged by invaders in AD 268, its scholars began to drift away. The Arab invasion of AD 642 removed the last remnants of scientific activity from the area that had formed the Eastern Roman Empire.

However, the Arabs themselves began to study the works of such eminent Alexandrians as the astronomer and geographer PTOLEMY *(see page 15)*, the mathematician Diophantus (lived *c.* AD 250) and the physician Galen (lived *c.* AD 170). To this they added scientific knowledge brought back from centres of learning farther east, including the system of numerals from India that we still use today. In the 800s and 900s a group of ARAB ENCYCLOPEDISTS arose that later provided the West with its most ample source of organized knowledge in the fields of mathematics, mechanics, chemistry and medicine.

Meanwhile, in the Christian West, many Greek ideas, including those of Aristotle, had become combined with the ideas of Christianity, mainly through the influence of St Augustine (AD 354–430). The great system of knowledge called ARISTOTELIANISM thus became an important part of Western tradition, an influence that dominated scientific thinking for well over 1,000 years.

The grip of Aristotle

Aristotle (384–322 BC) was the greatest of ancient encyclopedic thinkers and teachers. His system of knowledge embraced the theories of many earlier Greeks, so that we find in it the cosmology of Pythagoras, Anaximenes and Eudoxus (400s–300s BC) with its picture of a geometrically-perfect GEOCENTRIC UNIVERSE. This concept later became a fixed part of the Christian tradition. Also, the ATOMIC THEORY of Democritus, the medical knowledge of Hippocrates (450–377 BC) and, of course, the mathematical discoveries of many earlier and contemporary Greeks, were fused by Aristotle into his total explanation of the world.

Aristotle inherited from the Pythagoreans the idea of a world which could be properly explained only by certain ideas of perfection. These ideas were both geometrical and ethical. For example, the circle was a perfect geometrical

Below: An atlas of the Earth and the heavens by Andreas Cellarius, 1661. This is a late version of the idea of the geocentric, or Earth-centred, universe, which was held until 100 years or more after the death of Copernicus. From that time onwards the Ptolemaic system, as Cellarius refers to it on his atlas, had to give way to the Copernican system of the heliocentric, or Sun-centred, universe *(see page 31)*.

Reference

A **Acceleration, Law of,** was discovered by GALILEO about 1590. He rolled smooth metal balls down a groove in a wooden beam about 6 metres long. By carefully timing their descent, he established a mathematical relation between the distance descended and the time taken for descent. This experiment, together with Galileo's own PENDULUM experiments, did much to promote the use of modern SCIENTIFIC EXPERIMENT.

al-Kindi (died about AD 873) was an Arab philosopher who combined the views of Plato and ARISTOTLE. He wrote over 200 works on scientific subjects.

Arab encyclopedists were the caretakers of Aristotelian science in the later years of the Arab Empire. They also added scientific knowledge obtained from India and even from China. See AVICENNA; al-KINDI.

Aristotelianism is the great system of knowledge and ideas that chiefly influenced scientific development from the time of the ancient Greeks until well after the RENAISSANCE. It was an encyclopedic classification dealing with physics, biology, logic, philosophy and the humanities. Because of Aristotle's characteristically Greek fixed notions of the natural and the ideal, his system was responsible for holding back progress in physical science. However, his ideas on biology, based on accurate observation, were beneficial.

Aristotle (384–322 BC) was a Greek philosopher and disciple of Plato. His theories (ARISTOTELIANISM) were the basis of SCHOLASTIC thinking until after the RENAISSANCE.

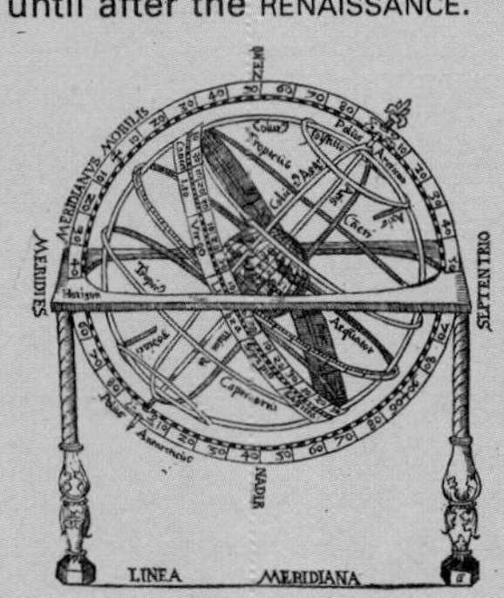

Armillary sphere, 1539

Armillary sphere was an astronomical instrument for demonstrating relative motions of planets. It comprised a number of metal rings, arranged in the shape of a celestial sphere, representing such aspects as the celestial equator, horizon, meridian and ecliptic. Armillary spheres were first made by the Chinese about AD 600.

Atomic theories began with the theory of Democritus (lived *c.* 400 BC), who had the original idea that there were material particles in the universe, the atoms, which partially filled the

figure, and so circular motion was also perfect and good. The notion of circular, therefore perfect, PLANETARY ORBITS dogged astronomy for 1,000 years because other pathways, even if more likely from the evidence of observations, were less perfect and good, and therefore less acceptable.

The assumption that nature behaves according to an ideal led Aristotle into a method of argument or explanation that is the opposite of what we now call scientific thinking. Modern scientists try to avoid influencing the conclusions of their observations or experiments by in any way assuming a favourable outcome. But Aristotle did the reverse in his physics.

He explained the behaviour of objects and beings in nature as the fulfilment of a purpose, that purpose being the achievement of perfection. This sort of explanation led Aristotle to conclusions, particularly those about MOTION, which seem very strange to us.

His general way of thought here is not analytical in the way of modern science, but TELEOLOGICAL in a way we now apply only to moral or ethical problems. But his idea that everything in the universe behaves purposefully was one which suited the early Christian Church very well, with its belief in a universe benevolently ordered and designed by the power of an omnipotent God.

The wrong-headedness of Aristotle's physics did not mean that he lacked the power to observe nature closely and draw scientific conclusions. Nor did he lack the imagination to build his observations into convincing general explanations of natural events. For example, after looking closely at rainfall and the natural evaporation of water, he offered the first scientific explanations in GEOLOGY AND METEOROLOGY.

Another major aspect of Aristotle's thought is also much more in line with modern science: indeed, it is one of its foundations. This is the idea of classification. Aristotle saw clearly that many aspects of nature could be placed into groups on the basis of similarity, and that this was a way of learning more about them.

Aristotle's method of classification is the one still most widely used in biology – even if it is more usually credited to the Swedish botanist Linnaeus (1707–78). A plant or animal is named scientifically according to what it is most like, so that all animals or plants which are very similar are placed together in the same genus (group).

Right: Most famous among Leonardo da Vinci's designs for flying machines is this helicopter. Its rotor is in the form of a large helix or screw which, when turned rapidly, provides lift. Leonardo is said to have been given the idea from a toy brought from China, which, when twirled by the hand, was propelled briefly through the air by the same lightweight screw mechanism. Although Leonardo abandoned many of his brilliant inventions half-complete, he is known to have tested out his flying machines. He had no success, however, because in his time, no man-operated flying machine could have worked. The lightweight yet very strong materials necessary were not then available. Nor had the principles of physics which explain flight been discovered.

This principle has been extended to place plants and animals in larger and larger groups, such as families, orders and classes, until the whole of the plant and animal kingdoms is included in the classification.

Dissenting voices

As a part of Christian belief after the time of St Augustine, Aristotle's ideas of the universe and of all motion taking place in it, had the great force of Church law behind it. But men of independent mind lived then, as now, among whom were some who disagreed with Aristotle's ideas.

Even soon after Aristotle, Aristarchus of Samos (310–230 BC) argued for a HELIOCENTRIC UNIVERSE in which the Sun, not the Earth, lay at the centre. He was largely ignored because this idea conflicted strongly with too many other Greek ideas, and it was not until the time of the RENAISSANCE that the Frenchmen, Jean Buridan (died about AD 1366) and Nicole Oresme (died about 1382) seriously reconsidered the notion of a Sun-centred universe.

Somewhat more fruitful was the disagreement of John Philoponus, who lived in the AD 500s. Though a good, even zealous, churchman, he could not believe Aristotle's explanation of the motion of a body as caused first by the initial push or thrust which starts the motion, and thereafter by the body following its natural or ideal path. Philoponus maintained rather that it

void. Epicurus (300 BC) and Pierre Gassendi (1592–1655) elaborated on this idea, but not scientifically. Isaac NEWTON (1642–1727) used the idea of atoms in his work, but John Dalton (1766–1844) proposed the first atomic theory.

Avicenna (980–1037) was an Arab philosopher and physician who wrote a major medical textbook, *The Canon*. He also studied logic, maths, chemistry and geology.

C

Causality is the principle that nothing can happen, or change, without a cause. Both the science of today and that of the ancient Greeks are causal. Modern science investigates a change to find its cause. Aristotle's science, on the other hand, presupposes an ideal order in the universe, towards which all changes tend.

Cosmology is the study of the universe as a whole, and thus includes both ASTRONOMY *(see page 13)* and the history and philosophy of the universe. Early cosmologies, such as those of the ancient Egyptians and the Christian SCHOLASTICS, were inextricably mixed with religion. Modern cosmology dates from the time of GALILEO and NEWTON and tries to be purely scientific.

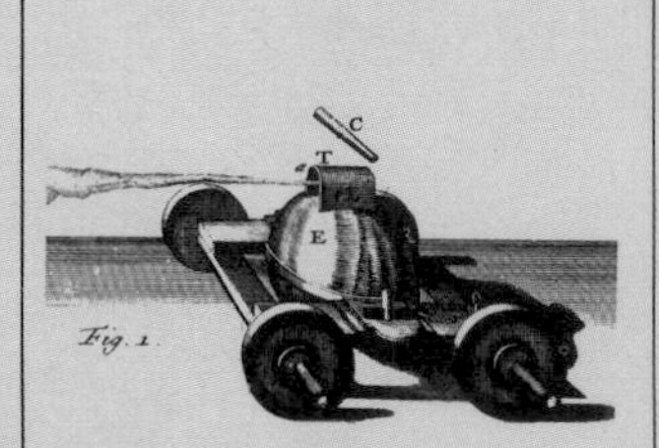

Newton's steam engine

D

Dynamics is the branch of mechanics that deals with the motion of objects under the influence of forces. As a pure science it is the invention of GALILEO. He was the first experimenter to make exact measurements of objects in MOTION, and then to express the nature of this motion accurately by mathematics. Isaac NEWTON followed and enlarged upon his work by discovering the scientific laws of motion.

E

Ether was, in ARISTOTLE'S physics, the substance, or element, that filled all the universe beyond the distance of the Moon. All below this, the Sublunary Sphere, was made up of the other 4 elements, earth, fire, air and water. Much later in physics of the 1800s, ether came to have another meaning. It was the hypothetical substance existing throughout space, which was necessary for the motion, or transmission, of light and other forms of radiant energy.

F

Force is either a push or a pull. The 4 basic forces in nature are: GRAVITA-

Right and **below**: This model of a spring-driven automobile is just one illustration of Leonardo da Vinci's inexhaustible flow of mechanical invention. The vehicle is steered by a tiller and is propelled by the action of two large springs, released alternately.

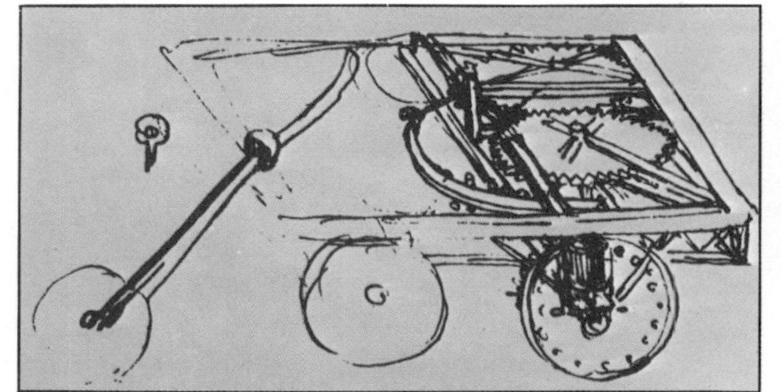

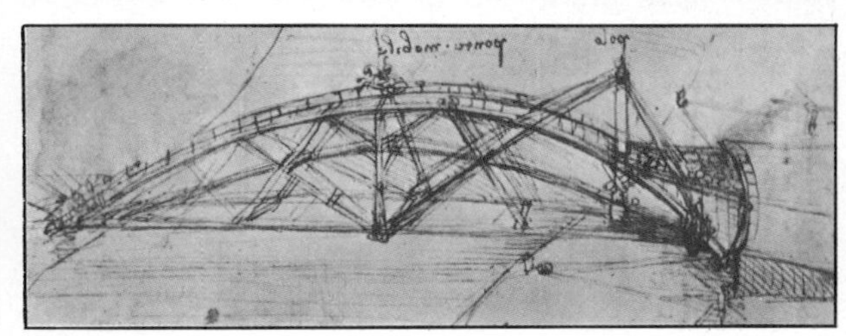

Left and **above**: Among Leonardo's many building designs is this two-level bridge, the upper level being reserved for pedestrians and the lower level for horsemen and vehicles. The same very 'modern' idea of separated traffic flow was incorporated by Leonardo in his plans for a city – which, like so many of his inventions, were never realised.

Right and **below**: Leonardo, as a military engineer in the pay of such generals as Duke Cesare Borgia, designed many weapons. Among the most prophetic was this fighting tank, which, had it ever been built, must have wreaked havoc among opposing armies of horsemen and foot soldiers. Its chief drawback was the lack of an engine, necessary to propel such a heavy vehicle effectively. In the event, tanks were not to appear on the battlefield for another four centuries, during World War 1.

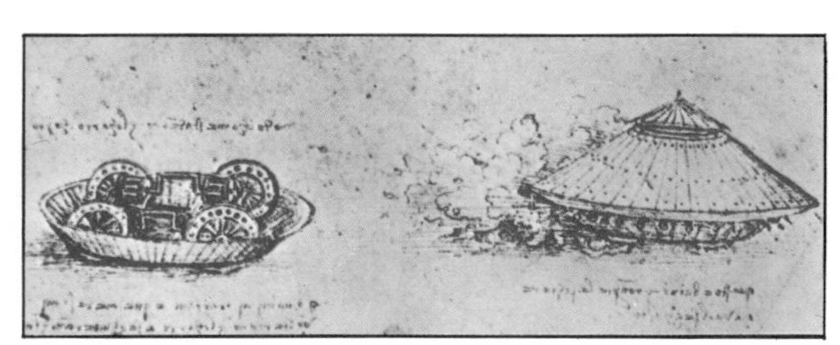

TION, electromagnetism and strong and weak nuclear forces.

G **Galileo Galilei** (1564–1642) was an Italian, and the first modern scientist. His achievements include laws of ACCELERATION and of the PENDULUM and the development of the TELESCOPE. He conceived the laws of MOTION that were later formulated by Isaac NEWTON. His astronomical observations led to a conflict with the Church, because he backed the HELIOCENTRIC view of the universe.

Geocentric universe was the idea of the universe chiefly held from ancient Greek to late medieval times. The Earth was at the centre and 8 outer concentric spheres were the homes, respectively, of the Moon, planets, Sun and stars. A ninth sphere was the *Primum Mobile* which moved all the inner spheres. A tenth sphere was the abode of God, who moved the *Primum Mobile.*

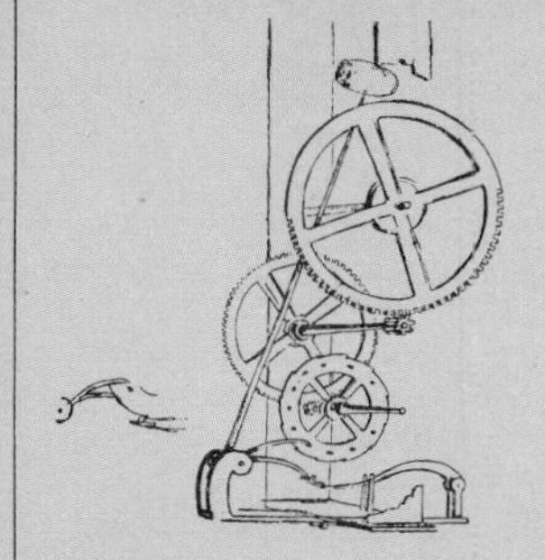

Galileo's pendulum

Geology and meteorology began, as sciences, with ARISTOTLE. In his *Meteorologica* he explained the origins of rivers by describing the natural cycle of evaporation and condensation of water. He also recognized water erosion.

Gravitation is the force of attraction between material bodies. Several major scientists of the 1600s maintained that such a force was necessary to hold planets in their orbits around the Sun. Isaac NEWTON demonstrated this mathematically in his law of universal gravitation published in 1687.

H **Heliocentric universe** was the idea of the universe with the Sun at the centre and the planets, including the Earth, revolving around it. In ancient times this view was held by a few Greeks, notably Aristarchus (lived *c.* 300 BC). It did not become popular until after the time of Copernicus (1473–1543). He, and later GALILEO, helped to destroy the idea of a geocentric, or Earth-centred, universe which had ruled thinking about COSMOLOGY throughout the Middle Ages.

Hypothesis is an idea or proposition put forward to explain facts, but which has not been shown to be true in all cases. In science, hypotheses are the working explanations which help progress but which await final proof

was something in the moving body itself which kept it moving.

This something later came to be called IMPETUS. Philoponus's idea was taken up first by Arab scientists, who eventually passed it back to such European Renaissance thinkers as Jean Buridan.

Mechanical ideas

Below these and other disagreements with Aristotelianism lay a single fundamental idea. This was the idea of the universe as a machine, obeying its own mechanical laws and not the requirements of a distant Greek ideal.

But this Greek ideal had become so entangled with that of Christianity, that medieval scholars, or SCHOLASTICS, were very reluctant to throw it out as false or irrelevant.

However, even if the idea of a mechanical universe could not be accepted, this did not mean that mechanical progress was discouraged. As European states became more stable, their trade increased and new ideas and inventions were accepted readily if they proved useful.

Sea-going Arab traders had advanced the science of navigation by the use of such instruments as the astrolabe. As a parallel development arose cartography or mapmaking, which required a very practical turn of mind but which broadened both man's physical and his mental horizons.

The rediscovery of Greek science included the works of ARCHIMEDES *(see page 12)*, which, together with Arab science, became a subject of study in the new European UNIVERSITIES. Among the scholastic thinkers themselves were men of a practical turn of mind, the best-known being Roger Bacon (1214–94) who thought of practical uses for optical lenses (OPTICS) and who had at least some notion of a SCIENTIFIC EXPERIMENT,

Above: A famous observatory built at Jaipur, India, by Maharajah Jai Singh II in the early 1700s. Stars and other heavenly bodies were sighted from gnomons, the triangular erections shown in the picture. Jai Singh's interests in science were wide-ranging. For example, he was also a capable mathematician. But his astronomy was more than tinged with astrology.

---

or disproof. An historical example is Ptolemy's *Hypotheses of the Planets* which explained planetary motions on the basis of many astronomical observations – and was later proved to be untrue.

I **Impetus** is the force with which a moving body continues in its path against resistance, for example that of friction. The idea of impetus was first employed to explain continued MOTION by John Philoponus (lived *c.* AD 500s). In doing so, he refuted ARISTOTLE'S idea of ideal motion, but Philoponus's arguments were not proved correct until the 1600s. In modern scientific language: impetus = MASS × VELOCITY.

M **Mass** is the quantity of matter forming any material body. Unlike weight, it remains the same for a body whatever the force of gravity acting upon the body. In ARISTOTLE'S system, there is no idea of mass or IMPETUS and it was this defect that led him and later scientists to put forward incorrect ideas about the motion of bodies.

Motion is, in modern physics, expressed in terms of FORCE, MASS, VELOCITY and ACCELERATION. But in ARISTOTLE'S physics anything that moved did so according to its proper place in the natural order of things. For example, a dropped stone fell because it belonged to Earth, and flames rose because they belonged to the zone of fire above the Earth.

Aristotle, drawn 1531

N **Natural philosophy** is a name for the sciences, particularly for physics, used as late as the mid-1800s. It reflects the influence of ARISTOTLE, who believed that things behave according to a natural and ideal order.

Newton, Sir Isaac (1642–1727), was an English mathematician and natural philosopher who discovered the laws of MOTION and proposed the principle of GRAVITATION. He separated white light into the colours of the spectrum and put forward the particle theory of light. He also made a long study of alchemy.

O **Optics** is the branch of physics that deals with the properties of light and the nature of vision. In ARISTOTLE'S optics, light behaved in an ideal way, according to geometry. This idea was

although his thoughts on this subject were still affected by mystical ideas.

Among the most remarkable of practical inventions of this time were those of a French architect, Villard de Honnecourt (lived about AD 1250), which include a screw-jack for raising heavy weights, a saw driven by water power, and, less convincingly, a PERPETUAL MOTION MACHINE. Indeed, architects and builders were among the most progressive engineers of the Middle Ages since they had the largest-scale problems to solve. In particular, they had the problems involved in the design and construction of the great Gothic cathedrals.

At a rather later date, Leonardo da Vinci (1452–1519) displayed astounding virtuosity as an artist, scientist and engineer. Among the more startlingly modern of his mechanical inventions were flying machines, a military tank and a diving bell. None of these could have been made to work, chiefly because Leonardo's inventiveness was truly ahead of his time. His inventions lacked the mathematical foundations of all truly scientific method. This was to follow, with the work of GALILEO, 50 years after Leonardo's death. Leonardo da Vinci heralded the SCIENTIFIC REVOLUTION but did not participate in it.

New ideas of the universe

Nicholas Copernicus (1473–1543) was the Polish astronomer who first put ideas about the universe on a new footing. He believed in a Sun-centred universe and placed within it the planets, at least those that were then known, in the order of distance from the Sun. He also demonstrated that the Earth was spherical and not flat as was then widely believed.

But Copernicus was still enough of an Aristotelian to believe in orbits made up of circles, and in a number of other ways he clung to the ideas of earlier astronomers such as Ptolemy. Copernicus lived before the invention of the TELESCOPE, which revolutionized astronomical observations. Yet it was another, rather later, astronomer, also lacking a telescope, who made the accurate observations necessary for the new HELIOCENTRIC theory.

This Danish astronomer was Tycho Brahe (1546–1601), who, working with such instruments as the quadrant and sextant, built up a catalogue of positions of stars and planets which was much the most accurate at that time.

Tycho Brahe disbelieved Copernicus's theory of a Sun-centred universe but his assistant Johannes Kepler (1571–1630) was much more

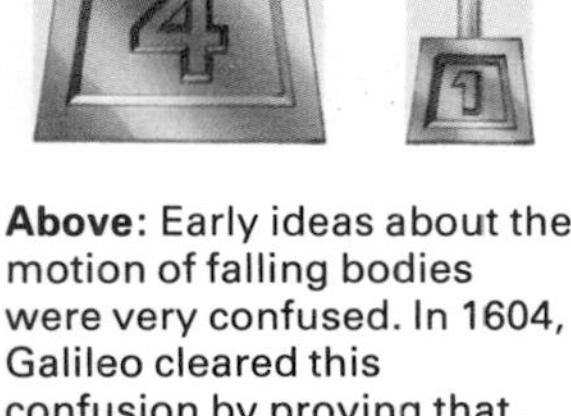

Above: Early ideas about the motion of falling bodies were very confused. In 1604, Galileo cleared this confusion by proving that, allowing for air resistance, all bodies fall at the same rate. Eighty years later, Isaac Newton proved that this falling motion was caused by gravity.

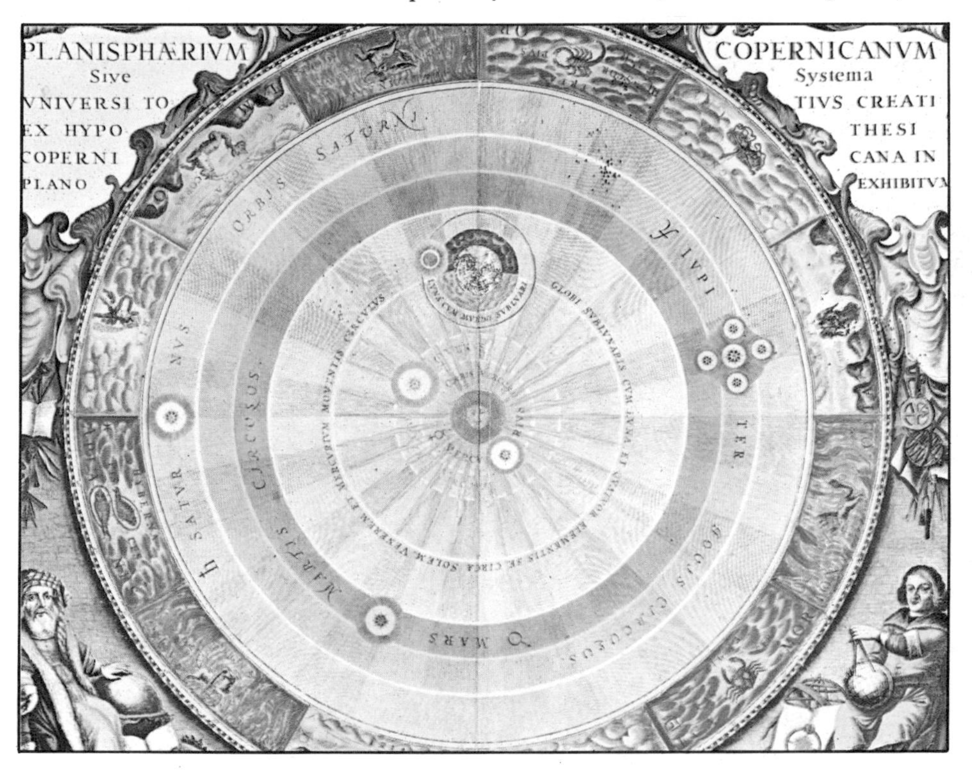

Left: The Copernican Sun-centred universe, as illustrated by Andreas Cellarius, 1661. Copernicus, in his *Concerning the Revolution of the Celestial Spheres,* 1543, proposed for the first time a detailed picture of the solar system, with the known planets and their moons correctly placed. But notice that the stars – represented by the signs of the zodiac – inhabit the outermost solar orbit: Copernicus lived before the invention of the telescope, and so could not have observed the remote independence of the stars from the solar system.

helpful, in that it presupposed that light travelled in straight lines. However, it was unhelpful in that it supposed that the study of geometry *determined* the study of optics.

P **Pendulum, law of,** was discovered by GALILEO in 1582 when he observed that the time taken for the swing of a pendulum is independent of the weight of the pendulum bob. This is another way of saying that lightweight objects fall at the same rate as heavy ones.

Perpetual motion machines are machines that, having started work, go on doing so forever without new supplies of energy. They were constructed by many engineers of the Middle Ages but never worked because perpetual motion contradicts a fundamental scientific law.

Perpetual motion machine

Planetary orbits were assumed to be made up of circles, a geometrically 'perfect' shape, until Johannes Kepler proved by calculation and observation that they were ellipses. However, these 'impure' orbits matched the idea, then growing, of a force of attraction between the Sun and planets and between the planets themselves. This force was finally formulated by NEWTON.

Plenum is space entirely filled by matter. The philosopher René Descartes (1596–1650) explained all motion in the universe by supposing this to be a plenum of material particles. Planets and other bodies moved in paths according to swirls, or vortices, in this plenum.

R **Renaissance, The,** was the great revival of learning and surge of artistic creativity which began in Italy in the 1300s and spread to other European countries during the following 200 years. Its scientific foundations were mainly Greek, both as a direct European heritage and also indirectly through the transmission of Greek science by later Arab scholars.

S **Scholastics** were those academic thinkers and teachers of the 900s to 1400s in Europe who followed beliefs of Christianity mingled with those of ARISTOTLE. Their dogmatism in science held back progress, as when their opposition forced GALILEO to refrain from publishing his discoveries.

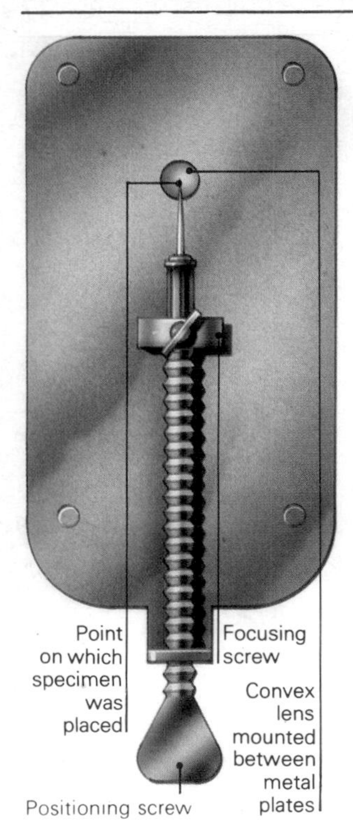

Above: The microscopes of Anton van Leeuwenhoek (1632–1723) were the best of their day. They had only a single lens, but this was very finely made.

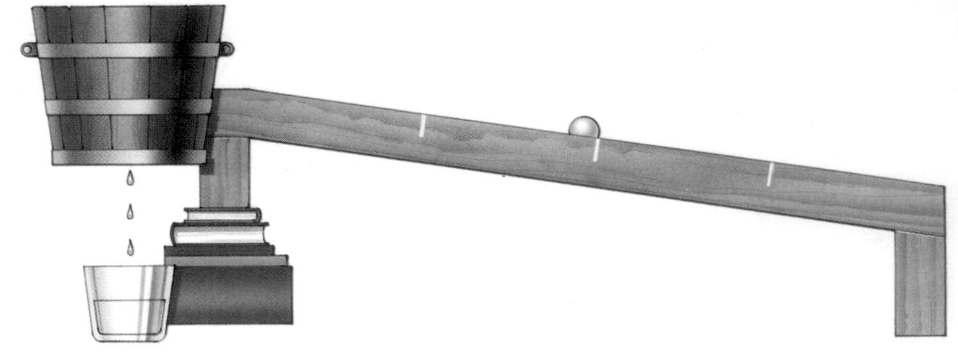

Right: In one of his most important experiments, Galileo rolled a ball down an inclined plane, timing its descent carefully with a water clock. By repeating the trial over various distances, he proved that the distance travelled was always proportional to the square of the time. This mathematical relation was a vital first step towards the scientific laws of motion, later formulated by Newton.

sympathetic to the new idea. Whereas Tycho was a great observer, Kepler was an excellent mathematician. Using Tycho's observations he checked in detail the Copernican picture, adding another vital element to it.

Much against his own belief in perfectly circular orbits, Kepler was forced to the same conclusion that Nicholas of Cusa (a German cardinal), 100 years earlier, had more intuitively supposed. This was that planetary orbits were *not* circular. In fact, Kepler proved that they were elliptical, a conclusion that was to have profound effects on future COSMOLOGY.

The first modern scientist

Without doubt, this title belongs to a slightly older contemporary of Kepler, the Italian GALILEO GALILEI (1564–1642). He earned the title by an independence of mind which was particularly admirable in his age of religious dogmatism.

Galileo's mind was both practical and mathematical. As an astronomer, he took the newly-invented telescope, improved it, and made astronomical observations which clearly contradicted the Aristotelian picture of the universe. For publishing his results he was condemned by the Inquisition (a court set up by the Roman Catholic Church to seek out heretics), made to recant, and more or less confined for the rest of his days.

Earlier in his career Galileo had made experiments in DYNAMICS even more fundamental to the future development of science. By these experiments he established laws of ACCELERATION and the PENDULUM. These are scientific because they depend only on the physical conditions of an experiment, in this case the MASS of a body and the time during which it moves. Previously, even such an acute mathematical brain as Kepler's had been diverted in its search for scientific truth by the powerful influences of religious and classical ideas. Later in the 1600s, an even greater mathematician, Isaac NEWTON (1642–1727) was to need Kepler's proof of elliptical orbits for his theory of universal GRAVITATION, and was to use Galileo's work in dynamics as the basis of his discovery of the scientific laws of MOTION.

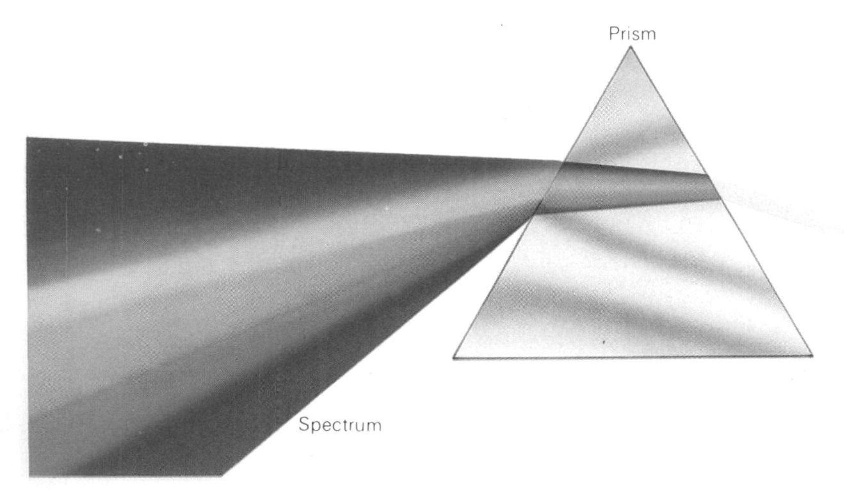

Right: Among Isaac Newton's great discoveries was that of the optical spectrum. White light, beamed through a slit into a prism, separates out into its member colours.

Scientific experiment is one in which the experimenter tries to get information about the world without in any way influencing the results of the experiment. This detachment permits such an experiment to be repeated under identical conditions, so that its results can be checked reliably.

Scientific Revolution began in the early 1400s, both with the rediscovery of Greek science and also with the questioning of Greek ideas which had come to be held dogmatically. The scientific revolution reached its peak with GALILEO (1564–1642) who established SCIENTIFIC EXPERIMENT as independent of other influences.

T

Teleology is the study of causes and purposes. In teleological thinking, such as that of ARISTOTLE in his physics, an object or being behaves in the way it does because it is its purpose to behave in this way. The aim of teleological science is thus to explain observations in terms of their natural purposes, rather than to investigate them.

Telescopes were invented in Holland about 1608, probably by Hans Lippershey. He is also credited with the invention of the compound microscope. GALILEO quickly improved on Lippershey's telescope and used his instrument to make the discoveries which finally destroyed belief in the medieval idea of the universe.

U

Universities began in medieval Europe as cathedral schools in which theology was chiefly studied. As more subjects for study were admitted, the schools enlarged until, by the late 1100s, they had become the universities of Bologna, Paris and Oxford.

V

Velocity of an object is specified by its speed and the direction in which it is moving.

A university doctorate being conferred, 1500s

Despite its brilliance, ancient Greek science contained various superstitious elements, which delayed the development of modern Western medicine until the 1500s and modern chemistry until the 1600s.

Healers and Alchemists

Magic and method

Like the medicine men of primitive tribes, physicians of the oldest civilizations depended largely on magical treatments for relieving pain and curing disease. Later, Greek physicians lived in a world less dominated by superstition and so could practise a medicine in some ways more like our own, using a greater number of specific treatments for particular illnesses.

Like any doctor today, a physician of ancient Greece, when examining a patient, would first attempt not to drive out devils but to make a MEDICAL DIAGNOSIS. Having identified the causes of disease, he would go on to treat it.

Similarly, a Greek scientist or philosopher, observing some change or effect in nature of the kind we would now describe as physical or chemical, would no longer put it down to the work of the gods. He would ask the question: 'What causes it?'

As physicians, naturalists and experimenters the Greeks used their powers of observation and commonsense to the full and so their medicine, biology and experimental science are in many ways the basis of these sciences today.

At the same time, Greek medicine and science remained consistent with Greek philosophy. In doing so they came into contact with ideas which strike us now as very unscientific but which were then so powerful that they held up real progress for a very long time.

Doctors, elements and humours

Hippocrates of Cos (lived *c.* 400 BC) was the most famous of the early Greek physicians and the one whose influence is most traceable in modern

Below: Kohl, a dark eye make-up powder, was an important cosmetic in Ancient Egypt. Various pots and tubes were used to store it, made from pottery, wood, stone, glass, alabaster and metal. The tubes could be double, triple, or even quintuple, as well as the single one shown. Other Egyptian cosmetic items include a stone palette, half a metre long, showing the victorious King Narmer overcoming his decapitated enemies.

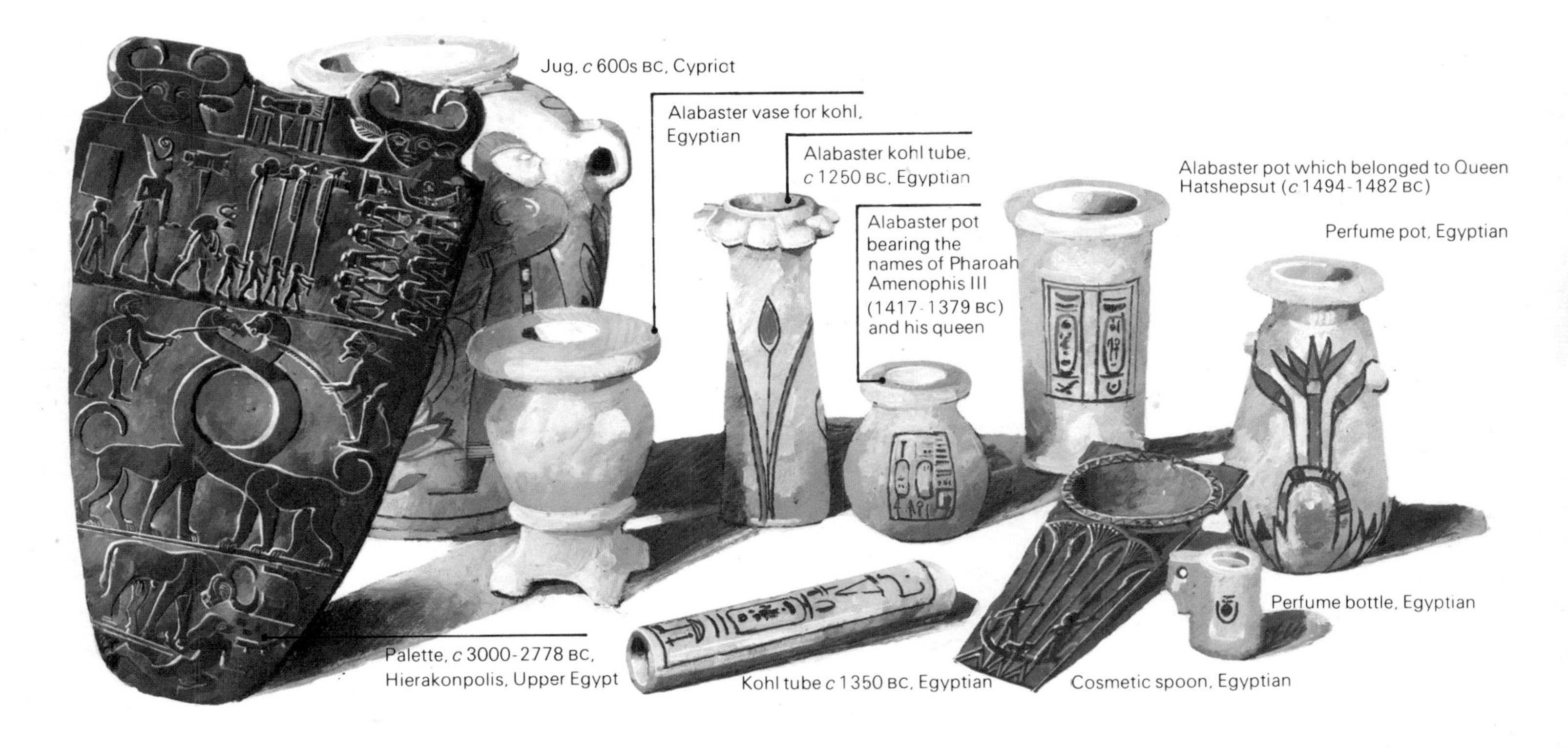

Reference

A

Alchemy was the chemistry of the Middle Ages. Alchemists were especially interested in the TRANSMUTATION of baser metals into gold.

Alembic was an early type of distillation apparatus used by alchemists. It comprised a bottle, usually of glass, in which substances were heated. This was topped by a tube, often of metal, which pointed downwards. Vapours condensed by DISTILLATION or SUBLIMATION in the metal tube and were collected in a flask.

Alkali or base is a chemical compound that is highly soluble in water and neutralizes acids.

Anaesthetics are substances that cause loss of sensation (local anaesthetics) or loss of consciousness (general anaesthetics). They were unknown in ancient times. Opium mixtures were prescribed for the relief of pain as early as 1150. Sometimes strong alcoholic drinks were given to patients awaiting surgical operations. The first true general anaesthetic was nitrous oxide or laughing gas, discovered by Sir Humphry DAVY *(see page 42)* in 1799.

Anatomy is the study of the structures of animals and plants. Human anatomy has been studied from the time of the ancient Egyptians, and Greek anatomists described, although not often accurately, the functions of organs such as the heart, brain, nerves and blood vessels. Modern anatomy began in the 1500s with the work of people like Andreas VESALIUS and William HARVEY.

Anatomy, drawn 1543

Antiseptics are substances that kill or inhibit the growth of harmful microbes. They can thus be used to ward off infection from rooms and buildings, and to sterilize infected areas. They were unknown before the 1850s because microbes had not yet been recognized as a cause of disease. Carbolic acid was the first antiseptic in general use.

C

Calx was a term used by early chemists for the ash or powder formed when a metallic substance was strongly heated in air. The

Right: This medical print of the early 1200s reveals the extent of medical ignorance in the medieval West. Without dissection, there could be no accurate knowledge of anatomy and physiology. Physicians were unable to diagnose most internal complaints and often could only pinpoint the sources of pain – which any patient could do well enough for himself.

Below: China was the home of the most advanced medical and surgical practice of ancient times. Medical treatment was largely based on herbal remedies. Acupuncture, another effective medical technique, was, and still is, more mysterious in its effects. The picture shows the heart meridian of the Chinese acupuncture system.

Below right: A drawing made in 1508 shows how elements, humours and their related zodiac signs were thought to influence various parts of the body. Such ideas derived from Galen (AD 129–199), most influential of Alexandrian physicians. This connecting of physiological with cosmological events held back medical science for 1,500 years.

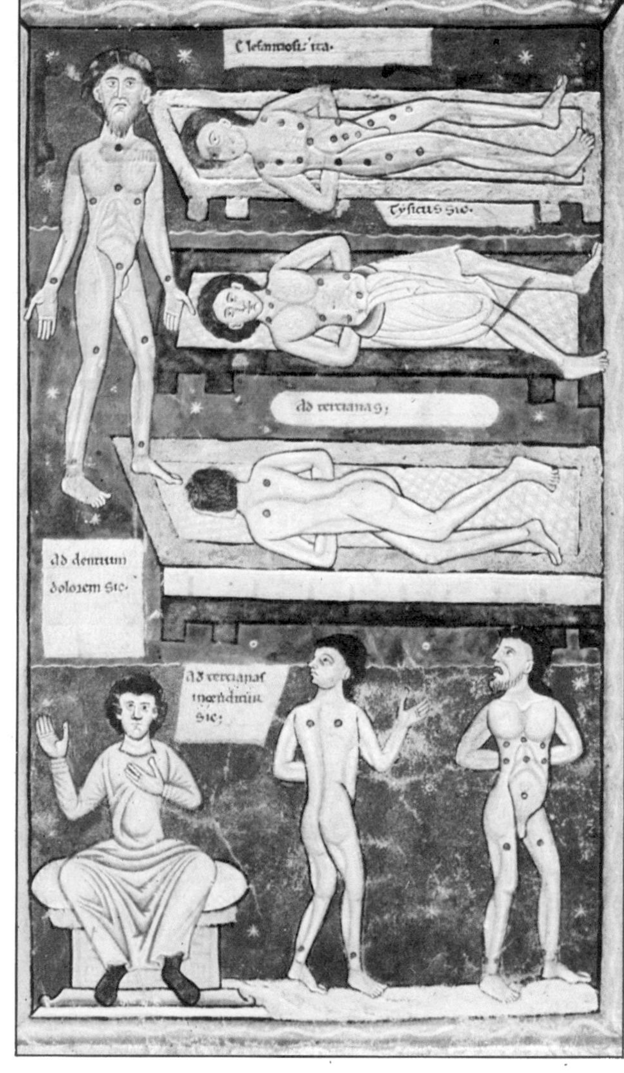

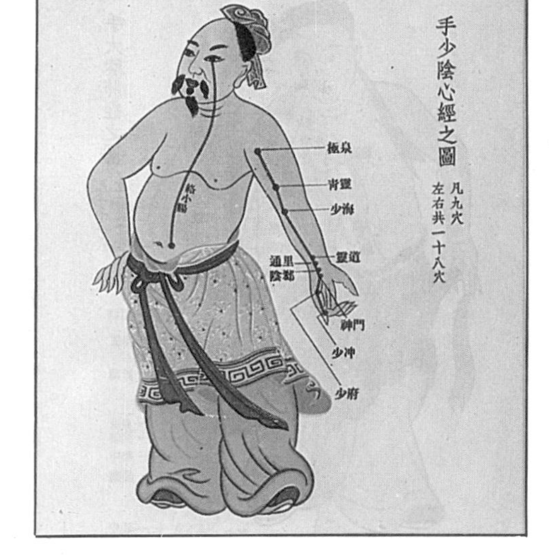

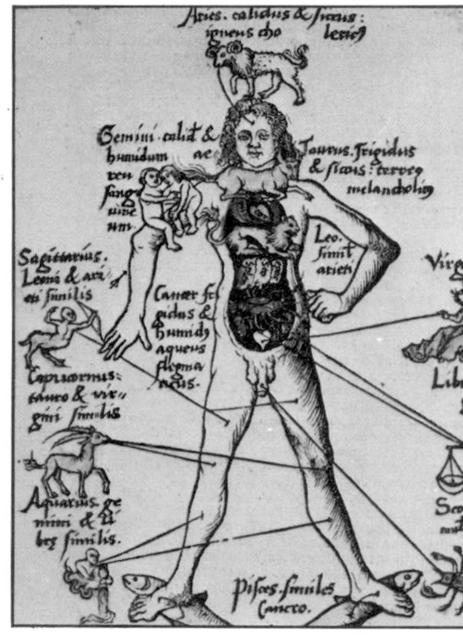

medicine. He and his school declared as their aim the cure of disease using the evidence of experience, that is, of repeated observations of patients. They rejected philosophical and religious ideas as unhelpful to medicine. The Hippocratic school, then, stood for the practical, empirical tradition in medicine. Yet before long, the powerful Greek urge towards fundamental explanations brought philosophy into Hippocratic medicine.

To explain the roots of sickness, health, and even personality, four HUMOURS, or fluids, were supposed to act in various organs of the body. In turn, each of these humours was associated in its activity with one of the four ELEMENTS. These, by their mixture and movements, made up all the substances and processes of nature.

This fusion of ideas satisfied the Greek desire for a harmony between man, nature and the cosmos. But it could only hinder the scientific progress of medicine, PHYSIOLOGY and chemistry. Indeed, the great authority of ARISTOTLE *(see page 27)* and of Galen, last of the great Greek physicians, perpetuated the doctrine of the four elements and the four humours into the 1600s.

Galen's study of ANATOMY was the most thorough of ancient times. His practice of MEDICAL DISSECTION led him to an equally thorough, but almost entirely mistaken, interpretation of how the body works. He described this not only in terms of the four humours but also of the three SPIRITS which he supposed to reside in various organs.

As only one example of mistaken treatments arising from these ideas, the practice of blood-letting continued almost until modern times as a cure for a fever. A raised temperature was related to the element, fire. This element corresponded to the humour, blood. Therefore, to remove blood from a patient was to cure his fever. Of course, a doctor of the 1700s or 1800s, applying leeches to his patients to draw blood, may or may not have known these original arguments for doing so!

Medicine in the East

After the capture of Alexandria by the Arabs, organized schools of medicine and other sciences virtually ceased to exist in Europe. However, the Arabs themselves, the inheritors of Greek science, attached great importance to medical

---

metal was supposed to have lost phlogiston *(see* PHLOGISTON THEORY*)*. We now know most calxes as metallic oxides, combinations of metals with oxygen from the air.

D **Distillation** is the process by which liquids are heated until they vaporize, the vapours being recovered by the process of condensation. Alchemists and early chemists, using ALEMBICS or stills, distilled many substances in order to purify them. Distillation is still widely employed as a process in the chemical industries.

A distillery, 1514

E **Elements** were supposed by the ancient Greeks to make up all the substances of the world. The 4 elements were, respectively, earth, water, air and fire, one above the other. Their order explained their motion, as they sought to regain their proper places. Thus, flames leaped upwards and stones (earth) fell downwards. The elements also represented qualities such as dryness, wetness, lightness and heat.

Elixirs were magical substances. Alchemists sought them for changing base metals into gold. The PHILOSOPHERS' STONE is an example. Healers sought the elixir of life to prolong life and cure disease.

Epidemics are large-scale outbreaks of infectious disease. Those caused by harmful bacteria were common until the 1900s when modern drugs, and better SANITATION and nutrition, helped to control or eliminate them. When a disease is at large in a population, waiting for a chance to break out into an epidemic, the disease is said to be endemic.

Essence is a word now used to mean a flavouring or aromatic compound which can be distilled or extracted, usually from vegetable substances. Among the ancient Greeks and the alchemists, the essence of a substance was its living principle or identity which persisted unchanged through chemical or other reactions.

F **Folk medicine** uses herbs and other natural substances for the treatment of disease. It was the only kind of medicine available in the West from the fall of the

knowledge, many of their major scientific figures, such as AVICENNA *(see page 28)* being physicians.

Like Greek medicine, that of Islam was confused with ideas from astrology and other unscientific sources. On the other hand, the Arabs were almost alone in recognizing the importance to health of SANITATION AND HYGIENE. They also discovered a number of useful drugs, and their anatomical studies of the eye later influenced both medicine and optical physics.

Chinese medicine at this time was also highly organized, including even a sort of medical welfare service. Like the FOLK MEDICINE of the West, it depended heavily on herbal remedies but it had investigated a far greater number of these and applied them in a much more detailed way. A Chinese medical treatise of AD 610, for example, lists more than 1,700 diseases, classified into 67 groups. Chinese PHARMACOPOEIAS, at a later date, detail many thousands of herbal treatments.

In surgery, too, the Chinese were well ahead of the West. As early as the AD 700s they were performing delicate SURGICAL OPERATIONS such as the removal of a cataract from the eye.

Rebirth of Western medicine

Western universities were founded as early as the 1100s, but for several centuries more, their scholars rigidly followed Greek ideas. Medical dissections were regarded less as scientific investigations but rather as demonstrations of the truth of Galen's work on anatomy and physiology.

Only in the 1500s, with the anatomical studies of Andreas VESALIUS (1514–64) and his followers at the University of Padua, did modern medicine begin. By careful dissections of animal and human bodies, this school of anatomists revealed many of the true structures and functions of the eye, internal ear, heart, lungs and reproductive organs.

Most famous of the students of the Padua school was an Englishman, William HARVEY (1578–1657). Galen had taught that blood ebbs and flows in the veins, under the influence of spirits, but Harvey discovered that the heart, a pump, recirculates blood through the arteries and veins of the body. By so effectively replacing a semi-mystical idea with a mechanical one, Harvey had helped to destroy the lingering influence of Galen and to promote the development of scientific medicine.

Biology and medicine

From ancient times, biological knowledge helped the progress of medical science. Aristotle's descriptions and classification of animals and plants both encouraged the study of animal anatomy and led to a tradition of useful botany,

Below left: Part of a page from Gerard's Herball of 1597, showing an illustration of the 'Prickly Indian fig tree'. The accompanying text describes its botanical details and medicinal value. Herbals, as books of natural medicines, were famous long before this time. Most accomplished of herbalists were the ancient Chinese. Their studies began in pre-Christian times and culminated in the 1500s with the Great Herbal of Li Shih Chen, which contained thousands of herbal remedies and treatments.

Below: An Arab chemist of the 1100s prepares perfumes. From early times times, both animal secretions, such as musk, and aromatic vegetable oils were used in perfumery. The long-established method of extracting plant oils is by steam distillation of petals or other parts.

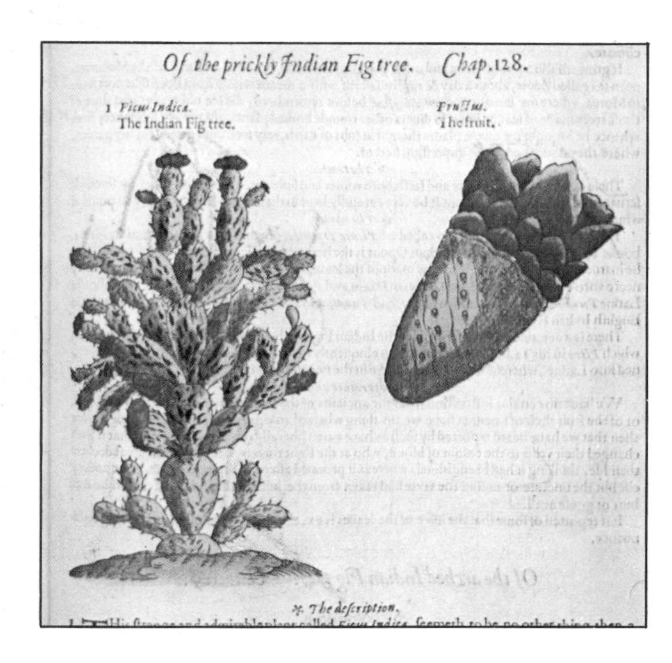

Roman Empire (AD 400s) to the rise of university medical schools in the 1200s.

G **Gunpowder** was invented by the Chinese at some time before AD 1000. It was introduced to Europe, probably by the Mongols, by about 1250. Its constituents were readily available: nitre from mines, charcoal from wood-burning, and sulphur from volcanic deposits such as those of Sicily.

H **Harvey,** William (1578–1657) was the English physician who discovered that blood circulates around the body.

Herbals were European books describing and illustrating plant life. They gave information about both botany and medicine. Dating from Roman times, they were later often written and drawn by Christian monks, who kept herb gardens and extracted medicines from their plants.

Gunpowder in fireworks, 1628

Hospitals were first built by the Romans for housing their sick and wounded soldiers. Formerly, patients were treated either in doctors' homes or in Greek temples, particularly those of Aesculapius. The first Christian hospital was that of the Emperor Constantine, about AD 300.

Humours were supposed by Greek physicians and their successors to determine both health and personality. The 4 types were, respectively, blood, yellow bile, phlegm and black bile. The humours were linked, in this order, with the 4 ELEMENTS: air, fire, water and earth. Medical treatment was based on restoring any upset balance of a person's humours, and thus also of nature's elements.

M **Medical diagnosis** is the recognition of symptoms of an illness and the identification of their causes. It dates from the time of Hippocrates (lived *c.* 400 BC), although ancient Greek doctors often assigned causes of illness to HUMOURS and SPIRITS.

Medical dissection was first practised by the ancient Greeks, to obtain knowledge of anatomy and physiology. This tradition lasted from Herophilus and Erasistratus (200s BC) until Galen (AD129–200). From about AD 1500, dissection of human and animal corpses was a regular part of the training of European doctors.

as recorded in HERBALS as early as the Dark Ages. In China, botany and herbal medicine were studied on a still larger scale over a period of 2,000 years.

With the invention of the microscope in about AD 1608 a new world of biology, hitherto invisible, was revealed. Among the living organisms that could now be seen were bacteria. Some of these were the agents of the EPIDEMIC diseases that periodically ravaged Europe and the rest of the world. For centuries afterwards, however, bacteria and other microbes were dismissed as interesting but lowly and unimportant forms of life. Only in the mid-1800s was their true nature discovered, and from that time dates the development of drugs for the control of bacterial disease.

Alchemy

The history of chemistry, like that of medicine, started with philosophical and magical ideas. Alchemists were early chemists who saw chemical and physical changes in substances as movements of a life force. For example, when they heated a substance strongly until it became an ash, they supposed that they had killed or driven off the life force of the substance, and we still speak of a 'dead' ash. Alchemists explained many other chemical reactions as the transference of life force from one substance to another. This belief led them repeatedly to attempt the TRANSMUTATION of base and noble metals and to search fruitlessly over the centuries for the PHILOSOPHERS' STONE.

Above left: Greek Fire being used as a weapon in a Byzantine naval battle. This highly flammable liquid was in use long before gunpowder. It originated with the various chemical mixtures, containing such substances as pitch, sulphur and tow, employed in war by the Assyrians and Greeks. Naphtha, a petroleum product, was later added, thus making a true liquid fire.

Above right: An alchemist in his laboratory in the year 1508. Although his interests and intentions were largely unscientific, he used equipment and methods which were important to the development of science and technology. For example, shown in the picture are an alembic or still for distilling liquids, and a furnace with bellows for melting, alloying and extracting metals.

Alchemy began in Europe with such figures as Zosimus, who lived in the AD 200s and drew his knowledge from the ALEXANDRIAN SCHOOL *(see page 12)*. In China, alchemy was practised on a scale equal to or larger than that of Europe, but its mystical ideas were those of Yin and Yang, the female (passive) and male (active) forces of the universe.

The Arabs, while accepting the magical ideas of Greek alchemy, were more interested in its practical methods, which they sought to improve. In the 800s and 900s, such Islamic scientists as Jabir, Rhazes and Avicenna prepared many substances in a purer state by processes such as DISTILLATION and SUBLIMATION. They also recognized particular kinds of chemical reactivity such as those of acids and ALKALIS (like alchemy, an Arabic word). The ability to carry out specific chemical reactions, and so to prepare chemicals in a pure state, earned the Arabs the title of fathers of the chemical industry.

Chemical doctors

The manufacture of chemicals such as nitre, for making GUNPOWDER, and soda, for treating cloth, was carried out on a large scale in medieval Europe. Yet as late as the 1800s, the teaching of chemistry remained separate from chemical industry.

Among the Arabs, chemical experimenters were also usually medical doctors, and this tradition persisted in Europe. A school of iatrochemists, or chemical doctors, arose to seek

---

P **Paracelsus** (1493–1541) was a Swiss alchemist and physician who opposed the theory of HUMOURS. He taught that diseases were specific and could be cured by specific remedies. His interest in ALCHEMY and minerals led him to introduce mineral baths and minerals as sources of medical treatment.

Paracelsus, drawn 1570

Pharmacopoeias are books listing medicines and their methods of composition. They were well used in ancient Chinese medicine, which employed a very large number of drugs derived from plants, and specified many thousands of medical treatments.

Philosophers' stone was a magical substance sought by alchemists. They believed that it could be used to change base metals such as lead into gold.

Phlogiston theory was invented by the chemists Johann Becher and Georg Stahl in the late 1600s to explain how substances burn. The theory stated that all substances which could be burned did so because they contained a 'fire principle,' or phlogiston. Thus, the ash from the combustion of wood or metal was formed by the wood or metal losing its phlogiston. This theory was generally accepted until 1774, when Antoine LAVOISIER *(see page 43)* showed that the process of combustion was a combination with oxygen, not a loss of phlogiston.

Physiology is the branch of biology dealing with the functions of the bodies of animals and plants and their various parts or organs. Human physiology began with the ancient Greeks but it remained largely unscientific. Its confusion with magical and philosophical ideas remained until the 1600s and the work of physicians such as William HARVEY.

S **Sanitation and hygiene** are modern developments. They have been encouraged by our knowledge of the ways in which harmful infections are transmitted. Before the 1800s few public health authorities existed and people were left to dispose of their own refuse and sewage. This encouraged the spread of disease.

Spirits were supposed by Galen, most influential of early Greek physicians, to provide vitality for the body's functions. The 3 types were made in the body both from food that was

Left: A chemical laboratory of the 1600s, when the beginnings of scientific chemistry were still a century ahead. But chemical industry flourished, as it had done since medieval times. In the foreground volumes and weights of chemical substances are being measured. At the back of the laboratory is a row of stills.

Below: Water-cooled stills of the type shown were being used in laboratories by the early 1500s. They condensed vapours at a much faster rate than the older, air-cooled alembics. Also by this time, the vapours of water and alcohol could be effectively separated from one another by the use of long fractionating columns.

new medical drugs with the aid of alchemy and chemical techniques inherited from the Arabs. The iatrochemists did, indeed, produce a few new useful drugs such as tincture of opium, but their activities were of most help to the development of experimental chemistry. In time, this new chemical knowledge was returned to medicine in the form of the revolutionary discoveries, ANTISEPTICS and ANAESTHETICS.

Modern chemistry only really began with the work of Robert Boyle (1627–91) who discovered the laws by which gases behave. He was also the first to recognize an element in the form we know it now, as the simplest kind of substance that takes part in chemical reactions. In the 200-year period after his death, chemistry rapidly developed into a true science, and the foundations of the modern chemical industries were laid.

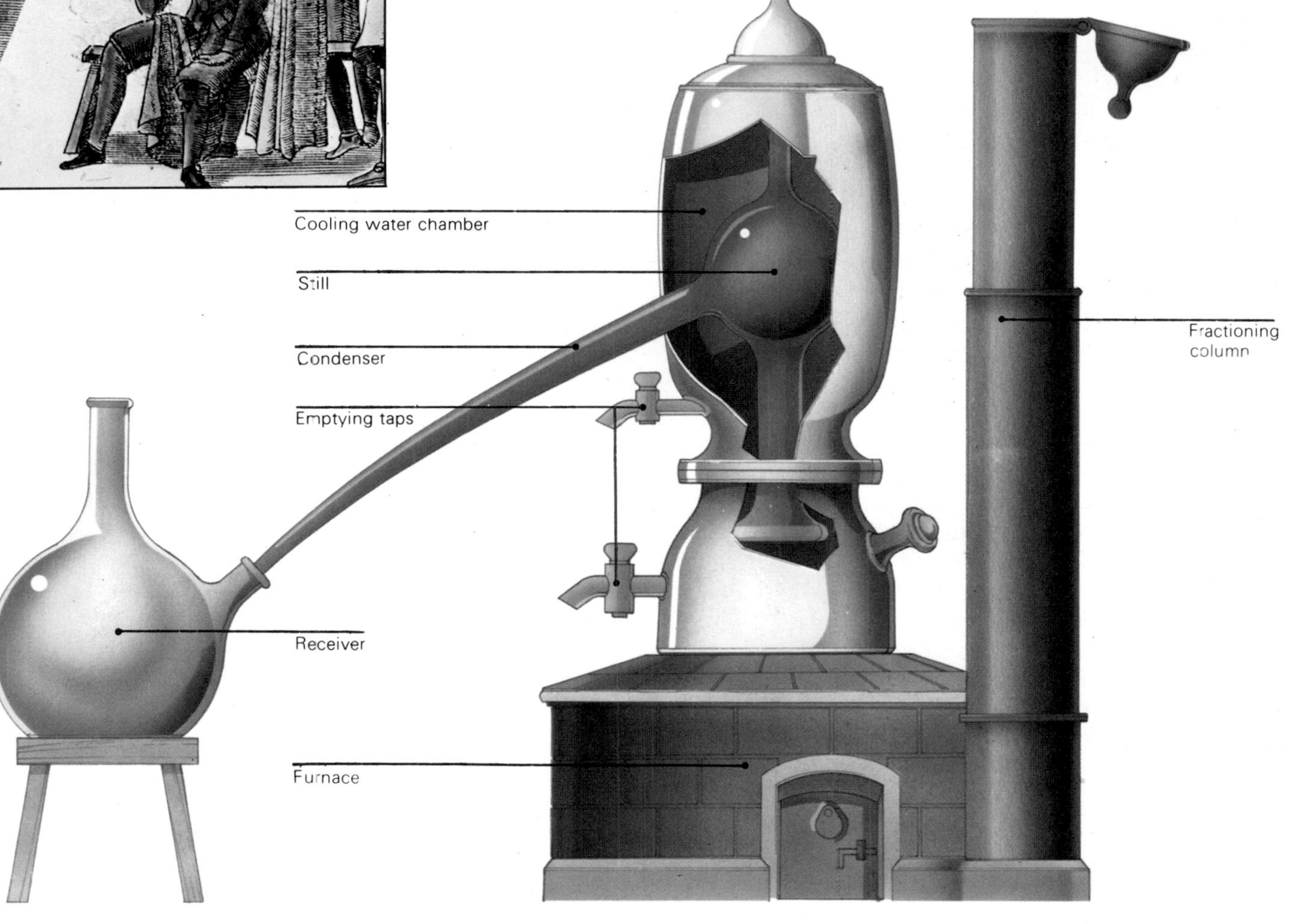

eaten and from air breathed. The vital spirit resided in the heart and lungs, and corresponded to what we now call spirit or soul. The natural spirit, in the liver, was responsible for growth while the animal spirit in the brain and nerves controlled bodily action.

Spontaneous generation was the idea, popular from biblical times, that lowly forms of life, such as worms, could arise from non-living matter. It was finally disproved by the French chemist Louis Pasteur in 1861, when he showed that even the simplest forms of life, such as bacteria, cannot arise spontaneously.

Sublimation is a change of physical state shown by some substances. When these are heated they become gases directly, without passing through a liquid state. Conversely, when their vapours are cooled these become solids directly.

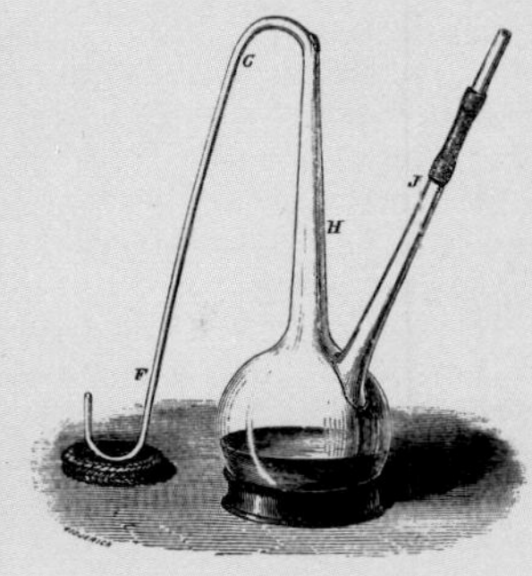

Pasteur's experiment

Surgical operations began in prehistoric times with trepanning (the cutting of a hole in the skull), probably with the intention of releasing evil spirits. The Greeks and Romans invented a number of surgical instruments, and the surgeons of ancient China used an even wider range of instruments to perform quite complicated operations. Modern operations began with the invention of ANAESTHETICS and ANTISEPTICS in the 1800s.

T

Transmutation is the changing of one thing into another. Alchemists long sought, but in vain, to transmute lead into gold. However, what they failed to do by magic, modern science can now accomplish with the aid of nuclear reactors. Any chemical element can now be changed into another in this way.

V

Vesalius, Andreas (1514–64) was born in Brussels and later, in Paris, became the greatest anatomist of his time. Both as a teacher and as a writer, he greatly furthered knowledge of the tissues and structures of the body.

From the 1400s, European scientists and inventors gained a new self-confidence, with broader horizons than ever before. New developments in science and technology culminated in the Industrial Revolution which began in the late 1700s.

Invention and Industry

Left: An illustration from the Luttrell Psalter of 1338 shows an improvement to the plough, the mould board. Although new to Europe, this originated with the Chinese more than a thousand years earlier. The invention ranks as one of the most important in history. By slicing and overturning the ploughed earth it allowed more air to reach the soil. This led to a rapid increase in soil fertility and crop yields.

The dominance of Europe

By the end of the 1200s the Arab empire had declined, passing the fruits of its scientific knowledge and industrial enterprise on to Europe. Europe by this time had achieved a certain stability with assertive feudal monarchies and was expanding rapidly. Trade and industry boomed, and modern business methods were born with the aid of the more efficient systems of arithmetic which were introduced from Morocco by the Italian mathematician Leonardo Fibonacci (1170–1240).

The 1300s, by contrast, brought disaster to Europe in the form of the Black Death and consequent famine, and the Hundred Years' War. There was little progress at this time in the fields of science and technology. This period, however, was only an interlude in the rise of European power. The Renaissance that began in about 1450, first in Italy then spreading to other European countries, expressed a new self-confidence which remained unshaken until our own century.

Domestic industry

As in many developing countries today, the industries of later medieval Europe were mostly of the cottage type, including such trades as those of the potter and the weaver. Until quite recently, the blacksmith's forge was an integral part of both urban and country life throughout Europe and a reminder of the long continuity and usefulness of this trade.

Below: Farmers grow their crops in different fields yearly in rotation to obtain high crop yields while maintaining soil fertility. By the 1400s, European farmers often used a four-course rotation, as shown. The root and clover crops served as animal feeds. Also, although farmers could not have known it at the time, the clover added nitrogen to the soil by fixing this from the air.

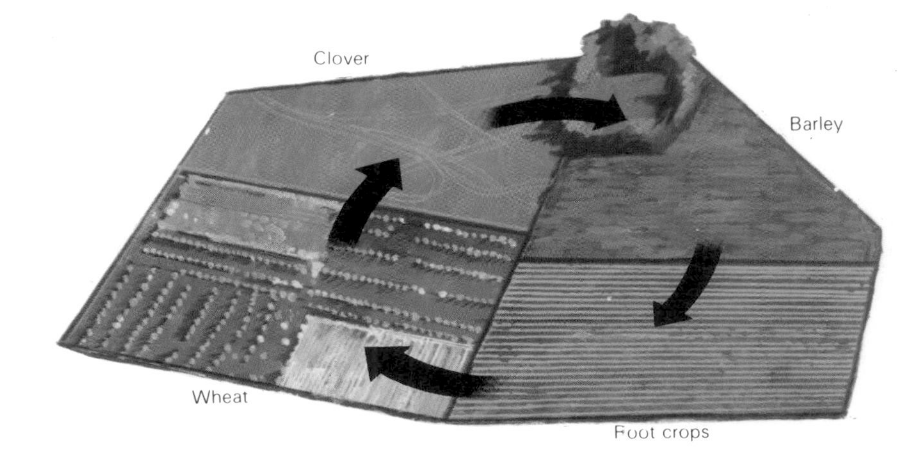

Reference

A **Alcohol** takes its name from the Arabic, *al Kuhl,* but was first made in quantity in Europe in the 1100s. During the plague years of the 1300s it was distilled in large amounts as a remedy, *aqua vitae.* Wood alcohol or methyl alcohol, an industrial solvent, was first distilled from wood in the 1830s.
Alum is the name of several salts, the commonest of which is potash alum or potassium aluminium sulphate. From ancient times alum was used for hardening leather. By the Middle Ages it was also being used as a MORDANT in cloth dyeing.
Ammonia has been used in leather tanning since ancient times, when it was obtained by concentrating urine. Its alchemical name, spirits of hartshorn, refers to the fact that it was also made by distilling the vapours of horns and hooves.
Artificial fertilizers were first made in the early 1800s by treating bones with sulphuric acid. The superphosphates so formed were later made on a much larger scale by reacting sulphuric acid with phosphate rock.

Mercury barometer, 1643

B **Barometers** are instruments that measure atmospheric pressure. The older type, invented by Torricelli in 1643, has a long tube, closed at the upper end, in which a column of mercury rises to a height which varies with, and so measures, the pressure of the air. Aneroid barometers have a coiled metal tube, exhausted of air and closed at both ends. This tube coils and uncoils with variations of air pressure, so moving a pointer on a measuring scale.
Bitumen is a hard, tarry substance obtained from asphalt lakes and seepages and from the distillation of wood and coal. From medieval times it was used for waterproofing ships' timbers, and later for surfacing roads. Distillates from bitumen were used as early as the 1600s as lamp oils, greases and paints.
Blast furnaces are tower-shaped, iron-smelting furnaces in which a blast of air is passed upwards to raise the temperature of heated iron ore until molten iron separates from the slag and runs out from the base, or

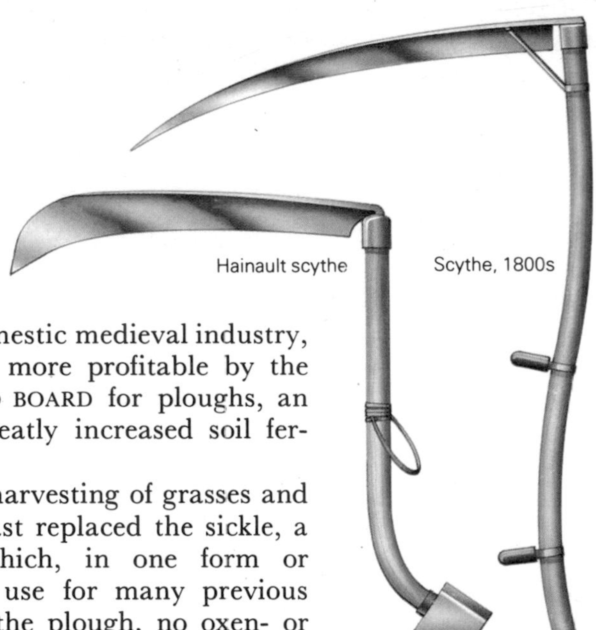

Above: Since Stone Age times, sickles of various kinds have been used for reaping. The scythe first appeared in the Middle Ages, as a sickle on a pole. The scythes shown here are more recent examples.

Farming, another domestic medieval industry, had been made much more profitable by the adoption of the MOULD BOARD for ploughs, an improvement which greatly increased soil fertility.

In the mowing and harvesting of grasses and cereals, the scythe at last replaced the sickle, a reaping instrument which, in one form or another, had been in use for many previous millennia. Aside from the plough, no oxen- or horse-drawn farm machine appeared before 1600, when a seeding machine, which drilled holes in the soil and fed seeds into them, first appeared in southern Europe. It is to be noted, however, that the Chinese had developed a hand-pushed, wheelbarrow-like seeding machine as early as 2800 BC.

European fishing, in contrast with agriculture, was a highly organized and centralized industry in late medieval times, and by the end of the 1200s brought in catches of more than 10,000 tonnes of fish annually. By the early 1400s, Dutch fishermen were towing behind their vessels drift nets more than 100 metres long. These swelled catches still further, particularly of herring. The fish caught in such great quantities were preserved by salting for long-term storage.

Timber-felling was a thriving activity in medieval Europe, providing domestic fuel and structural materials for building houses and ships. By the 1100s, over-exploitation of forests in Germany led to restrictions on timber felling. From this legislation, together with the gradually developing habit of replanting, grew modern forestry.

In the home, the women made cloth with the aid of the new spinning wheel, which had replaced the distaff and spindle. It was the first of a line of SPINNING MACHINES which gave rise to a mechanized textile industry in the mid-1700s.

Spinning and weaving also had great consequences for community health. The more widespread availability of LINEN allowed the wearing of underclothes which could be changed at regular intervals. This improvement in hygiene helped stamp out leprosy as an endemic scourge of Europe.

Among new fabrics for the richer Europeans was silk, introduced from China, which by the late 1200s was being woven on an industrial scale in water-powered mills. The voyages of European explorers in the 1400s and 1500s resulted in the introduction of other new consumer products such as the potato, tea, COFFEE and COCOA.

European agriculture broadened its base in the 1500s with the rise of market gardens, which employed new techniques of HORTICULTURE to produce a wider range of food vegetables and ornamental plants.

Cotton replaced flax as the principal textile for clothing only when it became cheaper as a direct result of the mechanization of the cotton industry in the mid-1700s. Yet even in the 1400s, the European textile industry employed sophisticated machinery, such as GIGMILLS which softened the texture of woven materials. Elaborate processes of cloth dyeing were used, taking advantage of the action of MORDANTS, particularly ALUM. Natural dyes such as saffron, indigo and

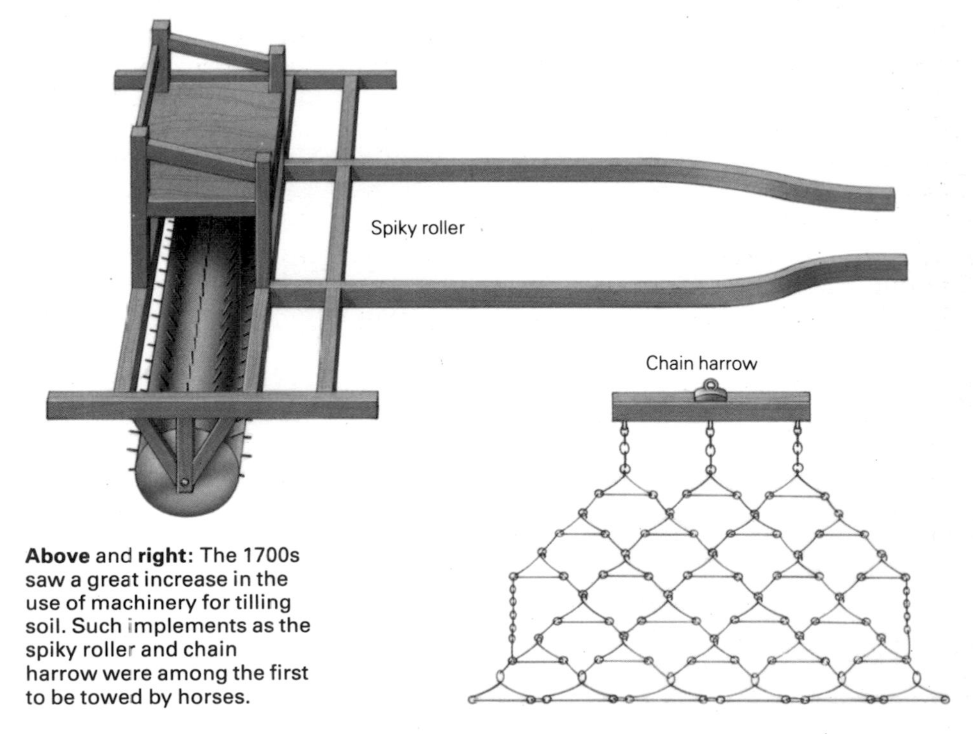

Above and **right:** The 1700s saw a great increase in the use of machinery for tilling soil. Such implements as the spiky roller and chain harrow were among the first to be towed by horses.

hearth, of the furnace. They date from the early 1600s.
Bleaching is the removal of colour or stains from textiles or other materials. Clothes were bleached by exposing them to sunlight or dilute acids, until the invention in 1800 of bleaching powder, or chloride of lime, through the work of the chemist Claude Berthollet (1749–1822).
Boring machines were invented in the mid-1700s to drill out the barrels of cannon. The drilling shafts of these early machines were turned by water power. Later boring machines were steam-operated and used also for drilling out the cylinders of locomotives.
Bottle kilns are pottery-firing kilns, well-described by their name. They were fuelled at first by wood, later by coal, but were inefficient and largely replaced in the late 1800s, with tunnel kilns.
Brass is an alloy of copper and zinc, sometimes together with small amounts of other metals. It has a yellow colour. The Romans made brass coins but brass alloys were used on a large scale only from the Middle Ages. Early scientific instruments were often made from brass. Muntz metal is a type of brass used from the 1830s to sheathe the bottoms of ships.

A brewery in London, England, mid-1800s

Brewing became a large-scale industry in England in the late 1700s, with the invention of pale ale and porter or stout.

C **Camera obscura** was a camera-like instrument invented in the 1400s. Light entered a pinhole at the front and was reflected by a mirror to project an image on to a screen. The camera obscura was used by astronomers for observing eclipses, and later by artists for sketching landscapes.
Cameras for photography evolved from the camera

Yarn twists as it slips off the end of the spindle
Spindle
Driving thread
Rolag (untangled mass of wool or cotton)
Spinner turns wheel by hand

Left: The Great Wheel was an early form of spinning wheel used for making cotton and woollen thread. Untangled raw fibres were spun on a spindle which was rotated rapidly by the movement of a large wheel, turned by hand.

Faller wire
Spindle
As each spindle turns, it twists fibres into a strong, even thread. The faller wire moves a treadle to wind thread on or off the spindle
Clamp
Bobbin

madder were mainly used as they had been in previous ages, but textile chemistry was gaining sophistication, as shown by the preparation of a dye by mixing green vitriol, or iron sulphate, with oak galls to make the intensely black iron stannate.

The spread of knowledge

With the invention of printing and of TYPE FOUNDING in the mid-1400s, knowledge through books became generally available for the first time in history. By 1500 as many as 40,000 editions had issued from the European printing houses, including a number of technical works. Among scientific books of a slightly later date are those on botany and zoology by the Swiss naturalist Konrad von Gesner (1516–65) and the French naturalists Guillaume Randelet (1507–66) and Pierre Belon (1517–64). These reflect an increased curiosity about nature stimulated partly by the reports and discoveries of the explorers.

De la Pirotechnica, written by the Italian metallurgist Vannoccio Biringuccio (1480–1539), surveyed the heavy industries of the Renaissance, including those manufacturing chemicals, metals and glass. In turn, this influenced the most important of all writers on early industry, the German mineralogist Georg Bauer (1494–1555) (who took the Latin name of Georgius Agricola) to write several books on medicine, chemistry and mineralogy, including the famous *De re Metallica,* published 1556. This contains many accurate and detailed descriptions and illustrations of metallurgical and mining activities, including the machinery of mine ventilation, drainage and haulage, and also deals with the economics of mining.

The spread of ideas and information through books, itinerant technologists and explorers resulted inevitably in the growth and diversification of industry. Glassmaking, for example, now not only provided exquisite stained glass for the great windows of cathedrals and other churches but also plain glass for the windows and mirrors of country houses, although a process for casting GLASS SHEET had to wait until the 1600s for its invention.

The demand for metals

Metals were being used more widely and variously, although not always with a clear idea of their particular identity. The 1500s saw applications of alloys of such metals as bismuth, cobalt, nickel and zinc without their separation into individual metals. Indeed, zinc, in the form

Above: In 1770 the British engineer James Hargreaves patented his spinning jenny. This was the first machine to spin many textile threads at the same time – up to 30 in later versions. It played an important part in the rapid development of the British textile industry during the final quarter of the 1700s.

obscura, which by the late 1700s had become a portable instrument with a lens in place of the pinhole. A lens shutter was added in about 1825.

Candles were made from beeswax by the Romans. Grease-soaked rushes, or rushlights, were used for lighting throughout early medieval times. Tallow candles with wicks of flax were invented in the 1200s and modern paraffin wax candles about 1850.

Canning and bottling began, as processes of food preservation, at the start of the 1800s. In 1809, Nicholas Appert in France heat-sterilized foods, then bottled and sealed them. In 1810, Peter Durand in England invented the similar process of canning.

Cannon were the earliest guns. They were probably invented by the Chinese but the first European cannon, bucket-like and cast from bell-metal bronze, appeared in the 1320s. Wrought-iron cannon, firing stone or lead balls weighing up to 300 kilogrammes, were in use by the 1400s.

Caustic alkali was used for

English cannon, 1500s

soapmaking as early as AD 350. Caustic soda and caustic potash were used in the Middle Ages both for this purpose and for cleaning textiles. Caustic soda was made by reacting soda with lime, and caustic potash by reacting wood ash with lime.

Charcoal had been used in metallurgy since the time of the Hittites, both as a fuel and in the making of iron and steel. By the AD 1500s it was consumed in large quantities as a fuel in the manufacture of cannon and shot. From this time supplies became scarce and it was gradually replaced as a fuel by coal.

Chronometer is a very accurate, portable clock, developed to help mariners fix their positions accurately when they made sightings on the Sun or stars. The most famous of early chronometers was that of John Harrison (1693–1776) in 1761, which was accurate to about one second per week.

Coal gas is made by heating coal without air. It was first burned as an illuminant in the 1760s but only became popular for house-lighting

Silkworms are caterpillars of the moth *Bombyx mori.* They have been cultivated in China at least since 2600 BC, for the delicate silk thread they spin to make their cocoons. Silkworms feed on mulberry leaves, which are plentiful in China. Silk was not produced in the West before late medieval times but it eventually became an important industry there.
Above left: Silkworm cultivation of the 1700s, including the harvesting of mulberry leaves.
Above: A woman of South-East Asia spools silk from cocoons. A single cocoon can provide nearly a kilometre of thread.

of its copper alloy BRASS, had been in use for at least 1,500 years before the pure metal became known through the invention of ZINC SMELTING in 1738. Mercury, however, was clearly identified by its extraction from its Peruvian ores in about 1550. It became, in the hands of Paracelsus and other iatrochemists *(see page 36)*, one of the chief principles of alchemy.

The mining of metal ores, particularly those of copper, iron and lead, had long been encouraged by increasing demands for metals by farmers, craftsmen, the Church and the army.

Bell-founding had for centuries consumed large amounts of bronze, since individual bells might weigh as much as several tonnes. In the 1300s bronze also came into demand for the casting of CANNON. Armour, weapons and tools accounted for most of the iron mined and manufactured in the 1400s.

Lead mining, once practised extensively by the Romans, started up again on a large scale in medieval times, when lead was used to roof churches and other large buildings. The new printing industry made further large demands for lead as a component of type metal, and the pistols and muskets of the 1400s needed large quantities of lead in the form of bullets, as well as iron for their own manufacture.

Several other metal ores exploited by the Romans continued to be worked in late medieval times by methods that were more or less traditional. These included ores of zinc, for brass alloys, tin, for bronze, the precious metals gold and silver and such minor constituents of alloys as arsenic, antimony and bismuth.

Two new alloys in use by the 1400s were solder, a mixture of tin and lead that melts at a low temperature and so can be used to join other metals, and PEWTER, which came to be used widely for utensils.

Iron smelting

The greatest increase in demand, however, was that for iron. By the early 1600s, the use of iron for cannon balls alone could amount to thousands of tonnes monthly during a war.

Both cannon balls and the cannon they were fired from were, by this time, made entirely or mostly from cast iron. Traditionally, the difficulty in manufacture of cast iron for sizeable applications had been that of heating large amounts of iron to its melting point, 1,539°C. Because of this difficulty the cast iron of ancient China and India came to the West only in the form of small ingots.

Iron casting began in Europe before the 1200s

---

from about 1800, when large retorts were designed in England to generate gas from coal.
Coal tar is made, like coal gas, by distilling coal without air. It is a rich mixture of carbonaceous substances which yielded many products of the early chemical industry, including dyestuffs and carbolic acid. It also found an important application in road-surfacing.
Cocoa is made from the seeds of the cacao tree. It was introduced to Europe by Spanish explorers of the New World in the early 1500s. First as eating chocolate, and later in the form of sweetened drinks, it became popular during the following two centuries.
Coffee came to Europe from North Africa and was a popular drink by the early 1700s. Coffee houses of the mid-1700s were meeting places of European men of ideas.
Coke is the hard residue of coal after it has been heated to remove volatile matter. It was first made in the mid-1600s by English brewers as a fuel for malt-drying. In 1709, Abraham DARBY first used coke for smelting metals, an event which, as much as any other, can be said to have started the Industrial Revolution.
Cultivators are agricultural machines, larger than ploughs, used for breaking up hard soil preparatory to sowing. Horse-drawn cultivators of the late 1700s and early 1800s included the harrow, which had many narrow iron spikes, and the heavier-built clod crusher.

A levelling cultivator, c.1850s

D

Daguerreotype was a very early type of photograph, first made in 1829 by the inventor and painter Louis Daguerre (1789–1851). It was a silvered copper plate on which the photographic image was developed with mercury, fixed with sodium thiosulphate, and toned with gold chloride.
Darby, Abraham (1677–1717) was an English

Left: Refining lead and copper in the 1500s, shown in an illustration from *De re metallica,* the single most complete record of medieval industry.

Below: A large blast furnace, in which iron ore is reduced to molten iron – pig iron – with the use of coke and limestone. The pig iron is then used to make steel.

Iron ore, coke and limestone
Valve
Pipes to carry waste gases
Furnace lining
Hot air from heater
Slag
Gas-fired heater
Molten iron

when furnaces such as the STUCKÖFEN were made large and hot enough to attain the temperature required to melt iron in bulk. However, the iron cast into moulds from these furnaces represented only about half of the metal in the ore, and contained so much slag that it was too impure for general applications. Even in the 1300s, therefore, most iron was still being made as wrought iron, by beating the metal out from its ores in the manner first made customary by the Hittite smiths of 1400 BC. Steel, as CEMENTATION STEEL *(see page 18)*, was also still being made in this fashion.

A further difficulty in making cast iron arose from the shortage of CHARCOAL as a furnace fuel, forests in many parts of Europe having become severely depleted. This problem was solved when coal took over from charcoal in iron smelting, but the change extended over several centuries, with the gradual developments that led to an increase in coal mining.

The manufacture of iron and steel in bulk began with the development of the BLAST FURNACE, which by the late 1600s was 10 metres or more in height, fuelled by coal, and provided with an air blast from large bellows powered by a water wheel or a windmill. Such a furnace could turn out several hundreds of tonnes of cast iron

---

ore smelter and manufacturer. He was the first person to make good-quality iron using coke, at his ironworks in Coalbrookdale, Shropshire. His grandson Abraham Darby (1750–91) built the first cast-iron bridge at Ironbridge in 1779.

Davy lamp was the first safety lamp to be used by miners and was invented in 1815 by Sir Humphry DAVY. An oil flame burning behind a metal gauze would indicate mine gases without risk of ignition.

Davy, Sir Humphry (1778–1829) was an English chemist. Among his many discoveries were the effects of laughing gas and the isolation of such metals as sodium and potassium by electrolytic means. His investigations into fire damp (methane and other gases) led to the invention of the miner's safety lamp, named after him.

E

Escapement mechanism is the central part of a clock that is regulated by the balance wheel or pendulum. It was invented by the Chinese, probably by Yi Hsing in AD 725.

F

Forge hammers were employed to shape hot metals from the early 1700s, when they were powered by water wheels. Nasmyth's forge hammer of 1839 was the first to be operated and controlled successfully by steam power.

G

Galvanism is the production of electric current by chemical means, as first described by Luigi Galvani (1737–98) in 1791, when he made a frog's leg twitch on contact with two dissimilar metals, both of which were partly immersed in chemical solutions. In fact, he had constructed an electric battery.

Nasmyth's forge hammer

Galvanizing is the process of coating steel with zinc in order to protect it from rusting. It was invented in 1836 in France, when steel was dipped into molten zinc. In modern galvanizing processes, zinc is deposited electrolytically on to steel.

Gigmills, machines for raising the pile, or nap, of textiles, were invented in the 1400s. Gigmills of the 1600s pressed cloth in close contact with teasel-heads, which raised the nap.

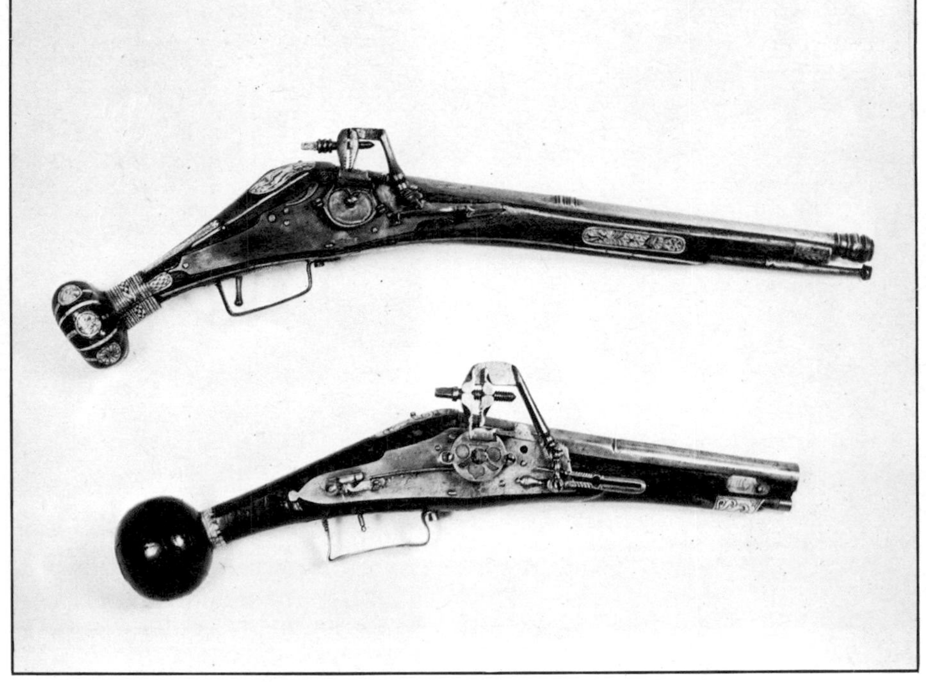

Above left: A farrier, or blacksmith, makes a horseshoe. The smith's craft became widespread in Europe with the breeding of different types of horses for farm work, travel and pleasure.

Above right: Two Venetian pistols from the 1600s. The earliest hand guns date from the 1330s.These were small cannon mounted on hand-held sticks, fired by a lighted match. In the 1500s various firing mechanisms were added, leading to the names matchlock, wheellock and flintlock guns.

yearly. Later blast furnaces of the 1760s also made quantities of PIG IRON, a form of iron which came to be favoured for remelting, moulding and forging for applications that demanded a resilience not often found in cast iron, which tended to be a brittle material.

In these later furnaces, COKE had replaced coal as a fuel, an event that signalled the beginning of the Industrial Revolution. As the result of the work of the ironmakers Abraham DARBY and his son and grandson of the same name, the first very large cast-iron structure, a bridge over the River Severn at the town now called Ironbridge in Shropshire, England, was erected in 1779. Within a few further decades, iron locomotives would be running on iron rails, and iron boats sailing the seas.

The usefulness of coal

The mining of coal as a fuel increased in Europe, stimulated by the shortage of wood, until by the end of the 1700s, coal and the coke made from it had become the energy basis of heavy industry. In the early 1800s, coal mining became somewhat safer with the invention of the DAVY LAMP and the demand for coal increased rapidly with the growth of industry and the railways.

Additionally, coal was beginning to be recognized not only as an industrial fuel, but also as a source of light energy, by the burning of COAL GAS, and of chemicals, by the distillation of coal and COAL TAR.

The extraction of organic chemicals from coal coincided with the growth of the science of organic chemistry. In our own century this has made possible the building of giant synthetics industries. The 1800s, however, had their own 'organics' industries, including those making RUBBER and the early plastic, celluloid.

Metals and machines

The large machinery of medieval times, such as that of the windmills and waterwheels used for drainage and for grinding corn, was for the most part made from wood, although iron or other metal bands might be used to strengthen wheels or other wooden structures. Iron bands were also employed in the construction of early cannon, to prevent these bursting during firing.

With the development and extension of its manufacture in the 1600s and 1700s, iron was used more and more in machine construction. It is still the only metal that can be employed for heavy duty applications, equally tough metals, such as cobalt and nickel, being too rare and expensive for these purposes.

Glass sheet was first made on a large scale in the 1600s in Italy. Molten glass was poured on to a casting table and rolled to an even thickness before it cooled and hardened. Two products of this new industry were windows and mirrors.

Gutenberg, Johannes (1400–68) is regarded as the inventor of printing. His presses used movable type. He made the first printed Bible, the *Gutenberg Bible,* before 1456.

H

Horticulture is the art or science of growing vegetables, fruit, shrubs and flowers, usually in market gardens. It arose in Europe during the 1500s, encouraging the cultivation of many new kinds of plants.

Hot rolling is a process in which a metal ingot is shaped by passing it through rollers at a high temperature. In the 1700s, pig iron was hot-rolled into bars in slitting mills. It was first used to make sheet steel in 1825, in south Wales, by rolling steel bars at 790°C.

I

Inks have been used for writing since ancient Egyptian times, and for printing since Chinese wood-block printing of the AD 700s. Inks were traditionally suspensions of soot in water with the addition of a little gum.

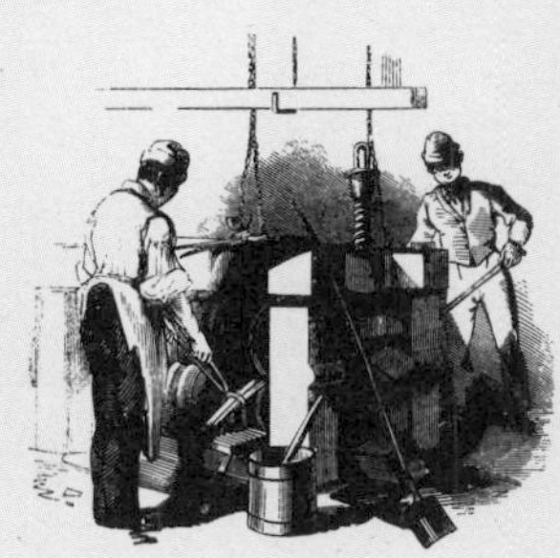

Rolling iron bars, c.1851

K

Knitting machines began with the stocking frame of the late 1500s. The first knitting machines to be power-operated were used for lacemaking, in Nottingham, England, in the early decades of the 1800s.

L

Lavoisier, Antoine (1743–94), a French chemist, is considered to be the founder of modern chemistry. He investigated many of the properties of oxygen and showed how it combined with hydrogen to form water. With other workers, he devised a system of naming chemical substances, which formed the basis of the present system.

Lead chamber process was the first large-scale process for the manufacture of sulphuric acid. It was invented by a Scottish chemist, John Roebuck (1718–94), and was central to the chemical industry from the 1770s until the early years of our own century. Sulphuric acid was made by burning sulphur or iron pyrites (iron sulphide) together with a

Above: In charcoal burning, small logs of wood were piled up as shown and ignited. They were then covered with earth to restrict entrance of air. This meant that they did not burn actively, but smouldered slowly, producing charcoal.

Above: Draining a mine in the 1500s. This illustration from *De re metallica* shows a waterwheel operating a ball-and-chain pump. The iron balls displace water in a pipe, thus lifting it to the surface. Such machines were not at all new. Both waterwheels and the Archimedean screw were used by the Romans.

Some of the new heavy iron machinery was needed for the fabrication of iron itself. By the early 1600s, this machinery included large water-powered trip hammers and tilt hammers for forging hot iron, and the 1700s saw the development of HOT ROLLING.

Early in the 1700s, steam engines began to be used as PUMPS in mine drainage. These engines were pioneered by Thomas Savery (1650–1715) and Thomas Newcomen (1663–1729). Many of them were very big machines indeed, Newcomen's first engine occupying a 10-metre-high engine house, and their long up-and-down pistons and broad cylinders were made of iron. John Wilkinson's steam engine (1776) was the first to be used to provide the air blast for an iron-making furnace. The reliability of some of these early steam pumps can be gauged from the fact that some of them were still working perfectly in the 1900s.

The armaments, agriculture and textile industries all made increasing demands for iron machinery in the 1700s. Among the heaviest of the machines of this period were those used for drilling out gun barrels. A very large example of these BORING MACHINES was invented for Boulton and Paul, makers of steam engines, by John Wilkinson in 1776; this had an iron steam-cylinder 125 centimetres in diameter.

Towards the end of the 1700s, such large, iron-made agricultural machines as CULTIVATORS and REAPING AND THRESHING machines began to come into demand. At about the same time, steam-operated POWER LOOMS, SPINNING MACHINES and KNITTING MACHINES were being installed to service a burgeoning textile industry.

Packaging, piping and protecting

The increased productivity and efficiency of iron and steel mills first stimulated and enriched other heavy industries. Later, it began to affect populations more directly, by helping to create a consumer society.

One of the first major events in this progression was the invention of TINPLATE which led to the development of CANNING as a process for preserving food. Another new process was GALVANIZING, by which steel sheet is protected from rusting and so becomes generally available as a building material. Galvanized corrugated iron is still the main roofing material in many developing countries.

More important still for the average citizen was the development first of iron, then of steel, piping for WATER PIPES AND MAINS and sewers and drains. The provision of clean domestic water, and the effective disposal of household sewage and other wastes, was to prove one of the greatest causes of increased community health and prosperity.

Chemical industry

Chemical experimenters in ancient times usually had only the haziest ideas of how and why they got their results. More often than not, they further complicated matters by referring causes to the realms of magic and philosophy.

This confusion is very understandable because

source of nitrogen oxides. The sulphur dioxide so formed was oxidized to sulphur trioxide by the nitrogen oxides. The sulphur trioxide was then dissolved in water to make sulphuric acid. This took place in large chambers lined with lead, a metal not attacked by sulphuric acid of the strength formed, 65–75%.

Leblanc process was the first major industrial process for the manufacture of soda, or sodium carbonate. It was invented by Nicolas Leblanc (1742–1806) in 1787, and used common salt as its starting material. This was reacted with sulphuric acid to make sodium sulphate, which was then mixed with limestone and coal. The mixture was roasted to make sodium carbonate, which was extracted with water.

Leyden jar was an early type of capacitor or condenser, a device for storing electricity. It was invented, both in Germany and Holland, in 1745. It comprised a glass jar containing water into which dipped an iron wire which was held in place by a cork. The exposed end of the wire was touched to a source of static electricity and an electrical charge was then stored in the jar, both its glass and its cork being good insulators.

Limelight was used briefly in the 1830s for lighthouses, but later and more famously for theatre illumination. It involved an early use of the arc-lamp, to heat a block of lime to incandescence.

Introduction of limelight on Westminster Bridge, 1860

Linen is a textile woven from fibres of flax, a plant which was widely cultivated in Europe of the Middle Ages. Linen was used not only for clothing but also for making paper of fine quality.

Lithography is a printing process invented by Aloys Senefelder (1771–1834) in the last years of the 1700s. It uses a flat plate, or stone, on which the lettering or design

Left: The town of Ironbridge in Shropshire, England, is named after its famous bridge. Built in 1779, this was the first bridge to be constructed entirely from metal. The area around Ironbridge was the birthplace of the Industrial Revolution.

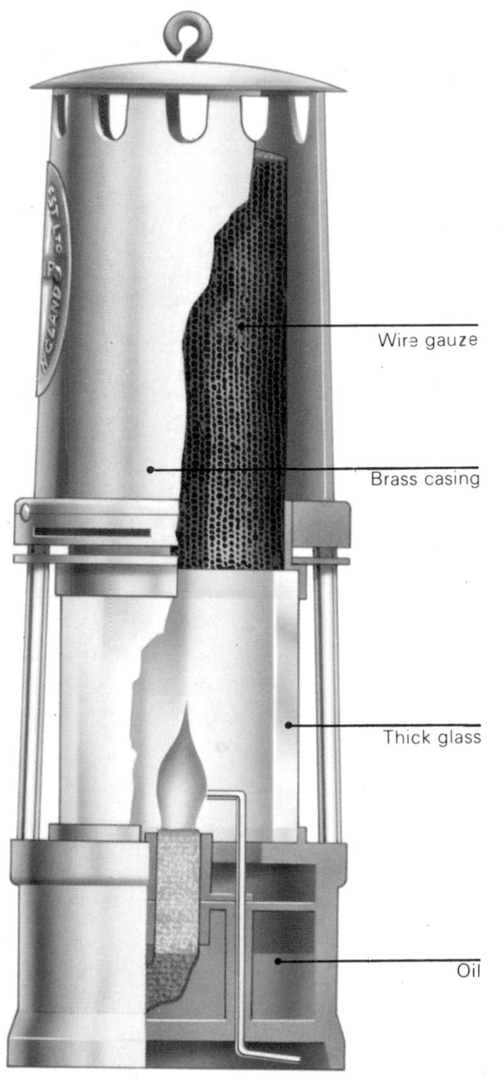

Left: A coal-miner's safety lamp of the type perfected by Sir Humphry Davy in 1815. The use of this lamp prevented firedamp (methane) explosions. The oil flame burned behind a metal wire gauze, which cooled the flame products so that they would no longer ignite firedamp. After the arrival of the electric lamp, the Davy lamp continued in use, to test for flammable gases.

the products of a chemical reaction are often utterly unlike the starting materials in appearance. Without a full investigation of reactions, which did not start until the late 1700s, a science of chemistry could hardly be expected to exist. Yet chemical industry existed even in ancient Egypt and Mesopotamia.

The Egyptians, for example, were experts in the manufacture of paints and pottery glazes. Among the pigments they employed for these materials was white lead, which they extracted from lead minerals with an acid substance,

Right: A precision machine of the 1700s, which is now considered a work of art. The machine was built in 1789 for cutting wheels for clocks, and is typically constructed of brass reinforced with iron.

is added with a greasy material. The stone is wetted and inked, when the grease absorbs ink but repels water. When paper is pressed against the stone the inked image is transferred to it.

M **Manures** are fertilizers of animal and vegetable origin. Animal droppings have been used to manure fields since ancient times. Seaweeds and potash from burned wood are vegetable manures used from the Middle Ages onwards. Industrial manures were available from the late 1700s as bone, horn, and leftovers from soapmaking.

Mordants are substances that help fix the colour to a textile during dyeing. The earliest mordants were ALUMS. In the 1800s tannic acid was used to mordant alkaline dyes, and metal hydroxides to mordant acid dyes.

Mould board is the curved blade of a plough that overturns the ploughed earth. In doing so it improves soil structure and aerates the soil, so promoting greatly increased crop yields. It was invented in ancient times by the Chinese but was used in European agriculture only from the 1000s.

N **Nitre** or saltpetre is a chemical salt, now called potassium nitrate, which was first used in large quantities to make gunpowder, and later in the LEAD CHAMBER PROCESS for making sulphuric acid. In the Middle Ages it was mostly extracted from nitrate-rich animal manures. Later it was made from nitrate minerals.

O **Ore concentration** is any process for enriching low-grade metal ores by increasing their metal content. It was invented in Sweden in the early 1700s when copper ores were

Nitre extraction, 1600s

floated and agitated in water. The heavier, metal-rich particles sank and were collected.

Oxygen was discovered by Joseph PRIESTLEY in 1774 and immediately was shown by Antoine LAVOISIER to be vital to the process of combustion. This demolished the PHLOGISTON *(see page 36)* theory of combustion, and prompted the growth of modern chemistry and the chemical industry.

P **Pewter** is an alloy of lead and tin, with smaller amounts of copper,

Right: Chemical industry began with the Leblanc process for making washing soda. The first stage is shown, in which common salt reacts with sulphuric acid to make sodium sulphate. Also produced is hydrogen chloride gas, which is led off and dissolved. Secondly, sodium sulphate is roasted with limestone and coal powder. This produces sodium carbonate, which is dissolved in water and then crystallized out.

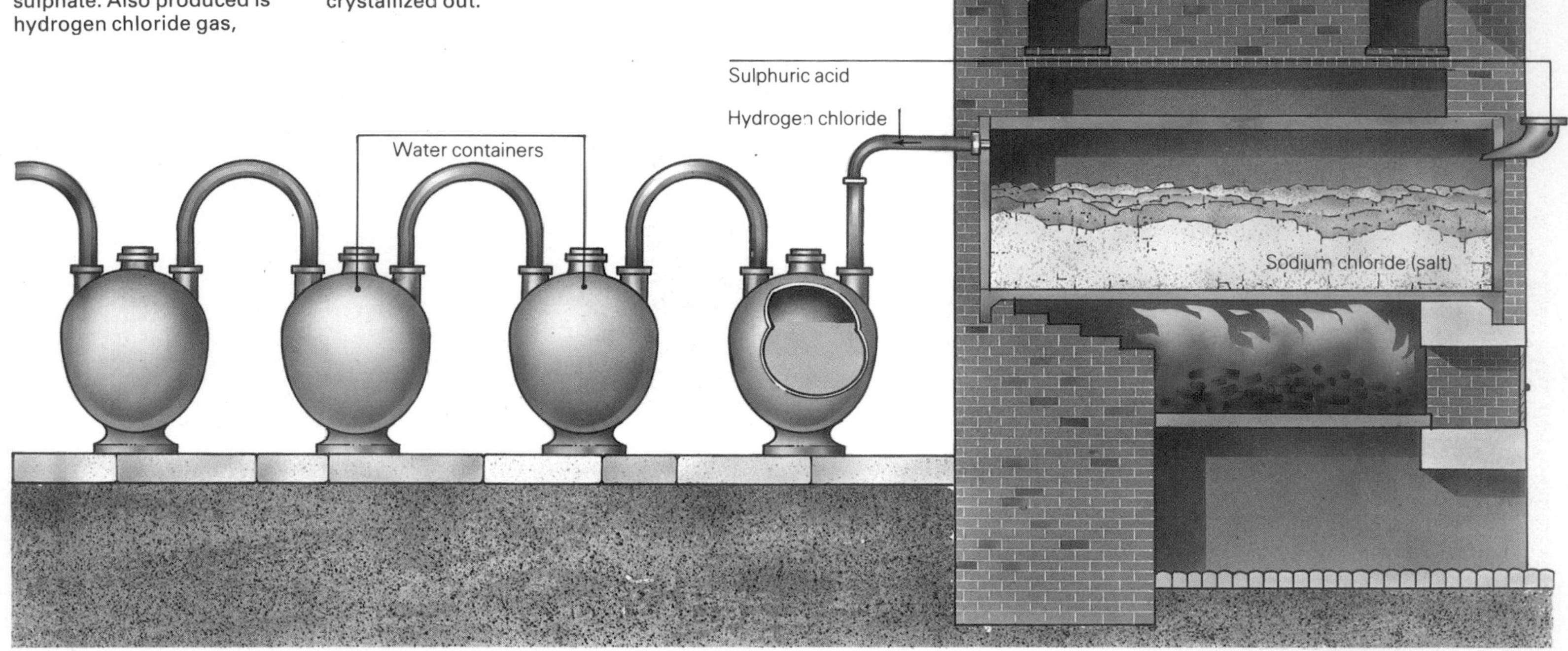

vinegar. They made this by a double process of fermentation which first produced ALCOHOL, then acid. These chemical and biological reactions, some quite complex in nature, had been arrived at by processes of trial and error, not by scientific knowledge. Such empirical, traditional procedures were the sources of chemical industry until relatively modern times.

Another chemical compound with a very long history is soda, still one of the most important products of the chemical industry. The Egyptians used soda as a glaze, a constituent of their artificial precious stone, lapis lazuli, and, like ourselves, as washing soda.

NITRE reveals one way in which it was obtained by its alternative alchemical name, saltpetre. This indicates that it is a salt (*sal*) extracted from rock (*petrae*). However, in the West it was more often extracted from MANURES. The most famous application of nitre is in the manufacture of gunpowder.

The Romans and Greeks generally lacked interest in chemical experiment for its own sake. However, their successors, the Arabs, not only performed many such experiments, but also manufactured such chemicals as acids, ALUM, AMMONIA, CAUSTIC ALKALI, NITRE, and soda on a larger scale than ever before. In their Spanish and North African colonies, and also through trade, the Arabs handed down their knowledge of and expertise in practical chemistry to the medieval West, thereby strongly influencing the formation of a European chemical industry.

Chemical manufacture in Europe in the 1300s was, therefore, a thriving scene. Besides industrial processes for making the chemicals already mentioned, people of the European Middle Ages had also inherited from their forebears the methods of SOAPMAKING, leather TANNING, glassmaking and textile BLEACHING and dyeing.

Household chemical manufacture also included that of rushlights and CANDLES for lighting and INKS for writing, and many a cottage and country household would engage in BREWING, making beer by fermenting grains of barley or wheat. A monastery might also possess the means for distilling such alcoholic beverages to make spirits and liqueurs.

Right: Wedgwood pottery ware includes some of the most refined products of the 1700s. The Jasper flower pot, 1769, is an early piece. The Jasper vase, 1785, has Venus decoration.

Right: Before the 1600s, plate glass was limited in size to the small pieces used in mullioned and stained glass windows. The plate-glass casting process shown here made possible large, single-sheet windows and mirrors.

zinc and bismuth. It has a silver-grey colour and is easily melted and cast into moulds. Pewter ware was made widely from the 1400s onwards.

Pig iron is iron cast into ingots called pigs. It is traditionally the main source of wrought iron. Production increased rapidly from 1790, with coke as a fuel.

Power looms for weaving date from the steam-operated loom of Edmund Cartwright (1743–1823) of 1787. Its invention was made necessary by the greater output of yarn from power-operated spinning machines, which had been invented in the 1760s.

Priestley, Joseph (1733–1804), an English chemist, discovered oxygen and a number of other gases including sulphur dioxide and ammonia. He also published works on history and politics.

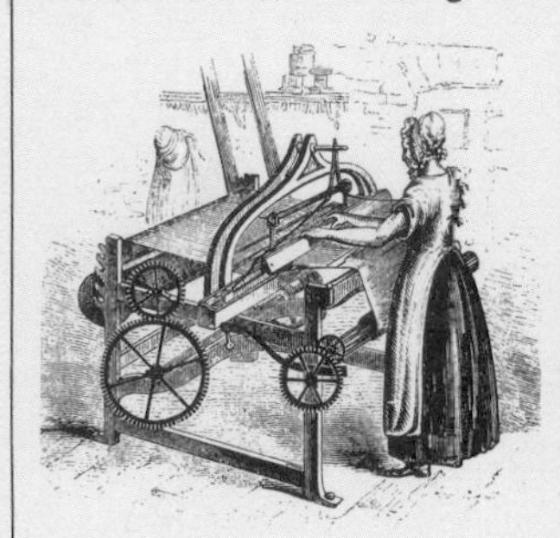
Cotton loom, c.1845

Pumps were invented by Alexandrian engineers. The Romans made practical use of these water-lifting devices in their aqueducts. In the Middle Ages, suction pumps were used to remove water from mines, and early windmills were employed to pump water for land drainage, particularly in Holland. The first working steam pump was that of Thomas Savery in 1698. Robert Boyle's vacuum pump of 1660 helped him discover the gas laws which were fundamental to the development of physics and chemistry.

R

Reaping and threshing were first mechanized, in northern Europe, from 1840 onwards, although reaping machines, pushed or pulled by horses, had been invented some 60 years earlier. Threshing machines of the mid-1800s were driven by steam engines.

Rubber became an important industrial material in the early 1800s. Natural rubber is made from a juice, or latex, which oozes from the rubber tree, *Hevea brasiliensis*. Seeds of this tree were brought from South American jungles to Kew Gardens in England, where they were raised and later used to start rubber plantations in Malaysia. In the 1820s, Thomas Hancock invented machinery for handling and forming rubber in large quantities.

S

Sawmills were early examples of mechanized

This scene, busy as it was, did not change significantly during the next four and a half centuries. The expansion and diversification of the chemical industry was, by and large, prevented during this time by the lack of scientific knowledge of chemistry.

At its simplest, this is the knowledge of practical quantitative chemistry – that is, of reacting known substances together in definite amounts in order to make definite amounts of other known substances. It really began in the late 1700s with the experiments of Joseph PRIESTLEY and his fellow workers in Britain, Antoine LAVOISIER in France and Carl Scheele (1742–86) in Sweden.

At the same time as these experimenters and theoreticians were establishing chemistry as a science, Nicolas Leblanc in France was inventing a new process for making that most time-honoured of industrial chemicals, soda. The LEBLANC PROCESS not only made soda on a truly industrial scale but also, because of its requirement for SULPHURIC ACID, stimulated another major process of chemical manufacture, the LEAD CHAMBER PROCESS. Although superseded in the 1860s by the Solvay Process which evolved from it, the Leblanc Process, more than any other, helped to found the modern chemical manufacturing industry.

Pottery and glass

These two ancient industries underwent rapid development following the adoption of coal as a fuel. As a result, kilns and furnaces could now be made larger and more productive, although, as was the case with the earlier BOTTLE KILNS of the pottery industry, they were often very inefficient.

A major pottery industry started up in the English Midlands in the 1700s, mass-producing both quality ceramics as table- and kitchen-ware, and bricks for houses, chemical retorts and furnaces.

The progress of glass manufacture was accelerated in the 1600s by the invention of processes for making GLASS SHEET and tube on a large scale. In the 1700s, optical glass was made to a new high standard, with revolutionary results in the manufacture of instruments using lenses. Mechanized mass production of glass bottles began in the mid-1800s.

industry, in Holland of the late 1500s, where log-cutting saws were driven by windmills. Water-powered saws were later widely used in Europe.

Tapping rubber, Brazil

Seed drill is an agricultural machine which drills multiple holes in soil, then feeds in seeds. The first mobile seed drill was invented by Jethro Tull (1674–1741) in 1701.

Soapmaking dates from ancient times, as evidenced by reports of the Roman, Pliny, who describes soap manufacture as the boiling of goats' fat with wood ashes treated with lime. This type of process, which makes soft soap, was prevalent in the Middle Ages, together with the manufacture of hard soap, in which fat was boiled with causticized soda.

Spermaceti is a white, waxy substance obtained from the head of the sperm whale and used principally for making cosmetics. It had a major earlier use for making candles, and was one of the most valued products of the whaling industry which started in the late 1600s.

Spinning machines began with the spinning wheel of the early 1300s. They became highly mechanized in the mid-1700s, with the invention by James Hargreaves (d.1778) of the spinning jenny in 1764; the spinning frame by Sir Richard Arkwright (1732–92) a few years later; and the spinning mule by Samuel Crompton (1753–1827) in 1779.

Steelyard is a weighing machine invented in Greek or Roman times and used ever since. It has a metal arm, hand-held or otherwise suspended from the middle. A weight slides along a graduated scale on the arm. Objects to be weighed are hung from one end of the steelyard and the weight moved along until the arm just balances.

Stuckӧfen was a furnace of the later Middle Ages in which iron was smelted with a charcoal fire, which was fanned by air from bellows sometimes worked by water power. This and later, taller, blast furnaces produced cast iron for the first time in Europe.

Sulphur was obtained in ancient times from volcanic deposits and used to make gunpowder and other pyrotechnic materials, and for alchemy. Later it was ob-

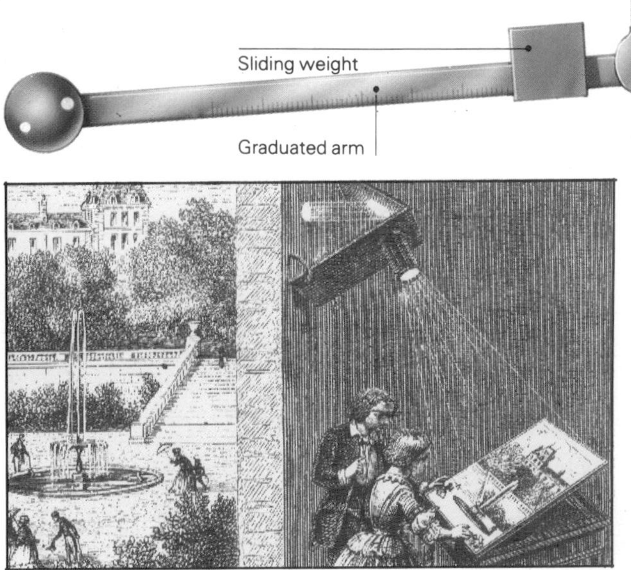

Sliding weight
Graduated arm
Fulcrum
Weighing pan

Engineering with precision

The earliest precision instrument might be identified as the weighing balance with which an Egyptian cosmetician, 3,000 years ago, would have weighed out fine quantities of materials. A more robust weighing machine, the STEELYARD, also has a very long history.

Most people, however, would think of the mechanism of a watch as the first precision instrument. Watchmaking has traditionally been associated with machine tools of fine quality, such as the lathes of the 1600s and 1700s on which the mechanisms of watches were made. The heart of a clock or watch, its ESCAPEMENT MECHANISM, had been invented long before this time, but only with gradual refinements in machine tools could accuracy and smallness be combined, as in the CHRONOMETERS made during the 1700s.

Precision instruments of another kind were in demand from the early 1600s, namely the telescopes and microscopes which were the first tools of the SCIENTIFIC REVOLUTION *(see page 32)*. As already indicated, the development of these and other optical instruments raced ahead in the late 1700s, when glass lenses were first made relatively free of faults that cause distortions and colour fringes. Lenses were also fitted at this time to the CAMERA OBSCURA, which for two centuries had been little more than a toy, but which now quickly gave rise to the development of the photographic CAMERA.

High quality glass was also made a part of such precision devices as the BAROMETER and the THERMOMETER, first of a long line of scientific measuring instruments. Brass was used for these until the 1800s when it was gradually replaced by harder-wearing iron and steel alloys.

The numbers and types of precision instruments and machines further increased with the harnessing of electrical power, signalled in the 1700s by such events as the discovery of GALVANISM and the invention of the LEYDEN JAR.

Above left: Early thermometers worked on the same principle as modern ones. The first to work well were made in Florence, about 1650. They were sealed glass vessels of elaborate shape. Inside, alcohol expanded and contracted according to temperature changes.

Above: The ancestor of the modern photographic camera was the camera obscura. In this device, light from a scene was projected into a darkened room, or box, through a pinhole or lens on to a screen. This late example from the 1860s is being employed for artistic purposes.

Above: The steelyard is a weighing machine with a long history, having been used since at least early Roman times. It can be held in the hand or suspended from a hook. Objects or materials are weighed by sliding a weight along a graduated arm until the arm just balances.

---

tained by mining and was employed mainly in the production of quantities of sulphuric acid.

Sulphuric acid became a very important chemical material in the late 1700s when it was required for the manufacture of soda by the LEBLANC PROCESS. Its own manufacture in bulk was accomplished first by the LEAD CHAMBER PROCESS and later by the contact, or catalytic, process.

T **Tanning** is a process in the manufacture of leather, for preserving skins and hides. In some form or other it has been practised since prehistoric times. Most processes have employed solutions of oak galls or other tannin-rich vegetable substances, in which hides are soaked.

Tanning goatskins, 1842

Thermometer is an instrument that measures temperature. The first to work properly were Italian instruments of the 1600s made of glass and containing alcohol as the liquid which expands and contracts with changes in temperature. Mercury-in-glass thermometers appeared early in the 1700s. They were invented by the German physicist Gabriel Fahrenheit (1686–1736) who also gave his name to a temperature scale.

Tinplate is mild steel rolled into thin sheet and coated with tin to stop it rusting. The tinning of sheet iron was invented in Germany in the late 1600s and tinplate was first used for cans and boxes in the early 1700s.

Type founding is the process of casting type metal, an alloy of lead, tin and antimony, into printing type. It was first practised by JOHANNES GUTENBERG about the year 1450.

V **Vulcanizing** is the process of hardening rubber and stabilizing it against tackiness by the addition of sulphur or some other vulcanizing agent.

W **Water pipes and mains** were often made of wood until the mid-1700s, when the first cast-iron water mains were laid in France. Later, water pipes were made from mild steel.

Z **Zinc smelting** was the invention of William Champion, who condensed zinc metal from its vapour in 1738. Previously, zinc had been used as a constituent of alloys but rarely or never in the pure state.

The Industrial Revolution has been marked by a tremendous explosion in scientific enquiry and technological progress, which has had a profound effect for good (and sometimes unfortunately for ill) on all our lives.

Making the Modern World

Prospecting and mining

With the increase in demand for metals in the final quarter of the 1800s came a world-wide search for metal ores, including those of the newcomers nickel, manganese, chromium and aluminium, which were beginning to have large-scale uses.

Underground ore deposits were proved by drilling to bring up cores of rock for examination, a technique used also in petroleum prospecting. In mines and quarries, ore-bearing rock was extracted using rock-boring machines and with high explosives, the first of which was DYNAMITE. Ore deposits nearer the surface were exploited by processes of OPENCAST MINING, which also came to be of major importance in the recovery of coal.

The age of steel

Cheap steel and the distribution of electric power together launched our age of modern technology. Of these, steel came first.

It had been made in Europe for centuries, first by the CEMENTATION process *(see page 18)* and later also by the crucible method of casting steel. Both these, however, were small-scale operations and as late as 1870, the production of steel was still far outweighed by that of cast iron and wrought iron. This situation was to change rapidly with two inventions, the BESSEMER CONVERTER and the OPEN HEARTH FURNACE.

The main problem in making steel had been the careful control of its carbon content. In the new furnaces, this control was achieved on a scale never before attempted, by blowing air not through the fuel but directly through or over a mass of molten metal. This burned away excess carbon and other impurities. At the same time, the air blast raised the furnace temperature because the combustible impurities had become, in effect, fuels.

At a somewhat later date, steel quality was

Above: The use of explosives for mining and quarrying was extended with the invention of dynamite and gelignite, which could be handled safely and be directed.

Below: Prospecting for petroleum. On land, geophysicists are generating seismic explosions. The echoes from these will help them to locate a salt dome with its associated petroleum deposits. At sea, a petroleum or natural gas field located in a similar way is proved by a test rig, which drills down to obtain samples from beneath the seabed.

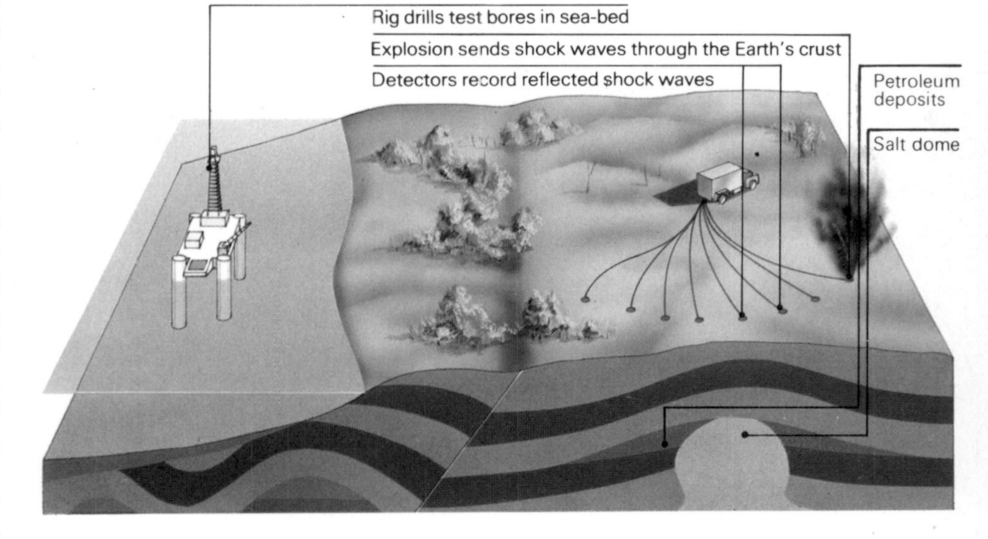

Reference

A **Accelerators, particle,** are large machines in which subatomic particles such as electrons, protons and neutrons are accelerated to very high speeds. They include linear accelerators, cyclotrons and synchrocyclotrons. They are used for research in particle physics, and in medicine for treating tumours.

Acoustics is the study of sound waves, particularly their transmission, reflection and absorption in buildings.

Amplifier is an electronic device for increasing the strength of electrical signals. Many scientific detector instruments include amplifier circuits.

Analgesics are drugs for relieving pain without causing unconsciousness. They include such drugs as aspirin and paracetamol.

Anodizing is a process for depositing a protective oxide coating on a metal, usually an aluminium alloy. The metal is the anode in an electrolytic process.

Antibiotics are drugs made by bacteria and fungi, which kill or inhibit the growth of other microbes. They include penicillin, streptomycin and the tetracyclines.

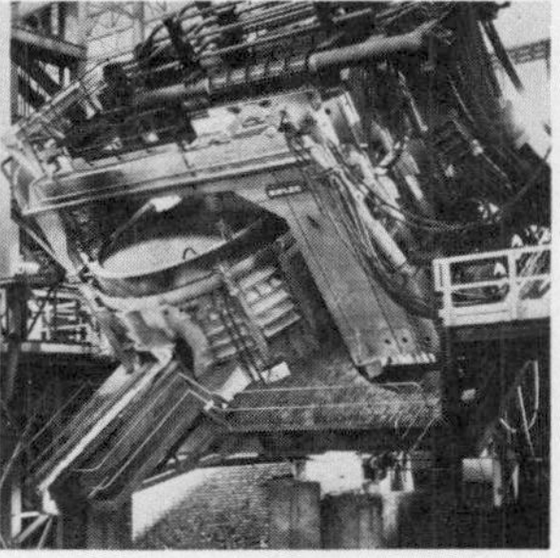

Steel pouring from arc furnace

Arc furnaces are employed mainly in metallurgy. Temperatures are raised by the passage of an electric arc between a large electrode and the metal or alloy being heated.

Auxins are plant hormones which promote growth of stems, buds and roots. They are used in horticulture for these purposes, and in agriculture as weedkillers, for their ability to cause lank overgrowth of unwanted plants.

B **Ballistic missiles** contain a mechanism which is preset to aim and guide the missile towards its target. Once it is in flight, however, its path cannot be changed like a guided missile.

Batteries, electric, provide an electric current by means of chemical reactions which take place in one or more electric cells. Accumulators, such as motor car batteries, have liquid-filled cells that can be recharged with electricity. Dry cells, such as torch batteries, contain a chemical paste and cannot be recharged.

Bessemer converter was a

further controlled with the development of special furnace linings, which reacted chemically with impurities such as silicon, sulphur and phosphorus, so further removing them from the metal. A wider range of ores could now be melted successfully, in quantities of about 25 tonnes, and converted to steel in a matter of minutes. From about 1875 onwards, mild steel, which is harder than wrought iron yet without the brittleness of cast iron, became cheaper to make than either of these. Over this period, world steel production increased 60-fold to reach about 30 million tonnes annually by the year 1900.

New types of steel were developed to meet the special needs of industry and the military. Steel was further hardened and toughened by such processes as TEMPERING and CASE HARDENING. Alloy steels, which contain substantial amounts of other metals as well as iron, were first manufactured for armour, and later, as STAINLESS STEELS, for more peaceful purposes.

To melt these special steels, electric ARC FURNACES were invented. In these and other modern steelmaking furnaces, oxygen is supplied from above by an oxygen lance. The capacity of a modern furnace may be several hundred tonnes.

Aluminium, the lightweight metal

Aluminium is, perhaps surprisingly, the commonest of all metals in the Earth's crust. For example, it makes up a large part of common clay, from which, unfortunately, it still cannot be extracted economically. The commercial production of aluminium from its oxide ore, bauxite, by a process of ELECTROLYSIS, had to await the coming of cheap electricity. The process began, on a small scale, in the 1880s, but aluminium was not a cheap metal until the 1930s, when aluminium alloys found their first large-scale applications in aeroplanes. Today, in the scope and variety of its uses, aluminium ranks second only to iron. It is a natural choice for any application requiring strength combined with lightness and high conductivity of heat or electricity.

Metal fabrication and protection

Factory technology also advanced quickly, to handle the new metals and alloys. Following World War I the use of the OXYACETYLENE TORCH for cutting and welding steel became widespread, but only after cylinders of gas had become available by the new technology of LIQUEFACTION OF GASES. Cheap electricity provided another popular method of welding metals, using the heat of the electric arc.

Special treatments of metals, for the sake of their appearance or preservation, included various METAL CLEANING OPERATIONS, ELECTROPLATING and ANODIZING. CORROSION PREVENTION became of great importance for metals exposed to atmospheric or chemical attack, and utilized new plastics materials as protective coatings.

More recently still, metal-forming techniques have come to include POWDER METALLURGY, used in the preparation of synthetic composite materials such as CERMETS and metals reinforced with non-metallic fibres. Such sophisticated materials are a necessary part of space-age technology.

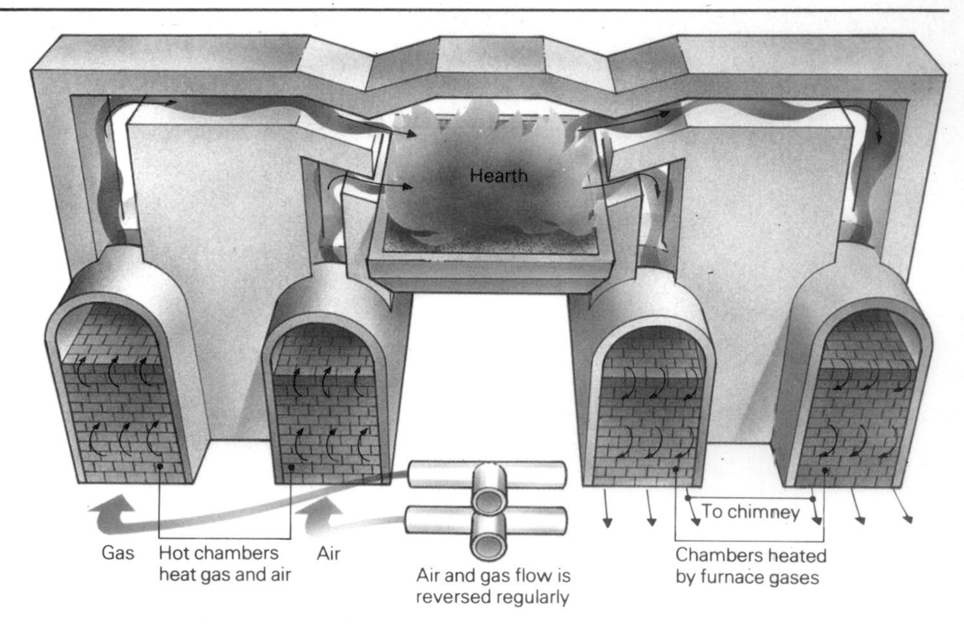

Above: Steelmaking by the basic open hearth process. Steel scrap is mixed and heated together with pig iron from a blast furnace (*see page 42*), limestone, and other materials. The furnace temperature is raised and controlled by alternating blasts of heated air and recycled furnace gases. Molten steel is poured or tapped off when all impurities have passed into the basic (lime) slag.

Left: A general-purpose, swing-centre lathe. Lathes for wood turning date from the earliest civilizations, evolving, like drills, from the bow.

large steel-making furnace, invented by Sir Henry Bessemer (1813–98) in 1856. It was barrel-shaped and could be tilted for pouring the steel. Hot air was blasted through pig iron, so burning off impurities and controlling the amount of carbon in the steel.

Bimetallic strip is a simple device used in thermostats and other temperature control mechanisms. It comprises two dissimilar metals, such as alloys of copper and iron, fastened together. When the strip is heated, one metal expands more than the other, so causing the strip to bend. It is used in this way as an electrical switch.

Biological warfare is the use of harmful microbes to cause disease in enemy populations. The enemy could be bombarded with missiles containing virulent bacteria or viruses to cause epidemic infections, or with their toxins, which would be expected to cause a very high rate of fatality.

C

Calculating machines date from the ancient abacus. The first mechanical calculator, with toothed counting wheels, was invented by Blaise Pascal (1623–62) for solving tax problems. The calculators of Charles Babbage (1791–1871) and William Burroughs (1857–98) in the 1800s were more elaborate mechanical devices. Modern calculating machines are entirely electronic in operation.

Burroughs' adding machine

Calenders are machines with rollers through which various industrial materials are processed. Rubber sheet is made in this way, and paper and cloth are provided with their finished textures.

Calorimeters are instruments for measuring heat, particularly that of chemical reactions. In the bomb calorimeter a known quantity of a fuel is burned in oxygen for the measurement of the calorific value of the fuel.

Cam consists of an arm on a rotating shaft, which engages with another component, so that the rotary motion is converted into a different sort of motion, such as an up-and-down motion.

Capacitor or condenser is a device which stores electric charge on two or more metal plates that are electrically insulated from one another. Variable capacitors, in which one set of plates can be moved to interleave with another set, have been

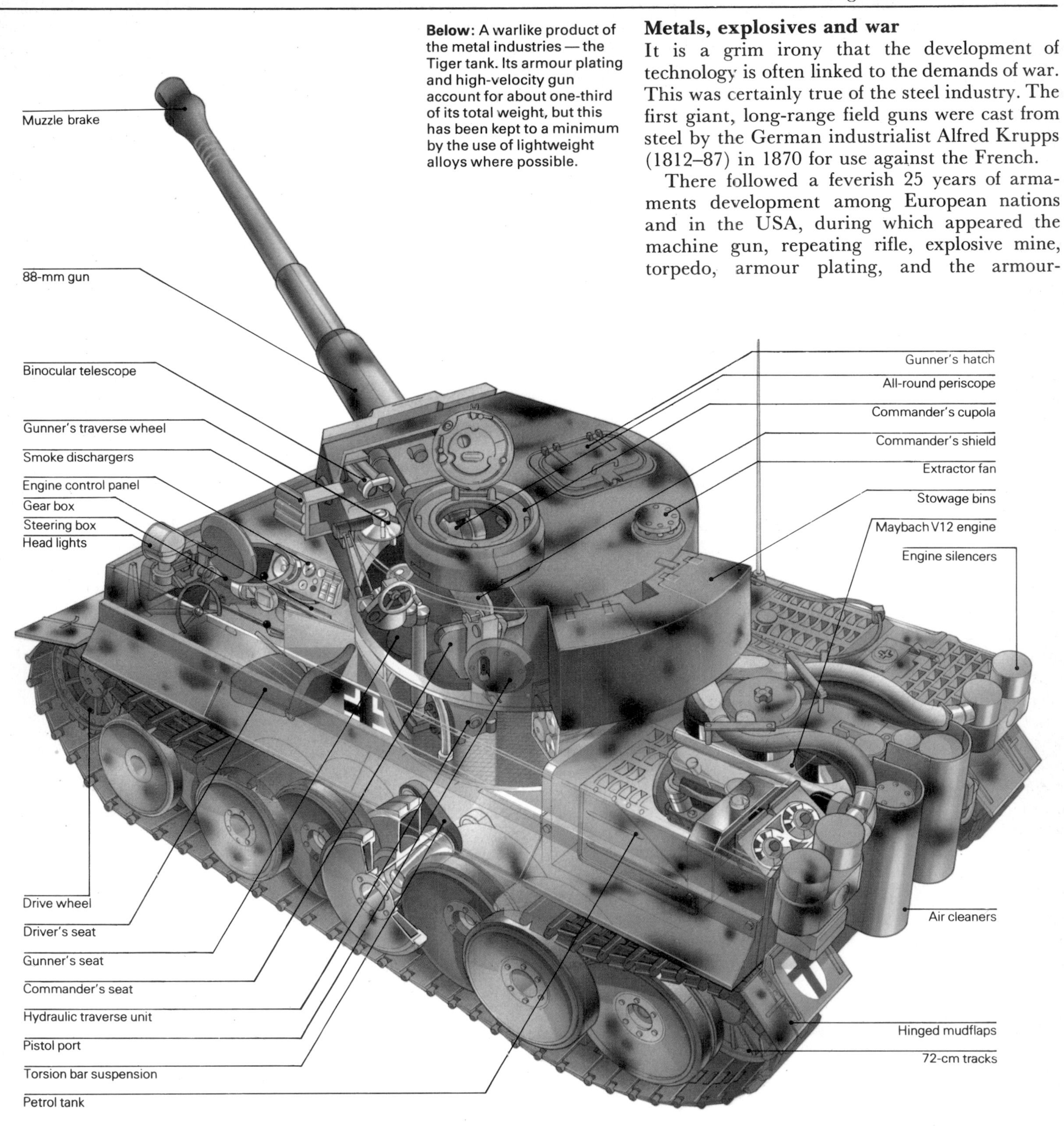

Below: A warlike product of the metal industries — the Tiger tank. Its armour plating and high-velocity gun account for about one-third of its total weight, but this has been kept to a minimum by the use of lightweight alloys where possible.

Metals, explosives and war

It is a grim irony that the development of technology is often linked to the demands of war. This was certainly true of the steel industry. The first giant, long-range field guns were cast from steel by the German industrialist Alfred Krupps (1812–87) in 1870 for use against the French.

There followed a feverish 25 years of armaments development among European nations and in the USA, during which appeared the machine gun, repeating rifle, explosive mine, torpedo, armour plating, and the armour-

widely employed as waveband tuners for radio sets.

Case hardening is the production of a very hard outer layer on objects of iron and steel. These are usually dipped in a bath containing molten cyanides and other salts, so that carbon from decomposing cyanide penetrates their surfaces to form hard iron-carbon compounds.

Catalyst is a substance that accelerates the rate at which chemical reaction takes place, without being consumed in the reaction. Catalysts are widely used in the manufacture of plastics, petroleum and industrial chemicals.

Catalytic cracking is the process of heating the heavier fractions from the distillation of crude oil together with a catalyst, usually metallic, in order to crack, or decompose them further into lighter petroleum fractions.

Celluloid was an early plastics material, invented in 1855 by Alexander Parkes and later used to make photographic film, toilet articles and toys. It is made from nitrated cellulose and camphor, and is dangerously flammable.

Cermets are composite materials made from metals

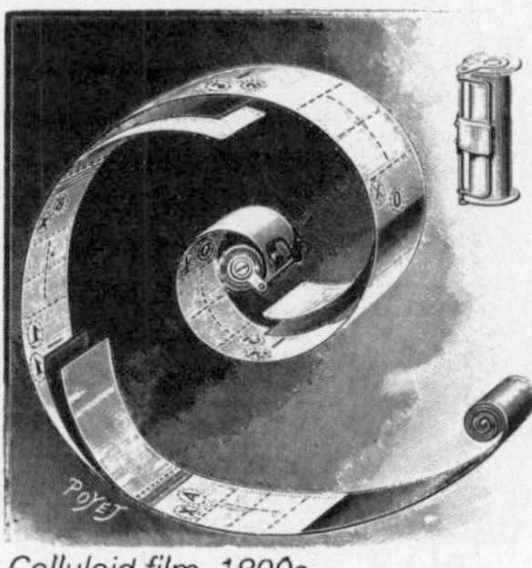

Celluloid film, 1800s

and ceramics by methods of POWDER METALLURGY. Their great hardness and resistance to corrosion and high temperatures finds them used as cutting tools and components of jet engines and nuclear reactors.

Chemical analysis includes qualitative analysis, in which the identities of particular elements and compounds in a mixture are determined, and quantitative analysis, in which their exact amounts are also found. For examples of the scientific methods and instruments used, *see also* CHROMATOGRAPHY, COLORIMETER and SPECTROSCOPY.

Chloroform and ether were the first general anaesthetics widely used for major surgical operations. Their use extended from the 1850s until the 1930s, after which chloroform was abandoned for this purpose as too toxic.

Chromatography is a versatile technique for separating out chemical substances from their mixtures. Its various processes all rely on the principle that the components of a mixture will be carried at different speeds through an absorbing

piercing shell, all of which consumed considerable quantities of mild or alloy steels.

Chemical explosives kept pace with this weaponry. Cordite, used as a propellant for bullets and shells in both world wars, was a mixture of nitroglycerine with another high explosive, guncotton, and petroleum jelly, all moulded into the form of thin sticks. The explosives packed into bombs and shells, some of more devastating power even than nitroglycerine, included TNT (trinitrotoluene), invented during World War I, and RDX (cyclonite), employed in World War II and one of the most powerful chemical explosives available.

However, the limits of destructive power had not been approached even with such explosives as RDX. The atomic bombs dropped on Japan by the USA at the end of World War II had an explosive force many thousand times greater still. By another great irony, the metallic explosives in these bombs were the same NUCLEAR FUELS employed in post-war years in power stations, for the generation of useful electricity.

Even now, the full limits of destructiveness had not been reached: H-bombs tested by a number of nations since the war have had an explosive power as great as that of 60 million tonnes of RDX. Nuclear warheads as powerful as this can now be fired halfway across the world in Intercontinental BALLISTIC MISSILES. The current military trend, however, is for smaller, tactical nuclear weapons in which warheads are carried in GUIDED MISSILES.

Urban technology

The modern city, in which most people of developed countries now live, is a product of recent technology. This fact is most obvious from the city's high-rise buildings and urban motorways. But it applies equally to what lies under the buildings and streets and is, perhaps, more taken for granted: not only underground railways, but also the systems of pipelines carrying fuel gas, purified water and sewage for disposal, and the network of conduits carrying the city's electrical supply. Often, the gasworks, water purification plants, SEWAGE TREATMENT plants and electrical power stations from which these

Below: Steel, which first became available in large quantities late in the 1800s, is the key material in modern building construction. It is employed widely both for main load-bearing structures and in the forms of rod and mesh for reinforcing concrete for floors, pillars and walls. Often, a tall building will consist mainly of these two materials together with glass for windows and lightweight aluminium and plastics for cladding.

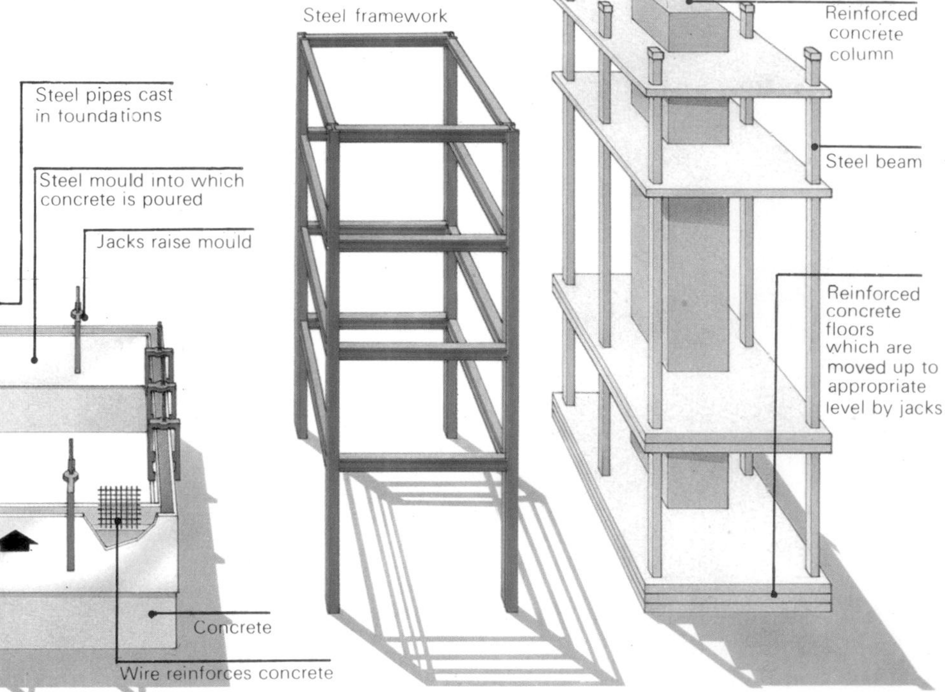

Above: The concrete walls of tall buildings are constructed by a continuous process of slip-forming. Liquid concrete, or slip, is poured into hollow steel shuttering. As the concrete sets, the shuttering is raised by jacks so that more can be poured.

Above: High-rise buildings are now often constructed by hanging walls from, and supporting floors on, a strong steel frame. Steel pillars and beams for this frame are raised from solid foundations by a crane which may be 100 or more metres in height.

Above: In the lift-slab system, floors of reinforced concrete are cast at ground floor level around steel pillars, then are raised to the required height by jacks. A large, central reinforced-concrete pillar adds strength to the building.

Below: Concrete piles to support large buildings may be driven as shown. A vibrator is used to drive the steel casing for casting the pile. This casing is afterwards removed and may be reused.

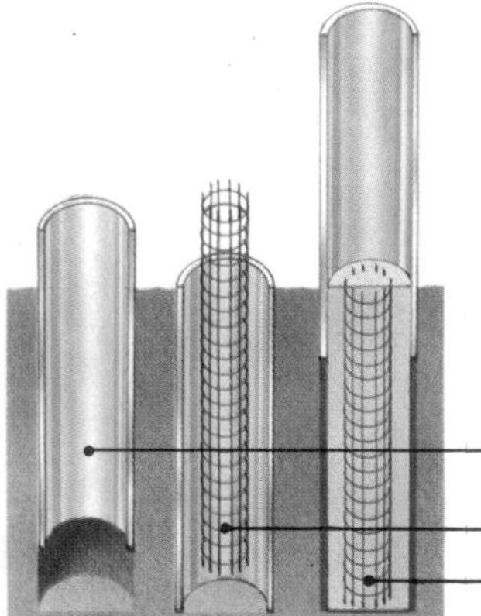

column by a flowing liquid or gas, and so be separate from one another. It is used for chemical analysis and for making small amounts of pure compounds.

Cinematography is the art of making motion pictures. It began with the invention, by Thomas EDISON in the 1890s, of a cine camera he called the kinetoscope, with which he took a series of still photographs in rapid succession. Commercial cinema projectors were in use by the early 1900s.

Coenzymes are organic chemical compounds related to vitamins. They combine in the body with more complex molecules called enzymes, activating these as biological CATALYSTS.

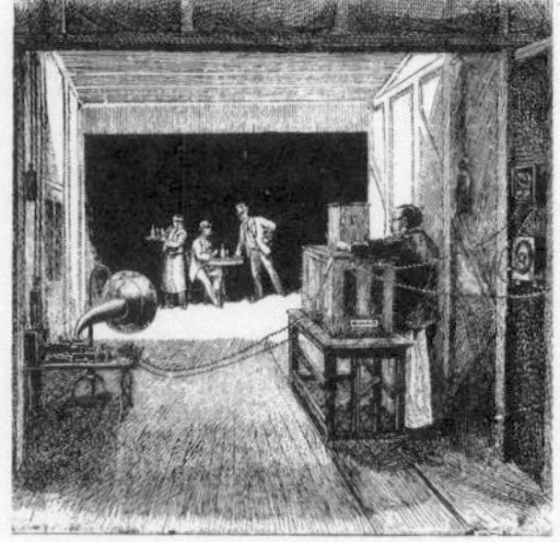
Edison's cinema, 1900

Colorimeter is a laboratory instrument used to measure the concentration of substances by the intensity of colour of their solutions. A sample solution of unknown concentration is compared with solutions of known strength until they match.

Combine harvesters are agricultural machines which cut and thresh cereal crops, separate and clean the grain and bind and deposit the stalks ready for collection. Invented in the 1890s, they were at first steam-powered, but the giant combines of today have diesel engines.

Compressors and blowers are machines that supply flows of air or other gases for industrial processes. From the heavy cylinders of a compressor, gas is delivered at high pressure. A blower supplies a greater flow at a lower pressure, by means of a fan or impeller.

Computers, digital, are electronic calculating machines. They perform laborious or complex operations in arithmetic very quickly and precisely. This happens in a unit called the central processor, which contains a memory bank. Information and instructions are fed in as a program, which, as it enters the computer, is coded electronically into a digital form. Answers are printed out and may also be shown on a screen.

Contact process is that now used industrially to make sulphuric acid. Sulphur dioxide gas is made by burning sulphur or a sulphureous mineral, mixed with air and brought into contact with a CATALYST, vanadium pentoxide. It oxidizes to sulphur trioxide, which dissolves in water to give sulphuric acid.

networks lead, are also within the city boundaries. All are developments of the last 150 years.

The first modern multi-storey buildings, the skyscrapers of American cities, began to rise in the 1880s. They also were products of cheap steel, their secret being a steel skeleton of great strength which evenly distributed the weight of the walls, so that great height no longer demanded inconveniently massive foundations.

High-rise blocks of today are built slimmer and more lightweight still. Often, their walls are largely of glass, held within narrow metal frames. Reinforced concrete, containing steel wire or bars, has become a favoured material for load-bearing structures, and aluminium and plastics materials are often chosen for non-load-bearing parts. During construction, complete sections of such buildings will be supplied to the site, having been made by processes of PREFABRICATION.

The birth of electric power

In 1831, Michael FARADAY demonstrated the first electric motor. This was a pivoted metal wire carrying an electric current, which, when Faraday placed it between the poles of a magnet, rotated about its pivot.

This demonstration also showed the way to electricity generation, for if the magnet moves and the wire is held steady, an electric current is generated in the wire. An electric generator, in fact, is an electric motor reversed.

Above: Modern urban roadbuilding, as in this London development, requires a range of engineering techniques which began only in the 1920s.

Right: The first electric light bulb to work satisfactorily was that of Thomas A. Edison, who patented his carbon-filament lamp in 1880. Filaments were made by burning strips of paper or wood. Earlier types, made by Edison and several other inventors of the 1800s had lasted only a few hours at best, but Edison's patented lamp had a burning life of several hundred hours. Of course, to make it useful, it needed a steady supply of electricity. This Edison also provided, by inventing a power station!

The first working electric generator, that of the French physicist Hippolyte Pixii (1808–35), appeared a year later. It had a hand-turned magnet which induced an electric current in a fixed wire coil. The rotation of a heavy magnet was inconveniently cumbersome, but all that was needed was for the wire and the magnet to move relatively to one another. In later, more efficient generators the wire coil was made to rotate in the field of a stationary magnet.

Other refinements were quickly added by a number of inventors. Generators were fitted with more powerful ELECTROMAGNETS, the electric current for which was supplied at first by electric BATTERIES and later, more logically, from the output of the generator itself.

By the 1860s such DYNAMOS, driven by steam engines, were being used to supply electricity for arc LAMPS in lighthouses, and by the 1880s their electric current provided a novel form of lighting for theatres and other buildings in Paris, London and New York. In 1882, Thomas EDISON opened

---

Conveyors are used in industrial processes wherever materials or products need to be moved about. Endless conveyors include those with moving belts, metal rollers and scrapers, overhead chains holding trolleys and cables holding skips. Powdery materials may be conveyed by fluidized beds and packaged goods are often conveyed by fork-lift trucks.

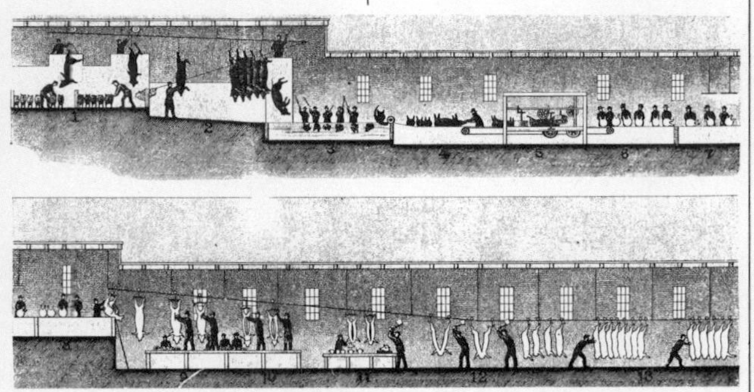

Early use of conveyor belt, Chicago, 1897

Cooling towers are familiar sights around power stations. Often 70 metres or more in height, they cool water from the turbo-generators. The hot water is pumped to the top and, falling inside as a spray, is cooled by a natural draught of air rising through louvres from below.

Corrosion prevention is often necessary for metals exposed to damp air or corrosive chemicals. Protective coatings for metals include special paints and sprays containing corrosion inhibitors. Large metal structures such as pipelines, bridges and ships can be protected by an impressed electrical potential which cancels out that caused by corrosion processes. This is usually called cathodic protection.

D

DDT is a powerful insecticide, employed successfully all over the world, during and after World War II, to kill insect pests. However, like other highly chlorinated hydrocarbons, it is toxic to animals. It is also very persistent, and for this reason is now used more sparingly.

Desalination is the removal of salt from seawater to make it drinkable. Desalination by distillation, using solar power, is a new development in some hot countries. Other methods use DIALYSIS.

Dialysis is the separation of dissolved substances by their selective passage, or diffusion, through a semi-

Right: The large water wheel has been a source of power for at least 2,000 years. It gave rise to the windmill, traditionally employed both for grinding corn and for pumping water. A much more recent power source is the solar collector, by which the Sun's rays are converted directly into useful electricity.

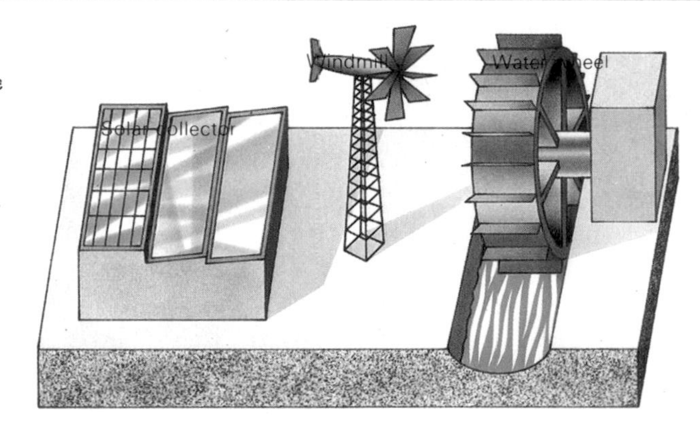

Right: In our large power stations electricity is generated from the chemical energy of fossil fuels such as coal and oil, from the nuclear energy of uranium and plutonium fuels, and, in hydroelectric power stations, from the potential and kinetic energy of water. In fossil-fuel and nuclear power stations, heat from the fuels is employed to raise steam at high pressure in a boiler or heat exchanger. The steam passes over and rotates turbine wheels which are coupled in large turbogenerators to the rotors of electric generators. The electricity output from these giant machines is at a high voltage, but needs to be stepped up much higher still, by transformers, before it can be distributed on the national grid.

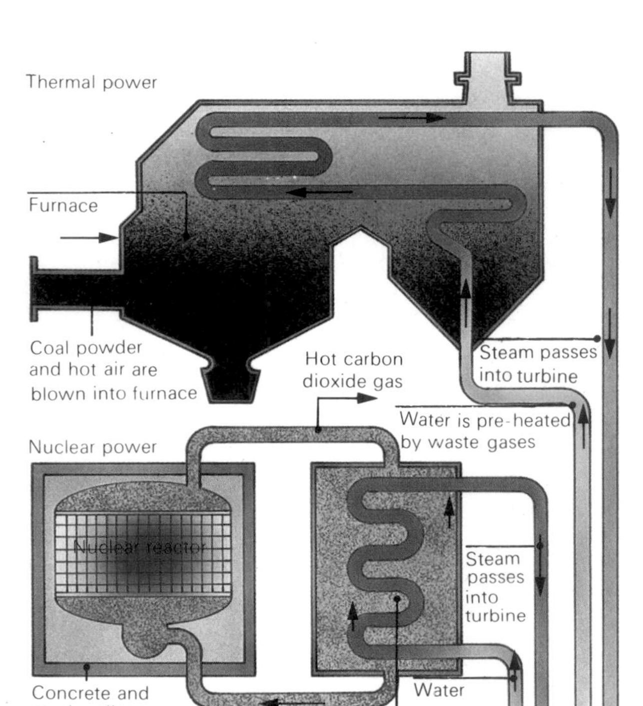

electric power stations containing several generators. These supplied electric current locally, partly for lighting up the incandescent filament lamp bulbs Edison had also invented.

Edison's generators, like most others before 1900, were driven by reciprocating steam engines. The first steam TURBOGENERATORS, ancestors of the giant machines now employed in modern fuel-burning power stations, were installed by the British engineer, Sir Charles Parsons (1854–1931), in 1888. A few years later a much larger power station at Niagara Falls, USA, began to generate electricity from turbogenerators driven by falling water. This was the first hydroelectric power station.

Modern electricity supply

The electricity generated by an early power station, such as those of Edison, would at most have sufficed to energize about a thousand modern domestic heaters, or ten times as many light bulbs. That is to say, its power output was about 1000 kilowatts. In contrast, the output of a modern power station is measured in hundreds, or even thousands, of megawatts, a megawatt being 1000 kilowatts.

Also, whereas Edison's generators made electricity as direct current, or dc (a form of electric current still used for such purposes as electroplating), modern turbogenerators make alternating current, or ac.

At the output terminals of a turbogenerator the alternating voltage is high, but still not high enough for power distribution. More energy is lost at low voltages and so very high voltages

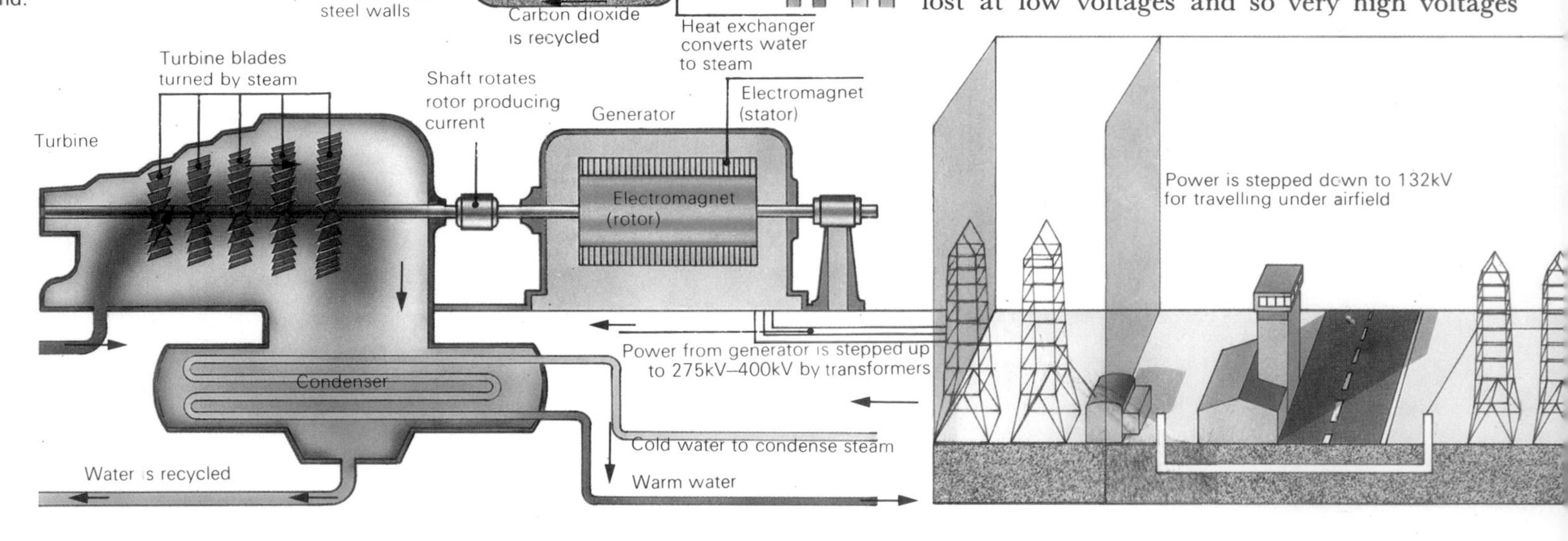

permeable membrane. It is used to purify solutions, as in medical dialyzers, which purify blood of nitrogen wastes, and seawater electrodialyzers, which remove salts.

Disinfectants are chemical substances used to kill microbes and prevent unpleasant odours caused by them. They began in the 1880s with the carbolic acid used by Joseph Lister as an antiseptic spray for operations.

Dyes, synthetic, began in 1856 with mauvein, a dye made from aniline, a coal tar product, by William Perkin. This was followed by other aniline dyes, the azo dyes, and eventually by dyes synthesized from petrochemicals. Nearly all synthetic dyes are aromatic chemical compounds, that is, their molecules contain benzene rings.

Dynamite is a high explosive invented in the 1860s by Alfred Nobel. He absorbed the dangerously explosive compound nitroglycerine in the inert, earthy substance, kieselguhr. The resulting material, made usually in the form of sticks, was safe to handle but could be detonated successfully for military and industrial purposes.

Dynamo is a machine which employs magnetism to make electricity. It ranges in size from such small machines as motor car dynamos, which provide direct current, to the giant generators of power stations, which provide alternating current.

E

ECG is short for electrocardiograph. This is a medical machine used to measure the activity of the heart, by small changes in its electrical potential.

EEG is short for electroencephalograph. This is a medical machine used to detect and record the electrical activity of the brain.

Edison, Thomas Alva (1847–1931) was an American inventor. His inventions numbered hundreds and included the telegraph and

Recording on the first model of Edison's phonograph

indeed are needed during long-distance transmission by metal cables. For this reason, the alternating voltage of the turbogenerators is stepped up to the required high level in TRANSFORMERS, before electricity leaves the power station in power lines of the distribution grid. These lines are slung safely on INSULATORS between tall pylons, or, in urban areas, buried underground.

The grid voltages of 100,000 or more, necessary for transmission, are far too dangerous for electricity users. Accordingly, before electricity is supplied to homes, offices and factories, voltages are greatly reduced by transformers in substations, to the familiar levels of about 240 volts or about 120 volts.

Sources of energy

The turbogenerators of power stations are turned either by steam from high pressure boilers, or by flowing water. The sources of this energy vary. The longest-established method for heating water to raise steam in power station boilers is the burning of coal. With the growth of the petroleum industry in the early years of our century, oil-fired power stations also began to be built, and in many countries these now rival or surpass coal-fired power stations in numbers and output. However, the fossil fuels coal and oil are in strictly limited supply in the exploitable regions of the Earth's crust. New sources of energy must be found to meet an ever-growing demand for electricity, or even to sustain supply at its present level.

One answer to this problem appeared during World War II, with the development of the NUCLEAR REACTOR. Nuclear power stations are now scattered widely across developed parts of the world, and nuclear fuels, if not in unlimited supply, can be expected to last out longer than the fossil fuels.

The grave drawback of nuclear power stations is that their waste products remain dangerously radioactive for the foreseeable future. Opponents of nuclear power programmes argue that the accumulation of these wastes could eventually poison the world.

Hydroelectric power, being pollution-free, is a much-favoured answer to the world energy crisis. Conventional hydroelectric power stations exploit either the potential energy of high lakes or the kinetic energy of large river flows. However, most countries have too few natural resources of this kind to be able to rely heavily on them for electric power.

This difficulty has turned the eyes of inventive scientists and engineers towards the oceans, as an almost unlimited hydroelectric power resource. A few TIDAL POWER STATIONS are already in operation. Wave energy machines are being keenly developed in a number of countries. A more distant prospect is the ocean thermal generator, a giant, floating power station which utilizes temperature differences between upper and lower parts of the oceans to move a fluid to generate electricity.

The heat of the oceans is one kind of geothermal energy. Another is heat in the Earth's crust, which may rise to the surface as hot water and steam. In a few places which have such hot

Below: Electricity is distributed widely from power station to factory, office and home by a grid system comprising a network of cables overground and underground. Electricity at the highest voltages is carried over the longest distances by the heavy cables of pylons, slung safely far overhead between porcelain insulators. This high-tension electricity is most suitable for wide distribution because power losses are minimized. At the consumer ends of the grid, high-tension electricity is stepped down in the transformers of substations to the lower and safer voltages required. Customer supply is then mainly by underground cable.

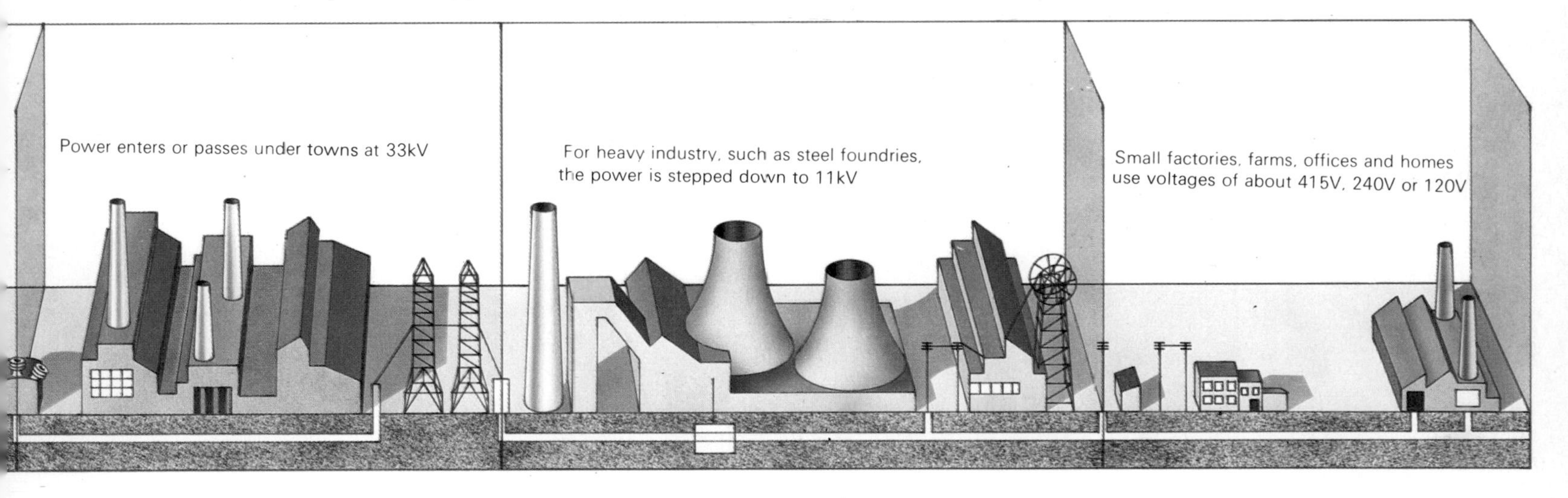

phonograph, as well as the first commercial electric light.

Elastomers are polymers such as natural and synthetic rubber which have elastic properties. This feature is explained by the ability of their long, polymer molecules to slide over one another when the elastomer is pulled or compressed, and afterwards to regain their original positions.

Electrolysis is the decomposition of an electrically conducting liquid, or electrolyte, by an electric current. Industrially it is used widely, as in electroplating and in the electrolytic extraction of metals, particularly aluminium.

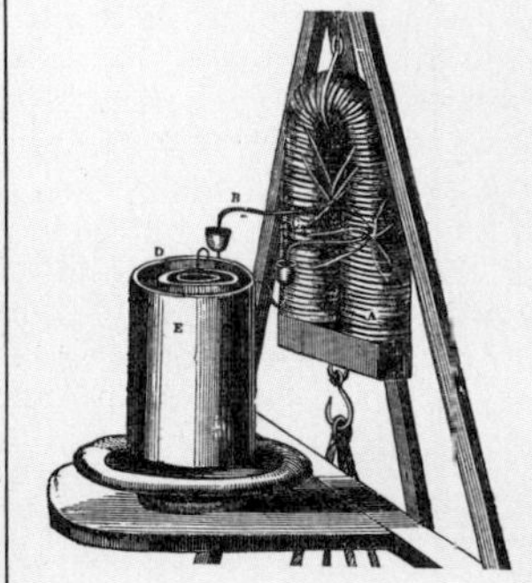

Electromagnet, 1835

Electromagnet has a soft iron core around which is wound a wire coil. When a large electric current is passed through the coil, a powerful magnetic field is generated around the core. The first electromagnets were made in 1825 by the English scientist William Sturgeon (1783–1850).

Electron microscope is capable of magnifying nonliving objects up to a million times. A beam of electrons is focused, usually magnetically, on the object. Electrons scattered by the object are detected on a fluorescent screen, on which the magnified image appears. This instrument was developed by various physicists during the 1930s.

Electroplating is the process of depositing a metal coating, usually on to another metal, by electrolysis of a solution of the metal being deposited. Cadmium, chromium, copper, nickel and silver are commonly electroplated on steel.

Enzymes are biological CATALYSTS, which enable most of the reactions in living cells to take place. Chemically they are proteins. They are prepared pure for biochemical analysis and synthesis. In the home, they are most familiar as constituents of 'biological' washing powders.

Excavators are large machines employed for earth-moving operations. They include bulldozers, power shovels, dredges and draglines.

Extruders are machines employed in the fabrication of metal, plastics and rubber strip and tubing. Material to be extruded is forced through a metal orifice of the required cross section.

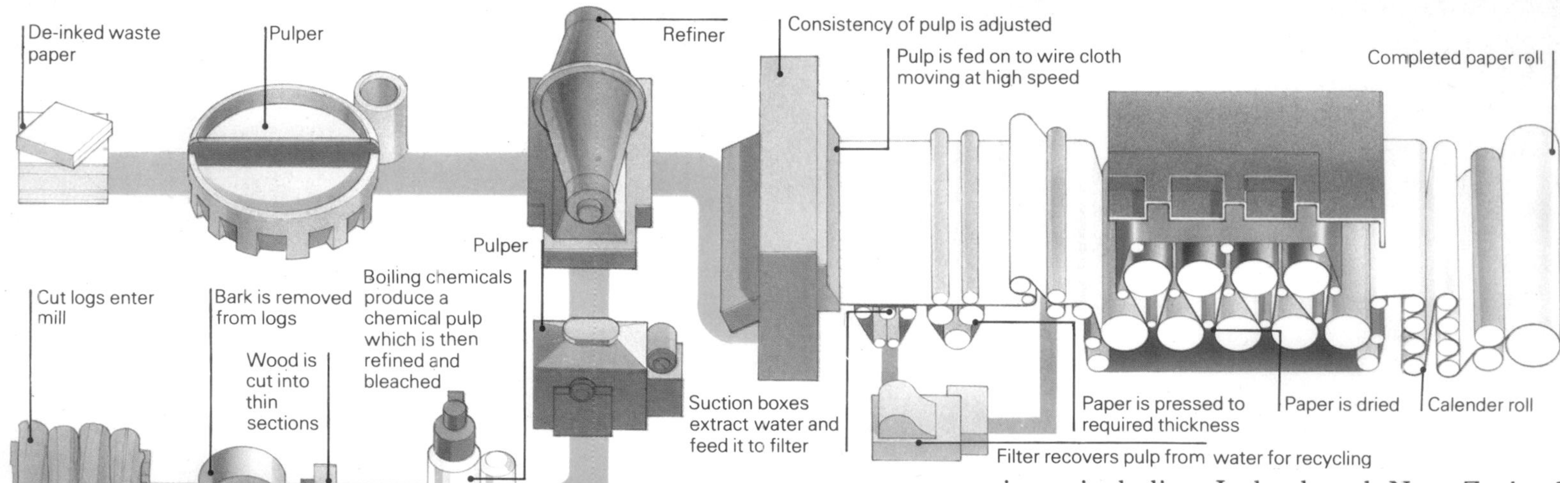

Above: Papermaking is now a continuous process in which raw materials, either fibrous wastes or timber, are fed in at one end, and the finished paper is rolled up at the other.

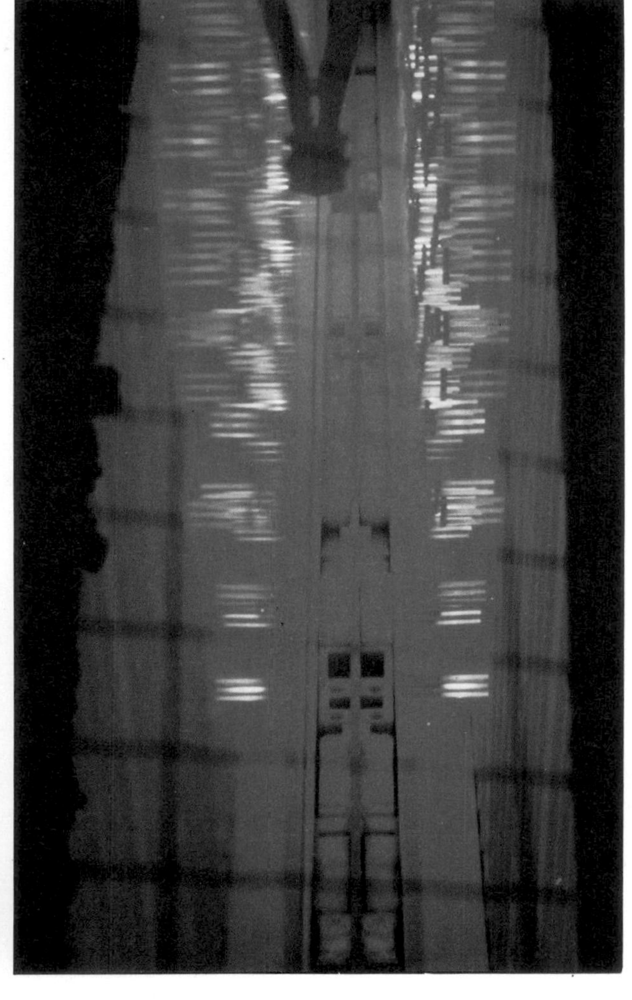

Right: The float glass process makes continuous sheets of plate glass. Molten glass is floated on a bath of molten tin, where it cools and hardens to an even thickness and a high polish.

Below: Modern ceramic ware is fired at up to 1600°C. This kiln is lined with a thin layer of almost pure alumina fibre, a specialized, efficient refractory material.

springs, including Iceland and New Zealand, geothermal power stations have been built to convert this heat into electricity. The extraction of useful heat from further underground is a future possibility.

The winds have long provided power for machinery to grind corn and to pump water. A new kind of windmill has recently been invented, which is made to generate electricity by the rotation of its huge, aerodynamic propellers. This type of generator is, however, more or less restricted to perpetually windy places, such as Jutland in Denmark, where the first of these windmill generators has been installed.

The energy of the Sun's radiation, as we receive it here on Earth, is another great, largely untapped power resource. Solar power stations are, as yet, undeveloped, but such devices as the SOLAR BATTERY and the SOLAR FURNACE point the way. Another distant possibility is thermonuclear power generation, which would tap the immense energy of nuclear reactions of the kind that cause the Sun and stars to shine. Man has already released this energy, with devastating results, as H-bomb explosions, but useful control of thermonuclear reactions presents difficulties that are now only beginning to be tackled.

Progress of traditional industries

With the rapid growth of consumer societies, the output of the pottery, glass, paper and printing industries increased greatly, and their technology grew more sophisticated.

As fuels for pottery, brickworks and glass kilns, coal and coke were replaced by coal gas and PRODUCER GAS and in our own century these gave

---

Metals are extruded either hot or cold.

F

Faraday, Michael (1791–1867), an English physicist and chemist, held the chair of chemistry at the Royal Institute from 1827. He developed the laws of electrolysis named after him, discovered electromagnetic induction and built the first dynamo.

Michael Faraday

Fibre optics uses glass fibres of high transparency to transmit light and images. Since the fibres are also flexible, they can be used to view objects around bends, as in the medical examination of lungs, stomach and intestine.

Fish farming is the cultivation of fishes for food. Over the past 30 years, tank-breeding experiments have been made in many places. In North American rivers, large stocks of edible fishes have been raised by farming methods.

Fluidized bed is a porous surface through which air or gas is pumped to keep a powdery material in suspension above the surface. It is used industrially for such processes as the roasting of powdered ores, and also as a method of conveying powders from place to place.

Freeze drying allows the long-term preservation of such products as foods and vaccines. The products are quick-frozen and at the same time most of their moisture is withdrawn in a high vacuum.

Fuel cells generate electricity directly using the energy of chemical reactions. They differ from electric batteries in that they are supplied continuously with fuels, usually gases. Their applications include powerpacks in spacecraft.

G

Genetic engineering is the science of altering, in a controlled manner, the genetic material of a living organism, that is, the giant molecules of DNA and RNA within its cells. It is practised in laboratories chiefly to breed new strains of microbes with special abilities, either for research or for more immediately practical purposes.

Geophysical prospecting uses the techniques of geophysics to locate natural reserves of petroleum and metals. These techniques include the sending of shock waves down into the Earth's crust, which, after they have been reflected from various

Left: Fluid-bed machines are used for processing and transporting powdered materials. Shown is a fluid-bed drier, in which warm air buoys up and dries powders for use as pharmaceuticals.

Right: Aerial view of a large chemical fertilizer factory. Fertilizers contain nitrogen, phosphate and potash which are the principal nutrients which plants need for effective growth.

way in turn to NATURAL GAS. Traditional pottery ware was joined by a range of ceramics compounded and processed scientifically for particular industrial duties. These notably include the glazed porcelain INSULATORS used for high-voltage electrical equipment. With the growth of the steelmaking and chemical industries, specialized firebricks came into demand for the lining of furnaces.

The mass production of glass became fully automatic, with large machines for the continuous drawing of glass sheet and tubing, and the moulding of glass containers. New, specialized products developed for use in other industries included GLASS FIBRE, foam glass for rigid thermal insulation, laminated safety glass and borosilicate glass which is resistant to thermal shock.

The technology of papermaking, like that of the mass production of glass, came to rely on large, continuous-process machines. At one end these accept wet, chemically-treated wood pulp. This is drained, squeezed flat and partly dried by heated rollers, and finished by such processes as watermarking and smoothing with a CALENDER, before being rolled up continuously at the other end of the machine.

Printing now employs many kinds of machines and processes. Printing type is assembled and composed into the required order by typesetting machines. The set type is then printed out on other machines. When very many copies are to be printed daily, as in newspaper production, both these processes may be controlled by a computer. In book production, special machinery is used for binding and covering.

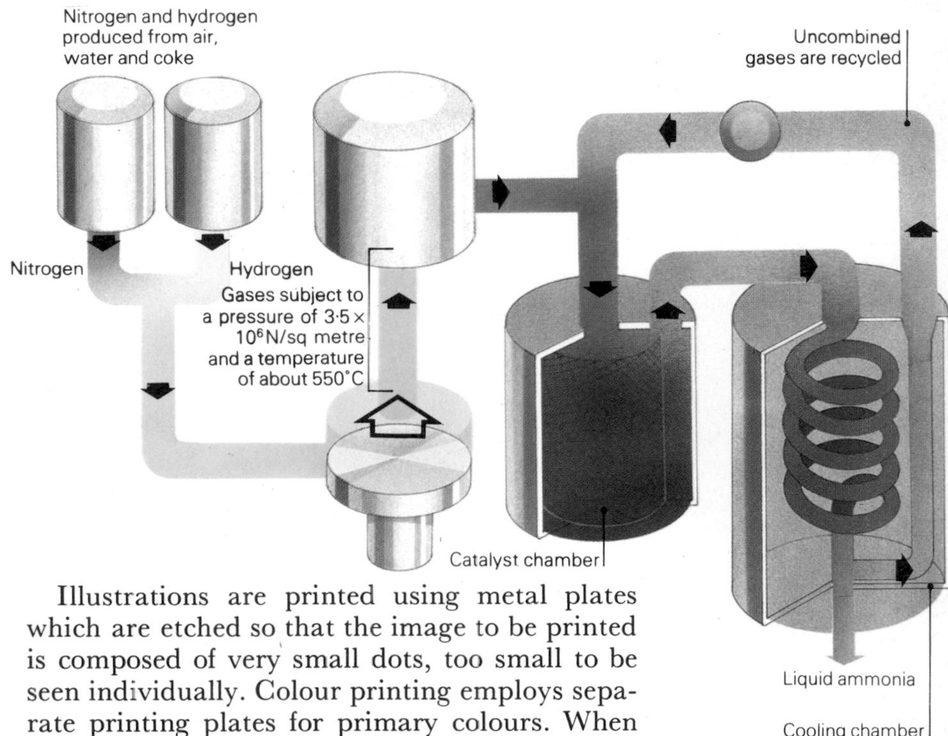

Above: Ammonia, one of the most important chemical raw materials, is synthesized industrially by the Haber process. Nitrogen (N) from the air is combined with hydrogen (H) from water, to make ammonia (NH_3). This reaction takes place in a heated, highly pressurized chamber, with the aid of a catalyst.

Illustrations are printed using metal plates which are etched so that the image to be printed is composed of very small dots, too small to be seen individually. Colour printing employs separate printing plates for primary colours. When these are printed together, the dots of different colours blend to the eye into a colour picture.

Chemical diversity

In the later 1800s, the chemical industry began to diversify into its present form. This started with the invention of more efficient processes for making chemicals of traditional importance, such as the SOLVAY PROCESS for soda and the CONTACT PROCESS for sulphuric acid.

The latter was an important early example of an industrial chemical process employing a CATALYST, although the catalyst used today,

---

strata, are detected and analyzed.

Glass fibre used for repair kits, or as glass wool for insulation, is made by blasting air or steam into molten glass, so converting it into fine filaments.

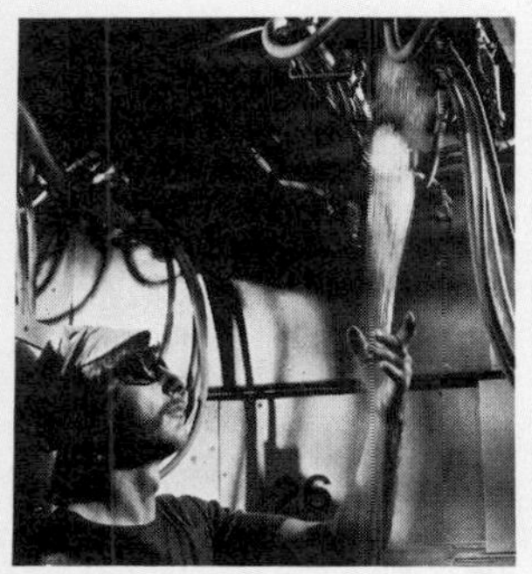

Making glass fibre

Guided missiles contain a mechanism which can be controlled remotely, so that the missile can be directed all the way to the target.

H

Haber process makes ammonia from hydrogen and atmospheric nitrogen, the reaction being catalyzed by iron compounds. It was invented just prior to World War I by the German chemist Fritz Haber (1868–1934).

Herbicides are chemical substances used to kill weeds and other unwanted plant life such as algae and fungi (algicides and fungicides). General weedkillers include simple inorganic salts such as chlorates and arsenates. More complex, organic herbicides are often more selective in their action.

Hormones are the chemical messengers of an animal or plant body, which stimulate growth, reproduction or other functions. Many are now made synthetically on a commercial scale by the methods of biochemistry and microbiology.

Hydrogenation is the process of combining hydrogen with chemical compounds, as in petroleum processing and the manufacture of fats from oils.

Hydroponics is the technology of growing plants without soil. Plant roots are immersed in solutions of nutrient salts, dissolved in carefully balanced proportions and concentrations.

I

Immunosuppressive drugs reduce the vigour of the immune reaction in the body. After transplant operations they make it less likely that the body will reject the transplant. At the same time, however, they increase the likelihood of infections.

Insulators, electrical, are materials of high electrical resistance which prevent the leakage of electricity from equipment or wiring. The insulators that support high voltage electric cables, such as those on pylons, are made of porcelain. Wire can be insulated with flexible plastic material and with rubber coatings.

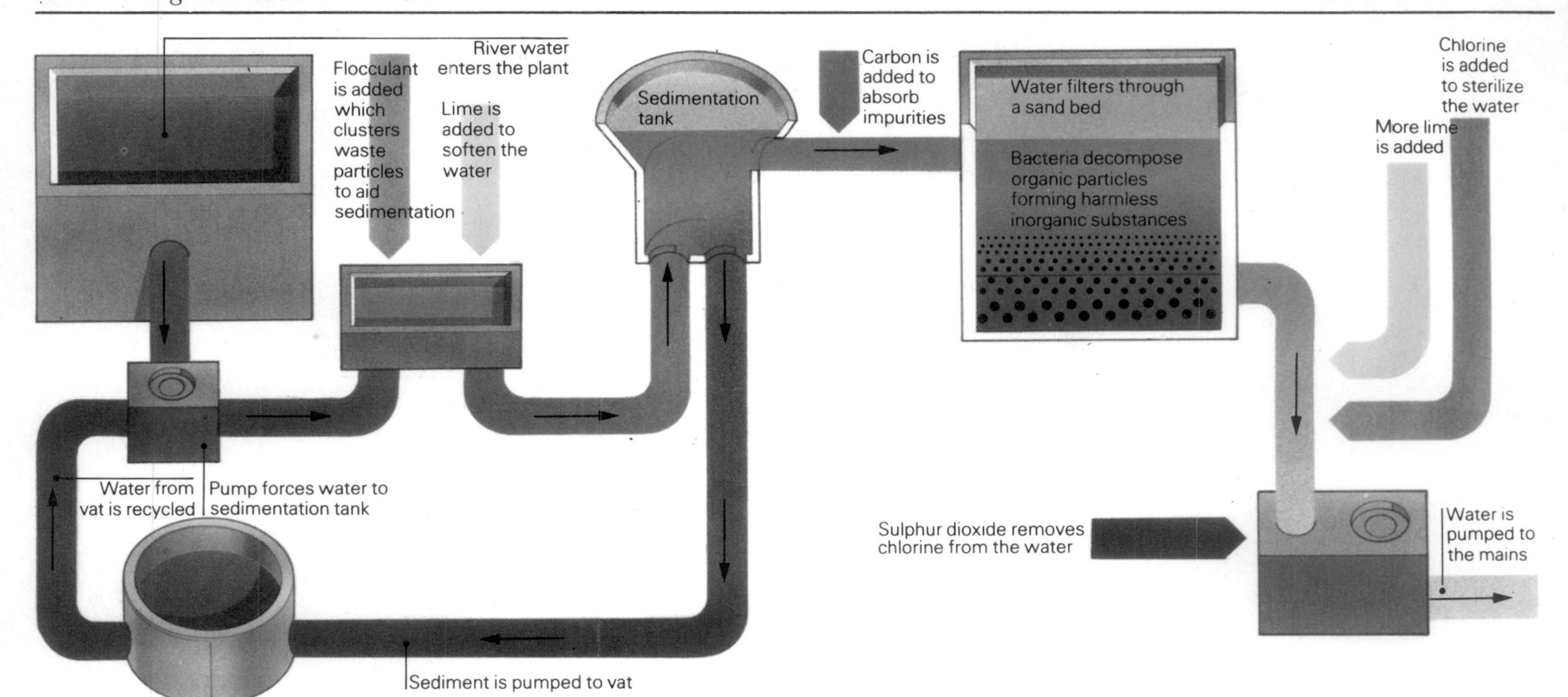

vanadium pentoxide, is a development of the 1900s. So too is the catalytic HABER PROCESS for the synthesis of the widely-used industrial chemical, ammonia; in turn, this development stimulated the production of a chemical compound with an even greater number of industrial uses, NITRIC ACID.

Electricity was put to good use in a number of industrial chemical processes of the 1800s, besides those of electroplating already described. These included the ELECTROLYSIS of brine to make chlorine gas and caustic soda, and the electrolysis of molten caustic soda to make sodium metal.

The above are major examples of inorganic chemical processes and materials. The progress of organic chemical manufacture took a new turn in the mid-1800s with the use of distillates of coal and wood as starting materials for the building-up, or organic synthesis, of more complex carbon compounds. Most diverse of these new organic materials were the synthetic DYES, but coal-tar chemicals later became important also for the manufacture of explosives, pesticides, drugs and synthetic polymers.

Together with the growth and specialization of chemical manufacture grew the science of chemical engineering, for the design, operation

Above: Water from our domestic taps has been purified in plants such as the one shown. Raw water from rivers or lakes is treated to remove all dirt, microbes and unpleasant smells and tastes. In some plants, very fine filters called microstrainers are used as well as sand filters, and ozone may be employed in the final treatment instead of chlorine.

Left: A copper mine and smelter, surrounded by an area made completely barren by pollution from copper wastes. Copper in more than minute concentrations is highly poisonous to plant life. Such pollution is also likely to persist almost indefinitely. There remain barren areas which are the sites of very early copper mining, dating from 6,500 years ago.

L **Lamps, electric,** date from arc lamps of the 1850s. The first successful filament lamps, similar to our own light bulbs, were invented by Sir Joseph Swan (1828–1914) and Thomas EDISON in the 1870s.

Laser is a device producing an intense beam of light of uniform width and great penetrating power. It was invented by several scientists in 1960. Its suggested uses include microwelding, long-range signalling and the production of thermonuclear energy.

Liquefaction of gases provides industry and research with a compact source of useful gases. Such gases as ammonia, propane and butane can be liquefied by pressure alone. Air, oxygen, nitrogen and hydrogen cannot, and are liquefied in special equipment in which they are alternately pressurized and expanded, each expansion being accompanied by a drop in temperature. Carbon dioxide is condensed not as a liquid, but as a solid.

Evacuating light bulbs, c.1883

Lubricants are materials that reduce friction and are used industrially to prevent excessive wear. They include many oils, greases and solid substances such as graphite compounds.

M **Margarine** consists mostly of vegetable oils, which are made more solidly fatty by blending or by HYDROGENATION. It was first made in the 1860s, mostly from groundnut oil.

Mercerizing is a process, invented in 1850 by John Mercer (1791–1866), of treating cotton thread with caustic soda to give it a silky lustre. A similar process was later used in the preparation of rayon from wood pulp.

Metal cleaning operations include pickling, in which metal parts are freed from grease and oxide films by immersion in an acid solution, before electroplating. In blast cleaning, metals are exposed to an abrasive stream of solid particles which remove dirt.

Microelectronics uses electrical circuits made very tiny, or microminiaturized, so that very many of them can be placed on a single silicon crystal. Assemblies of such circuits are fitted in modern computers and they are also the basis of miniaturized control devices.

N **Napalm** is a blend of petroleum with a

and day-to-day management of chemical plant. Industrial chemical processes, like those of the laboratory, can be described by such names as oxidation, nitration and electrolysis. In chemical engineering these and many other *unit processes* are carried out by a number of well-defined unit operations. These operations involve the use of specially designed equipment including handling machinery such as pumps, COMPRESSORS AND BLOWERS, various types of CONVEYORS, grinders and filters, and reaction chambers for such processes as combustion, absorption and evaporation.

The progress of manufacture on a large chemical plant is monitored constantly by recording instruments, which may also control processes automatically at preset levels of temperature, pressure and concentration. Although many chemical processes are nowadays completely automated in this way, the works laboratory usually still plays its part in the QUALITY CONTROL of the final products, by the CHEMICAL ANALYSIS of samples taken at various stages of the process.

In such plants as the above, manufacture can be continuous, 24 hours each day, for the bulk production of chemicals. These may be finished products in themselves, such as artificial fertilizers, or they may be starting materials or chemical intermediates for other manufacturing processes. Other chemical products, such as drugs and pharmaceuticals, are needed in far smaller amounts and are made by batch processes.

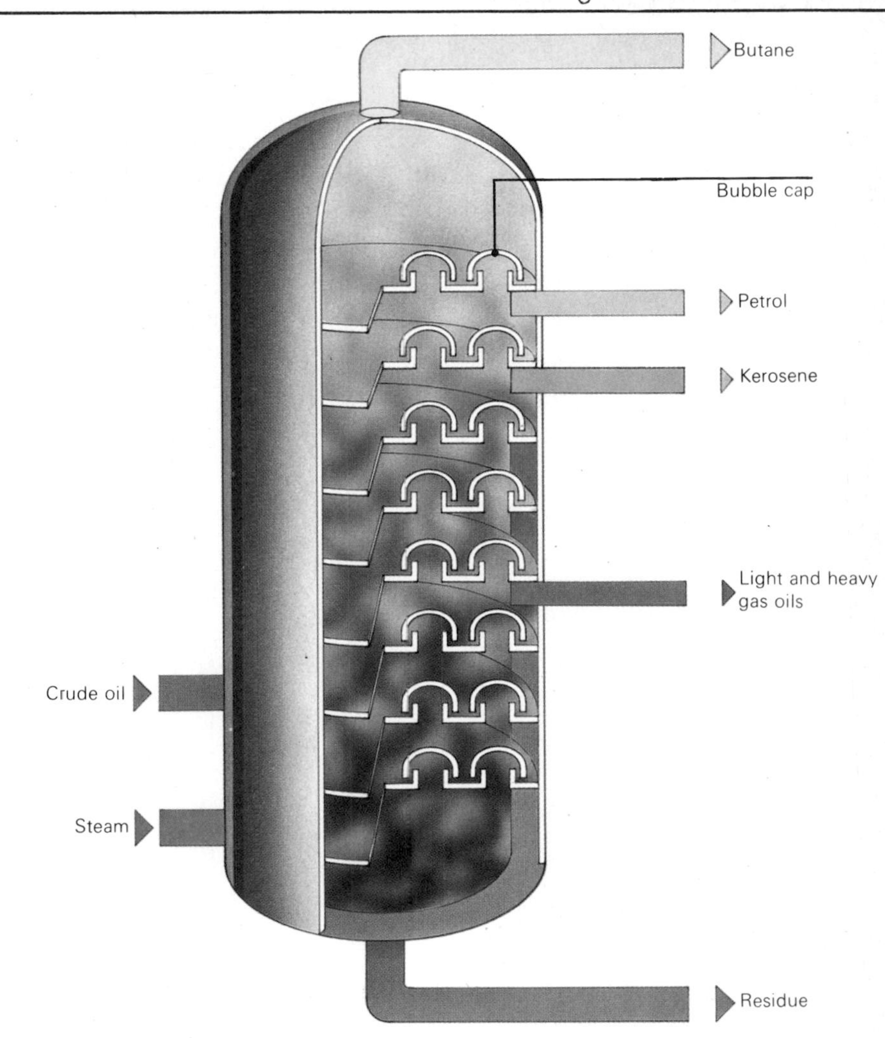

Above: Fractionating towers are used in industry for the continuous distillation of petroleum and other liquid mixtures. Fractions, or components of different densities and boiling points, are separated out at various points on the tower. The lightest in weight pass out at the top end as gases, and the heaviest condense out at the bottom.

Petroleum and its chemicals

Industrialists of the 1800s recognized petroleum as a valuable natural material. Drilling for petroleum started up on a large scale in the USA in the 1860s. Later, deeper drilling there and in other countries was encouraged by a number of developments. These included the availability of cheap steel for making derricks and drilling rods, and of industrial diamonds for tipping drilling bits to cut more easily through hard rock.

They also included the growth of the science of geology. In the search for oil, geologists located likely areas for drilling, then examined drilling cores brought up from deep underground to find minerals and fossils associated with petroleum deposits. GEOPHYSICAL PROSPECTING is a more recent development for locating oil deposits, which can then be proved, and later exploited, by drilling. In recent years, great new reserves of oil and natural gas have been located by these methods under the North Sea.

Petroleum refining is now one of the world's largest industries. Crude oil, as extracted from the Earth's crust and transported to the refinery, is there separated by distillation into more useful, lightweight fractions. The most familiar products of the fractionating towers are the hydrocarbon fuels, petrol (gasoline), kerosene and diesel oil. The quantities of these and other lightweight fractions are increased by further processes of CATALYTIC CRACKING.

As already indicated, crude oil is a rich mixture of substances and its distillation yields not only fuel oils but also a large number of other useful materials. Some, such as carbon black,

---

sodium or aluminium soap, to make a highly flammable, sticky jelly. As a weapon, it was chiefly used in the bombs and flamethrowers of the US forces during the Vietnam war.

Narcotic drugs are used medically to relieve pain and to induce sleep. They include alkaloids such as codeine, procaine and morphine. Opium and its derivative heroin are alkaloid narcotics that are the main causes of drug addiction.

Natural gas has become a major fuel in recent years, particularly for northern Europe. It is found associated with petroleum reserves. Chemically it is about 85% methane, the rest being ethane and other hydrocarbons.

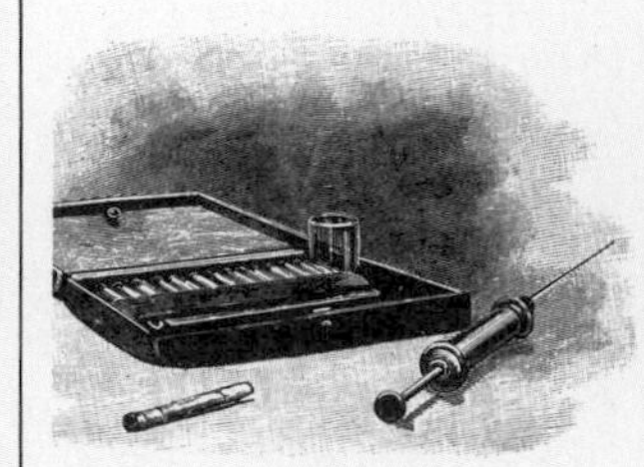

Morphine syringe and ampoules

Nerve gases are odourless, tasteless but speedily fatal. They were invented by German scientists during World War II but never used. Chemically they are organic phosphorus compounds.

Nitric acid has many important uses in the manufacture of explosives, dyes, fertilizers and plastics. Its large-scale manufacture became possible with that of ammonia, by the HABER PROCESS, in 1914.

Nuclear fuels are those consumed in nuclear reactors in the production of nuclear energy. They include uranium 235 and plutonium 239, either of which is likely to be mixed with the less radioactive uranium 238.

Nuclear reactor generates heat and radioactivity by the controlled breakdown of the nuclei of atoms of nuclear fuels. In nuclear power stations the heat is used to produce steam to operate turbogenerators for the generation of electricity. The dangerous radioactivity is absorbed by thick metal and concrete shielding.

O

Open hearth furnace was invented by Frederick Siemens (1826–1904) in 1856 for the large-scale production of steel. Its special feature was the recycling of hot waste gases to preheat the air blast. This both increased the furnace temperature and made steelmaking more economical.

Opencast mining is the extraction of coal or metal ores from seams near the surface. Giant excavators, particularly draglines, are used for this purpose.

Oscilloscope is an analytical instrument much used by electronic engineers. It displays patterns of electrical

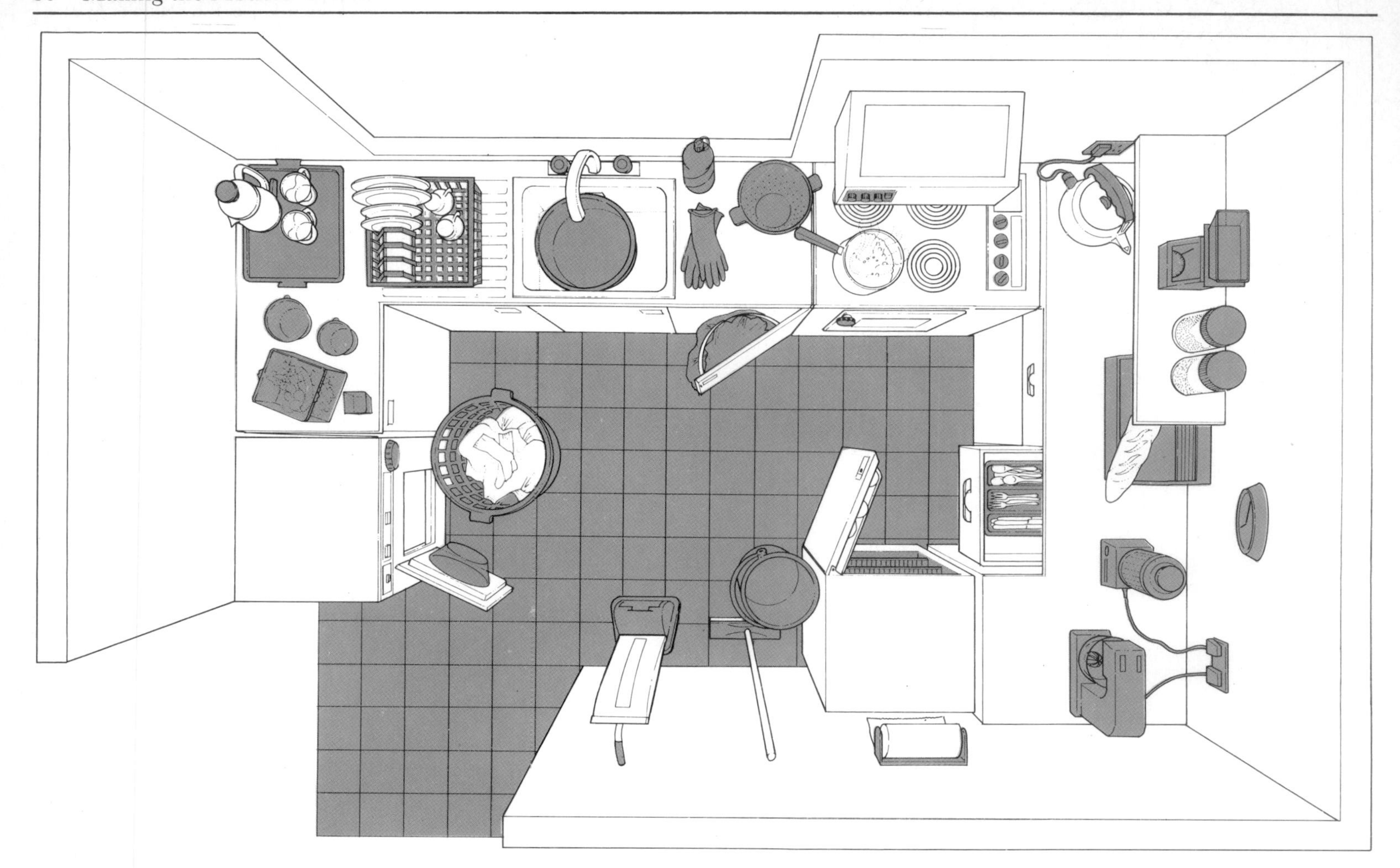

Above: This modern kitchen shows how hard it is to imagine an everyday life without plastics. Fifty years ago, most of the objects now made from plastics would have been made of wood. But even then, it is likely that a few early plastics would have been in evidence. For example, moulded formaldehyde resins being used for knife handles and for electric plugs and sockets.

ammonia and sulphur, are inorganic in composition, but most are organic compounds or mixtures. These last include heavy hydrocarbon materials such as asphalt; oils and greases used as LUBRICANTS; many kinds of industrial solvents; and, perhaps most important of all, hydrocarbon gases such as methane and ethylene, which are consumed in vast quantities by the synthetic polymer industries, and for the production of other PETROCHEMICALS.

Biochemical industry

Chemical weapons against disease began with the DISINFECTANTS of the 1800s. This period also saw the discovery of pain-killing drugs such as NARCOTIC DRUGS and a few milder ANALGESICS, together with the hospital anaesthetics CHLOROFORM AND ETHER.

The great majority of synthetic drugs are, however, inventions of the 1900s. SULFA DRUGS, which revolutionized the treatment of many diseases caused by bacterial infections, first appeared as recently as the 1930s, and ANTIBIOTICS, which abolished most of the remaining bacterial infections from developed countries, were mainly post-World War II discoveries.

Since 1945 the biochemical industry has grown and diversified rapidly, until it now rivals in number of products, if not yet in their bulk, the whole of the rest of the chemical industry. Hundreds of new drug preparations become available to doctors yearly. Among these are the synthetic HORMONES, best known to women in the form of 'The Pill', and IMMUNOSUPPRESSIVE DRUGS, both of which act in an intimate and complex way to affect the control systems of the body.

Many new biochemical products, including the antibiotics, are extracted from the cells of microbes by the methods of industrial micro-

---

waves on the screen of a cathode-ray tube.

Oxyacetylene torch burns acetylene gas in oxygen to give an intensely hot flame. It is used for cutting and welding steel.

P

Pasteurization is a process of heating milk quickly or gently to free it from harmful bacteria without impairing its taste. The process was invented in the 1860s by Louis Pasteur.

Petrochemicals are chemical materials made from petroleum products or natural gas. They now include a very large range of plastics, synthetic rubbers, solvents and chemical intermediates.

Photoelectric cell contains a semiconductor material which, on exposure to light, emits electrons. This effect is put to use in such devices as burglar alarms, the exposure meters of cameras and solar batteries. The first photoelectric cells, made in the 1880s, employed selenium as the light-sensitive materials but those of today usually employ silicon.

Powder metallurgy involves processes of heating and compressing metal powders in moulds until they form solid objects of considerable strength. Tools and gears are among products made in this way. It can also be used with mixed powders of metals and nonmetals, to make such composite materials as CERMETS.

Prefabrication is the manufacture of modular units for buildings so that these can be erected quickly and inexpensively on site. It developed during and after World War II, with the urgent need for more houses and the availability of new, versatile plastics materials.

Printed circuits are electrical circuits deposited as metal films on to plastics or other non-conducting surfaces, and etched into shape.

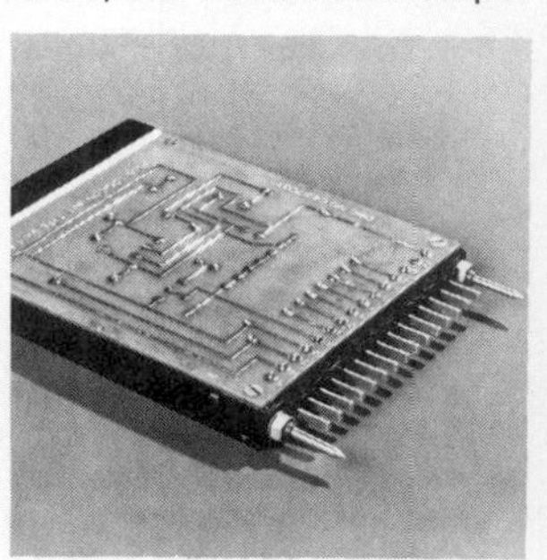

Printed circuit

They were important developments in the miniaturization of electronics but have been largely replaced by much smaller MICROELECTRONIC devices.

Producer gas was used from the 1850s onwards for heating pottery kilns and other furnaces. It is made by blowing air through red-hot coke, and chemically is about 29% carbon monoxide, the rest being mostly nitrogen.

Program is a set of data and instructions for processing in a digital computer. It is written in a programming

Above: Plastics can be called the characteristic new materials of the mid-1900s. For very many purposes they have replaced the wood and metals used 50 years ago, and unlike those natural materials, need never be in short supply so long as sufficient energy is available for synthesizing them.

Below: This bowl is made from the thermoplastic material polypropylene by a process of injection moulding. Propylene, the monomer gas, or raw material, which is polymerized to make the plastics material, is itself a product of the petroleum industry.

biology. Further examples are ENZYMES, which have a growing number of industrial uses, and vitamins, important as food supplements in maintaining a healthy diet.

More generally, microbes are coming to be regarded as useful in their own right, both as food for animals — so-called single-cell protein — and as microscopic machines which can be made to perform very special chemical tasks, including the disposal of organic wastes such as oil spills. The versatility of microbes can now be increased by new methods of GENETIC ENGINEERING, although this branch of science has become somewhat controversial for its connections with the dire possibilities of BIOLOGICAL WARFARE and the breeding of new, artificial viruses.

Synthetic polymers

Plastics materials, as the stuff of our food containers, insulating tiles and clothing fibres, we now take for granted. Yet the first plastics, such as CELLULOID, were invented only a century or so ago, and most plastics of the 1800s were primitive materials compared with the many different kinds we can choose from today, most of which are discoveries of the mid-1900s.

Raw materials for plastics manufacture are supplied mainly by the petroleum industry, as organic gases or other relatively simple organic chemical compounds. These are known as monomers. For example, the gas ethylene is the monomer of the well-known plastic material polyethylene or polythene, and the monomer vinyl chloride, itself manufactured mainly from ethylene, is the monomer of the plastic polyvinyl chloride, or PVC.

In polymerization processes, the simple molecules of the monomers link together, with the aid of a catalyst, to form long, chain-like polymer molecules. Unlike their monomers, these polymers are chemically very unreactive, or inert, solids. This chemical stability is, for their many and varied applications, their most valuable feature. Yet it can also be a disadvantage: plastics do not rot and, being manufactured in vast quantities and thrown away in almost equally large quantities, they cause increasing pollution problems.

Physically, plastics differ considerably in their flexibility and behaviour when heated, which leads to the classification of a plastics polymer as either a THERMOPLASTIC or a THERMOSET. These differences are explained by the numbers and types of chemical bonds, or cross-links, that form between one polymer molecule and another during polymerization. In general, heavily cross-linked polymers are stiffer and less easily deformed than those with fewer cross-links.

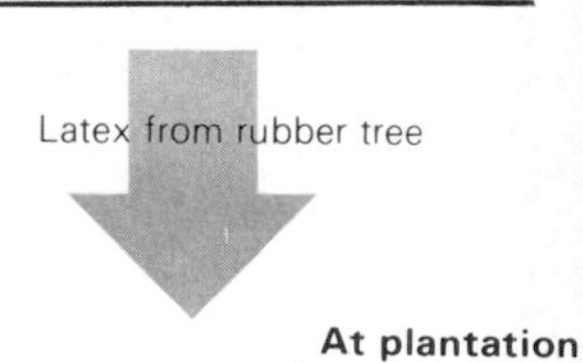

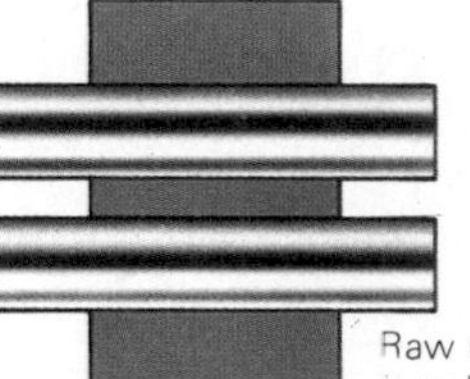

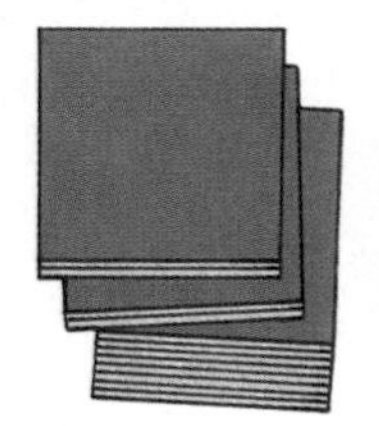

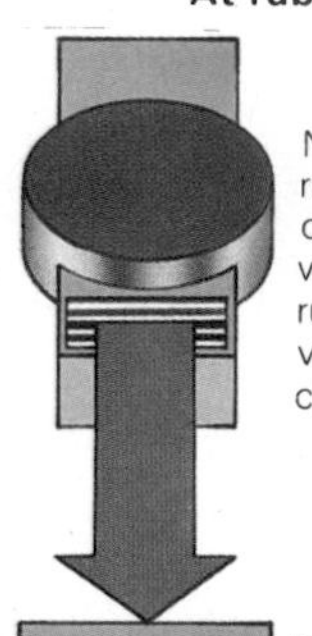

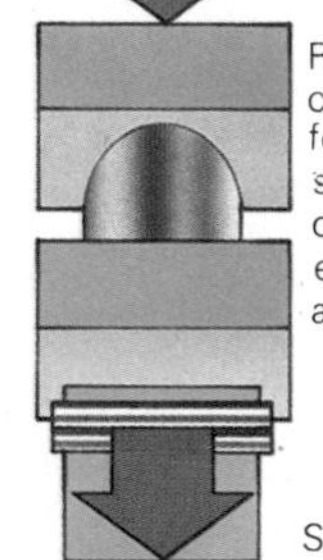

Above: Natural rubber, obtained from the rubber tree, is processed on the plantation into forms suitable for export. When it arrives at the factory, it is blended, or compounded, with other substances before being processed to the desired shape.

language, which enables the instructions to be expressed in a strictly logical sequence.

Prosthetics is the branch of surgery dealing with the development and fitting of artificial replacements for parts of the body which are for any reason missing or defective. Modern prostheses include highly engineered joints for limbs and hips, power-operated limbs, heart valves and pacemakers.

Q

Quality control is the maintenance of required standards of products in a factory. It includes chemical analysis and physical testing of products in works laboratories, and on-line inspection of products by workers or automatic instruments.

R

Radioisotopes are chemical elements that emit radioactivity. Some occur naturally and others are made in nuclear reactors and nuclear explosions. Because their radioactivity is easily detected, many are used as tracer elements to follow the course of complex chemical reactions such as those of the body. The penetrating radiation of radioisotopes such as cobalt 60 is employed in medicine to kill tumours.

Refrigerators began with the ice boxes and ice cellars used to preserve food in the mid-1800s. The heat absorbed by the expansion of the vapour of a cooling liquid within a closed piping circuit, cools the inside of a modern refrigerator. These machines date from about 1880, when the cooling liquid was ammonia. Domestic refrigerators today use cooling liquids called freons.

An ice store, 1861

Rubber, synthetic, was first made in about 1900. The first useful types were made by the polymerization of butadiene, a petroleum product. This 'buna' rubber was made on a large scale by Germany in World War I. Since then, many other types have been synthesized

The physical properties of another class of polymers, the ELASTOMERS, are also to be explained by the behaviour of their polymer molecules relative to one another. Better known as synthetic RUBBER, these are very versatile polymers. To name only a few of their uses, some are major components of tyres, others form the corrosion-resistant linings of chemical tanks, and yet others, the silicone rubbers, are chosen by surgeons for their inertness as internal prostheses (*see* PROSTHETICS) and for their stability at high temperatures they are often used by spacecraft designers.

Machinery for processing polymers includes EXTRUDERS, and calenders similar to those employed for papermaking. Objects made from thermosets are shaped by injecting the plastics into moulds before they have set hard. The machinery employed to make synthetic fibres imitates the spinning gland of a spider. A polymer is melted or dissolved, then forced through the small holes of a spinneret, after which it solidifies into fibres either by cooling or by contact with a chemical solution which precipitates it.

Agriculture and food

The steam-powered mechanization of farming in the 1800s resulted in great increases in farm production. In our own century this process has continued, with the development of such sophisticated machinery as the giant COMBINE HARVESTERS of cereal farms and the automatic milking machines of dairy farms. Modern mechanized farming systems also include highly-equipped facilities for the intensive rearing of farm animals in large numbers; many thousands of chickens may be raised together in a single battery, for the production of eggs or meat. Farm production has also benefited immensely from the scientific development of insecticides such as DDT and of many kinds of HERBICIDES, for the control of pests of various kinds.

With our greater understanding of the genetic, and other biochemical, processes of life has come a greater ability to alter plants and animals to our own requirements. A major example is the breeding of a highly productive rice strain, which has since been grown to alleviate food shortages in many parts of the world. The use of antibiotics to fatten pigs and other farm animals is a more

Above: Cereal farms today are often vast. The machines for harvesting grain, combine harvesters, have also grown larger. These machines carry out the whole harvesting process of cutting, threshing, separating grain and binding stalks.

Right: In recent decades, a major development in stock farming has been the widespread use of intensive cultivation methods. Most familiar of these is the chicken battery, in which tens of thousands of birds, raised for eggs or meat, are penned together in a single building. Farmers argue that only by such methods can they satisfy demands at reasonable prices.

---

and their use now far outweighs that of natural rubber.

Semiconductors are materials having an electrical conductivity which is normally very low but which may increase rapidly with exposure to heat or light. These peculiar electrical properties have led to the use of semiconductors in transistors and photoelectric devices. Examples are the elements germanium, selenium and silicon.

Sewage treatment began in the 1890s in London, with the chemical precipitation of sludge, which was then dumped at sea. In modern sewage plants sewage is purified biologically by the action of bacteria and protozoa. The dried sludge may be used as fertilizer and the purified water is recycled, usually to rivers.

Sewer, 1845

Solar battery is an assembly of photoelectric cells, each of which contributes a small electric current when the battery is exposed to sunlight. Its principal use has been to provide power for spacecraft.

Solar furnace is a device which focuses the Sun's rays with mirrors on to a furnace. This can be used to melt metals, or to raise steam in a boiler for electric power generation.

Solar roof heater is used to provide domestic hot water. Mounted in the roof behind a glass pane, it absorbs the heat of the Sun's rays through a blackened metal surface, and water in nearby piping takes up this heat.

Solvay process is the industrial method for making soda, or sodium carbonate, which succeeded the Leblanc process in the 1860s. First ammonia, then carbon dioxide, are bubbled into brine (a solution of common salt). The sodium bicarbonate so formed is heated to make sodium carbonate.

Spectroscopy is the study of the atomic spectra of chemical elements and compounds. It utilizes such instruments as spectroscopes, spectrometers, spectrographs and spectrophotometers. These have wide-ranging applications in chemical and biochemical analysis.

Stainless steels are iron alloys which resist tarnish-

disputed issue, many experts believing that this could lead to the spread of drug-resistant disease bacteria. Food resources of the future are expected to be increased by methods of FISH FARMING and HYDROPONICS, both of which demand an extension of our knowledge of biological engineering.

Food packaging and processing has undergone a revolution beginning in the 1800s with CANNING *(see page 40)* and with such processes as PASTEURIZATION. In our own time, plastics packaging, and such food preservation methods as FREEZE DRYING, have become commonplace.

Precision instruments

The 1900s can properly be described as the age of scientific instruments. Only with the growth of electronics, which began in the final quarter of the 1800s, did modern scientific instruments become possible. A host of such devices is now available for use in the factory and the home, the laboratory and the medical centre.

Electronic instruments are essential in industry for the control of processes and for the measurement and recording of their data. In industrial laboratories, routine testing of samples and research into new products both employ a wide range of instruments which are wholly or partly electronic in their operation. Important examples of analytical instruments are the chromatograph *(see* CHROMATOGRAPHY), OSCILLOSCOPE and spectrograph *(see* SPECTROSCOPY).

Among the first of large, electronic precision machines were X-RAY MACHINES, first developed in the early years of our century. They are now used both in industry, for such purposes as the detection of flaws in metal welds, and more familiarly, in medicine, for the diagnosis of internal abnormalities. The penetrating radiation of RADIOISOTOPES is utilized both in medicine and in industrial detectors. Electronic instruments such as the Geiger counter and the scintillometer have been developed to detect and measure radioactivity in these and other applications.

In hospitals, soon after the X-ray machine came the even more elaborate ECG and EEG machines. These detect minute electrical variations within the body, so allowing physicians to make diagnoses of heart and nervous disturbances. More recent still are body-scanning machines which detect internal abnormalities by the heat these emit, or by their reflection of very high-pitched sound waves. ULTRASONIC GENERATORS have uses also in metal finishing operations, and in navigation.

Electronic devices using light waves have also found both industrial and medical uses. Of these, instruments employing PHOTOELECTRIC CELLS have been in use for the longest time, but LASERS

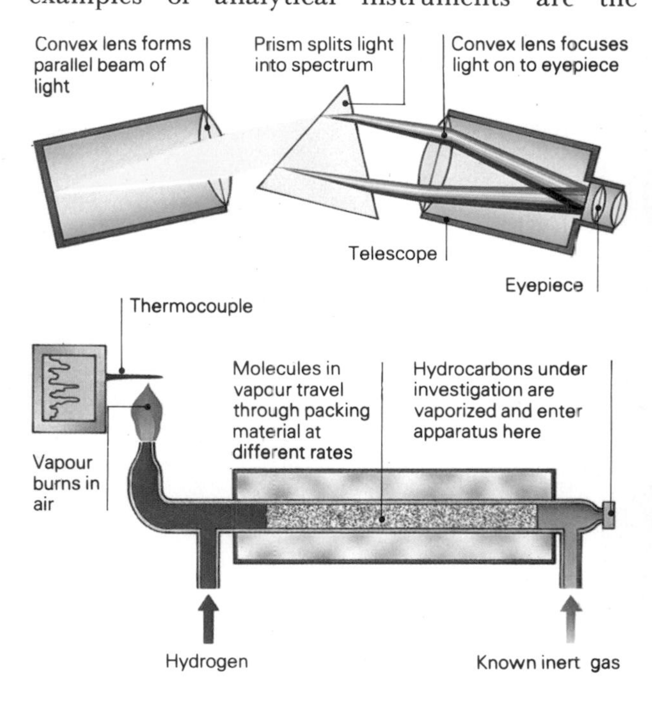

Middle left: A simple spectroscope. Light from a luminous object, such as a star, is viewed at the eyepiece as a characteristic number of spectral lines.

Bottom left: A gas chromatograph. The chemical mixture to be analyzed is fed into one end as vapour. A flow of inert gas carries along the components at different rates, so that they reach the outlet at different times. Here they are burned by the addition of hydrogen gas. A thermocouple connected to a recorder measures the heat generated, which is proportional to the amount of each component.

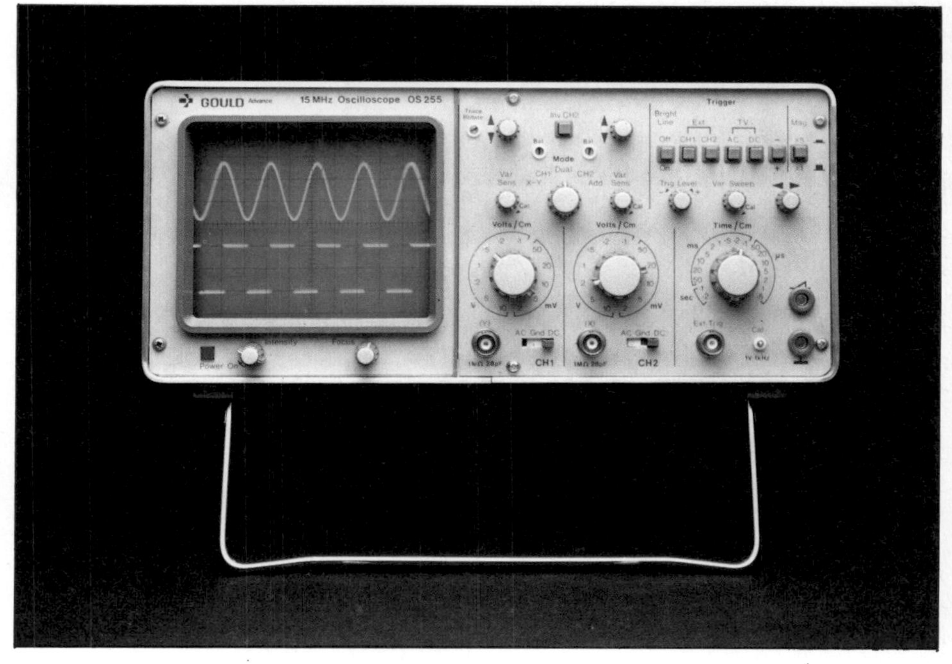

Below: The oscilloscope is a scientific instrument widely used by electronic engineers for recording and analyzing the wave-forms of electrical signals. It is also used to examine light waves or other types of electromagnetic radiation, which can be changed, or transduced, into electrical waves. Wave-forms are displayed as shown, on the screen of a cathode-ray tube similar to that of a television.

ing and corrosion and so are used to make cutlery and chemical equipment. Besides iron they contain chromium, usually together with nickel and sometimes also molybdenum.

Sulfa drugs or sulphonamides are powerful antibacterial drugs, first made, synthetically, in the 1930s. They revolutionized the treatment of many bacterial infections but after World War II were largely succeeded by the antibiotics, which were generally less toxic to patients.

Superconductors are metals and alloys which, when cooled to temperatures near absolute zero, lose their resistance to the passage of electric current. This promotes their use for ELECTROMAGNETS in which very large electric currents produce magnetic fields of great strength.

T

Tempering is a process of reheating and fast cooling, or quenching, steel and other alloys, in order to produce crystalline changes within the alloys which increase their strength and hardness.

Thermocouple is used to measure temperature. It comprises two wires of dissimilar metals, welded together at one end. When this end is heated, a voltage develops between the wires, which is proportional to the temperature, and is measured instrumentally.

Tempering steel, 1844

Thermoplastics are plastics that soften when heated. They include all plastics spun into textile fibres and most of those used to make domestic utensils.

Thermosets are plastics that set hard by a process of chemical cross-linking, and once set cannot be softened by heating. Bakelite, invented in 1905 by Leo Baekeland, was among the first of useful plastics. Epoxy resins, employed for hard, strong cements and metal coatings, are among more recent thermosets.

Tidal power stations convert the energy of the tides into useful electricity. A tidal barrier is erected across a river, containing turbo-generators through which flows the water both of ebb and of flow tides, to generate electricity. The longest-operating of such power stations is that on the River Rance in Brittany.

Transformers are electric devices for raising and lowering voltage. The princi-

and FIBRE OPTICS are important new developments. The microscope is an instrument that has been developed 'beyond light'. The ELECTRON MICROSCOPE can now be made to magnify objects up to a million times.

Technology in daily life

As ordinary citizens or consumers, we own or use a large share of modern technology. Indeed, this has been developed largely to meet the demands of consumer societies. The motor cars and motorways which are such familiar parts of the modern world are examples of consumer technology, as are many of the products and machines described above.

Technology in the home began in the 1800s with such inventions as the vacuum cleaner and the REFRIGERATOR, and mechanization of the office with the invention of the TYPEWRITER. Somewhat later, family entertainment began to be revolutionized with the invention of CINEMATOGRAPHY.

The idea of a home or an office as a machine for living or working in was extended by the development of central heating, and cooling and ventilating systems, a recent advance in heating devices being the SOLAR ROOF HEATER. The last is an example of what has come to be known as *alternative technology,* the general idea of which is the use of modern technology to exploit our natural resources, without using these up completely, while causing the minimum of pollution. The garden greenhouse is an old-established device of this kind, as is the compost heap, which helps to recycle food wastes. A mechanized version of the compost heap, the biological digester, is now to be seen in many suburban gardens.

More obvious products of modern technology are, perhaps, the pocket calculators that so many of us now use. CALCULATING MACHINES have a long history, but pocket calculators are of very recent development, the most advanced being miniaturized electronic computers. Large, centralized digital COMPUTERS increasingly affect our daily lives, preparing bills for the gas and electricity we consume, and identifying our personal documents in order to renew our driving licences and so on. Computers and other MICROELECTRONIC devices are products of the fastest-growing of all industries.

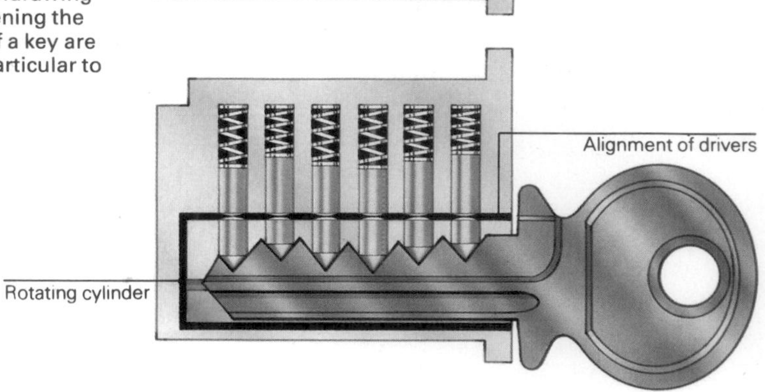

Right: The Yale lock, invented by Linus Yale in 1861, is still the most convenient and widely-used device for securing house doors. Inside the lock, metal pins project downwards into a cylinder, holding this and the connected door catch firm. When the key is inserted, its teeth raise the pins so that the cylinder is free to rotate, withdrawing the catch and opening the door. The teeth of a key are cut to a pattern particular to its lock.

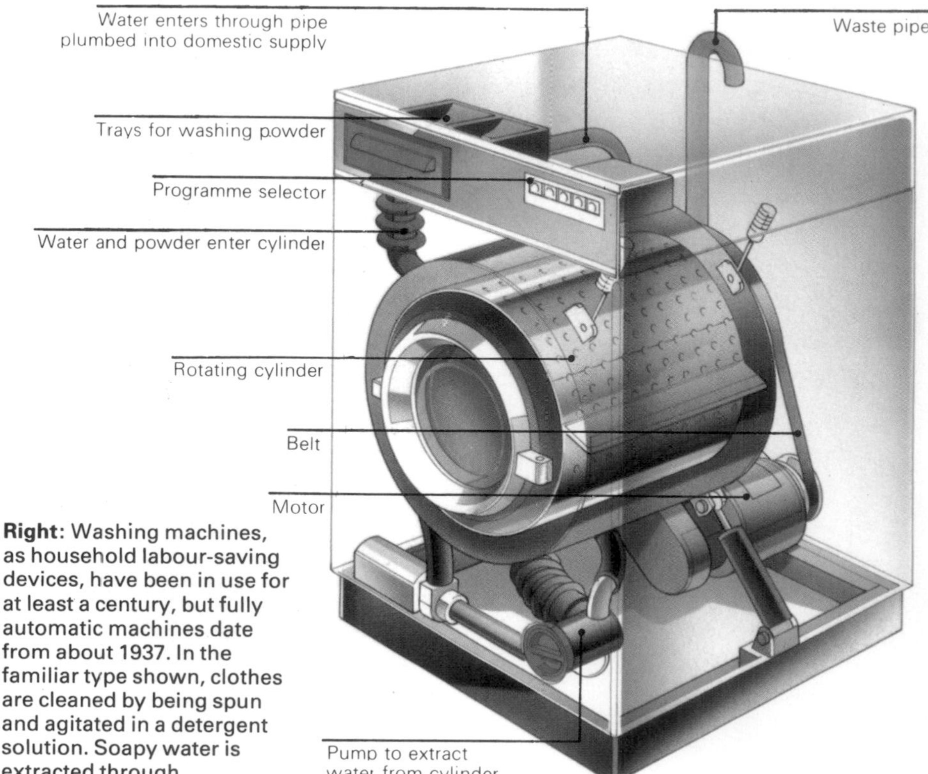

Right: Washing machines, as household labour-saving devices, have been in use for at least a century, but fully automatic machines date from about 1937. In the familiar type shown, clothes are cleaned by being spun and agitated in a detergent solution. Soapy water is extracted through perforations in the spinning cylinder.

---

ple of their operation was first demonstrated by Michael FARADAY in 1831. A major use is in the distribution of electric power, in which large transformers are employed to raise and lower voltage to and from the grid.

Turbogenerators are machines that generate electricity in power stations. The generator is a large dynamo producing alternating current. In hydroelectric power stations the generator is powered by a water turbine, and in other power stations by a steam turbine.

Typewriters were invented at several times during the 1800s, but the first practical machines were constructed by Christopher Sholes in 1867. These later became the first models of the Remington company.

Typewriter, c.1890

U **Ultrasonic generators** emit sound waves of very high frequency, inaudible to the human ear. These ultrasonic waves are utilized in medical and industrial research and have day-to-day applications for cleaning metals and textiles. Sonar equipment on ships generates ultrasonic waves for depth-finding.

V **Vacuum tubes, or radio valves** were used generally in electronics until the 1960s for the control and amplification of electric current. Since that time they have been replaced by more compact transistors.

W **Water gas** is made by blowing steam through red-hot coke. *(See also* PRODUCER GAS.) It is a mixture of hydrogen, carbon monoxide and carbon dioxide. It was once a major source of industrial hydrogen but this is now obtained from petrochemical gases.

X **X-ray machines** date from the discovery of X-rays by Wilhelm Röntgen in 1895. Today, their uses are widespread and various. In medicine, they are employed for radiological diagnosis and treatment. Their industrial applications include the detection of flaws in metals and other forms of non-destructive testing. In research they are employed for studying the structures of atoms and crystals.

Transport through the Ages

Mark Lambert

Introduction

Travel was once slow, uncomfortable and often dangerous. As a result, it was usually undertaken only when it was absolutely necessary. Transport by land and water became important as early civilizations arose. But it was the Industrial Revolution which triggered off the most rapid developments in transport, because industrialization would have been impossible without cheap and easy methods of moving bulky loads of raw materials, manufactures and foods around the world. The extent of the recent transport revolution is exemplified in the fact that, 200 years ago, it took a month or more to sail from Europe to the Americas while, in 1977, the Anglo-French supersonic airliner *Concorde* took only 3 hours and 42 minutes to fly from London to Barbados. **Transport through the Ages** tells the fascinating story of how the application of scientific principles has enabled inventors to discover ways of carrying products and people around at high speeds in comfort and safety, even beyond the Earth into space.

The wheel was one of man's most brilliant inventions. In the field of transport, it led to the development of war chariots, wagons for hauling supplies and, later, coaches for carrying passengers.

Chariots and Coaches

The early evolution of the wheel remains obscure. Possibly it developed from the use of tree trunks by Stone Age men as rollers to transport the massive stones we now find in structures such as Stonehenge and in megalithic (large stone) tombs. Gradually, men would have worked out that the central part of the roller was unnecessarily thick. From this idea it was a short step to making solid wooden wheels jammed firmly on to long axles. Such wheels could be fitted underneath a sledge with pegs to keep the axles in place. Drawings on clay tablets show that the Sumerians had these wagons in 3250 BC.

The oldest wheels that have been discovered intact come from tombs dating from 3000 to 2000 BC. They were made from three pieces of wood clamped together with cross-struts. At the same time, people learned that such light wheels could be made to turn on axles that were fixed to the wagon. The advantage of this was that the wheels on either side could turn at different speeds. Therefore, the wagon could turn corners more easily, without the outside wheels skidding and scraping along the ground. Pegs in the ends of the axles prevented the wheels from falling off. The Sumerians also found that they could prevent wooden wheels from wearing down by studding them with copper nails or by fitting bronze or leather tyres to them. When the tyre wore out, it could easily be replaced without having to make and fit new wheels.

These early wagons were simple in construction and were generally pulled by slow-moving beasts, such as oxen. However, as well as farming, the people of the early Mesopotamian civilizations were much concerned with warfare.

Below left: In the Stone Age people probably used logs as rollers to transport heavy weights.

Below: The development of the wheel. At first, solid wheels were probably cut from logs. However, such wheels break easily and stronger wheels were therefore constructed from several pieces of wood. Finally, spoked wheels combined both strength and lightness.

Reference

A **Arcera** was a Roman, 4-wheeled passenger wagon designed to be used for the sick. It was probably very uncomfortable and it was abandoned in favour of the litter — a bed carried on the shoulders of slaves.

B **Barouche** (1800s) was a 4-wheeled open carriage with a high driving seat.

Berlin coaches (1600s–1800s) were the first really successful light coaches. They had C-shaped steel springs and leather straps to help support the body. The bottom of the body was curved and bounced up and down instead of swaying from side to side. The front wheels were smaller than those at the rear, which allowed the coach to turn sharply without scraping the wheels against the body. The first Berlins only held 2 people, but later versions carried 4.

Undersprung step-piece barouche, 1851

Brougham (1800s) was a popular 4-wheeled carriage drawn by 1 horse. It was designed originally by Lord Brougham for his own use.

Buggy (1800s) was a 4-wheeled vehicle used in America. It was light, easily driven and durable, and its body was supported by 2 elliptical steel springs.

C **Cabriolet** (1800s) was a 2-wheeled carriage drawn by 1 horse. It had been developed from the CHAISE and was used for hire (hence 'cab').

Chaise (1700s) was an open 2-wheeled carriage drawn by 1 horse. It had wooden springs and a folding roof. In America it was known as the Shay.

Char, or Curré (1300s), was a 4-wheeled wagon used by queens and duchesses in France. It had been very popular during the 1200s, but in 1294 Philip II forbade its use by the wives of the bourgeoisie. Its body rested directly on the axles. Fixed

Above: Solid wheels are still used in some countries. The wheels of these Vietnamese carts have been constructed in much the same way for thousands of years.

The Sumerians had learned to control wild asses (onagers) and used them to pull their war chariots. Then it was discovered that horses could pull chariots even faster. This encouraged people to try to make their wagons lighter and, in about 2000 BC, the spoked wheel was invented somewhere in eastern Iran. At first, such wheels had only four SPOKES surrounded by a thick rim. However, the number of spokes was soon increased, therefore allowing the rim to be made thinner and lighter, without reducing the strength of the wheel. The use of spoked wheels spread rapidly into Mesopotamia. The warlike Assyrians used them not only for their chariots, but also for gigantic, turreted battering rams.

The early Sumerian chariots were four-wheelers and, although the wheels ran independently, they were difficult to steer round corners. Two-wheeled chariots did not have this problem and were lighter and faster. The Egyptians, who received the spoked wheel and the horse from Mesopotamia in about 1600 BC, used two-wheelers to great advantage in battle.

By this time the RIMS of wheels were being covered in copper. This is a fairly soft metal and, like the leather rims used by the Sumerians, quickly wore away. However, in northern Europe in 700 BC, the Celts were using iron rims for their wheels. They made an iron hoop, slightly smaller than the wheel. When this was heated, it expanded until it could be slipped over the wheel. Cooling the iron 'tyre' caused it to contract and grip the wheel tightly. So the tread of the wheel was made tougher, and the whole wheel became stronger.

The Romans, with their vast empire, needed fast communications and this encouraged them to build good roads and to improve the available means of transport. They copied many of the Celtic ideas and the best Roman wagons had iron-tyred wheels with metal-lined hubs. The cornering ability of a four-wheeled wagon was also improved by placing the front axle on a pivot so that it could be steered. In addition, the Romans also used a double shaft instead of a single pole. This meant that light carts could be pulled by one horse rather than two.

After the decline of the Roman Empire, European roads were neglected, but travel continued to increase. Horses were fitted with iron shoes and harnesses that did not strangle them. Wagons with a form of suspension appeared. The body of the wagon was built as a separate box and hung on chains or leather straps from four posts attached to the chassis. This reduced the jolting, but unfortunately some

to the front axle there was either a pole (if the horses were to be attached in pairs) or two shafts (if the horses were to be in tandem). It was mainly built of wood, but many of the fittings were of iron and the whole thing could weigh nearly 1,000 kg. As a result, the char needed 4 or 6 horses to pull it.

Concord coaches (1800s) were the famous stage-coaches of the American West. They were fast and very durable.

Conestoga wagons (1700s), also known as prairie schooners, were used in America to transport freight along the Santa Fé trail and settlers to Oregon and California. The bed of the Conestoga was about 5 metres long and the rear wheels were nearly 2 metres high. The hoops, which were covered in white canvas, stood over 3 metres from the ground. Up to 6 horses, mules or oxen were needed to pull it.

Curricle (1800s) was a 2-wheeled carriage drawn by 2 horses.

D

Diligence was the name given in England in 1783 to mail coaches. These were BERLIN COACHES and they could travel at a speed of 11 km/hour while carrying mail and 4 passengers, but no luggage. In France the name Diligence was given to a POST-CHAISE drawn by 3 horses abreast and driven by a postillion.

Dished wheels appeared during the Middle Ages. The top of a dished wheel could be angled outwards, while the hub remained vertically above the point where the rim touched the ground. Thus the body of the wagon could be built wider, while the axle remained short enough to bear the weight.

G

Gig (1670) was a light, 2-wheeled cart with a high chair in which the driver used to sit.

H

Hackney coaches (1600s) were 4-wheeled vehicles used for hire.

Hansom cabs were 2-wheeled carriages used for hire. Designed by J. A. Hansom in 1834, they superseded the CABRIOLET. A hansom cab had a square, enclosed body with a seat for the driver on the roof.

L

Landau (1700s) was a 4-wheeled carriage with a folding hood. The name Landaulet was given to smaller versions of this vehicle.

Hansom cab, 1887

passengers were made 'wagon-sick' by the swaying motion. Despite these few improvements travelling by wagon remained slow and uncomfortable for several centuries. Most people preferred the faster, pleasanter method of travelling on horseback, particularly when swift communications were needed.

In the 1500s the first coaches were built in Hungary. These resembled large bath tubs on wheels, but by the early 1600s coaches more familiar to us were being built in Germany. Some of these had glass windows and steel springs. Coaches could now go faster, but poor roads often prevented them from travelling quickly. Eventually, the invention of macadamized roads *(see page 93)* in 1810 led to the development of STAGECOACHES, the fastest of which were the mail coaches. Passengers found these much more comfortable because they had laminated spring suspension — the kind still used in some cars today. Their wheels had bolt-on hubs so that faulty or broken wheels could be replaced swiftly.

Below: A reconstruction of a Celtic chariot. In about AD 500 the Celts used to bury chariots with their owners and many metal parts have been excavated from such graves. Pictures on coins provide an idea of what such chariots looked like. A grave stone in northern Italy shows a long chariot with 2 semi-circular pieces on one side – as shown here. These were probably used as handholds by the warrior and his driver. The chariot had spoked wheels held on to the axles by linchpins. The trace reins were attached to the axle.

Left: An English stagecoach of the 1800s. First-class passengers rode inside, and second-class passengers rode on top with the coachman. The bolt-on wheels had iron rims and carved oak spokes.

P

Phaeton (1800s) was a high, 4-wheeled carriage drawn by 1 or 2 horses. It was noted for its ability to turn corners sharply and its drivers earned the reputation of being reckless.

Post-chaise (1743) was a 2-wheeled carriage suspended on leather braces and drawn by 1 horse. Travelling post meant that the driver and horse could be changed at stations every few kilometres. Later post-chaises had 4 wheels.

R

Raeda (reda) was a 4-wheeled Roman cart taken from the Celts, used for long-distance travel.

Rim of a spoked wheel is the outer part that makes contact with the ground. When spoked wheels were first invented the rim was made of a single piece of wood bent into shape using steam. Later rims had several pieces of wood (felloes) slotted into each other, bound with a metal tyre.

Phaeton, c.1857

S

Sedan chairs (1600s and 1700s) were enclosed chairs for 1 passenger only, carried by 2 men. Although they received much competition from the increasing number of coaches, they were more comfortable and so were used until the early 1800s.

Spokes are thin pieces of wood or metal that join the RIM of a wheel to the hub. The first wooden spokes were invented about 2000 BC when someone realized that an apparently thin stick will not break if a weight is applied along its length. This allowed a wheel to be both strong and light.

Stagecoaches (1800s) received their name from the fact that long journeys were broken at staging posts, usually inns, along the route. The coachmen, horses and passengers could get rest and food before continuing their journey.

W

Whirlicote was the English name for the CHAR.

Early sailors faced great dangers when their fragile craft were swept beyond the sight of land. But, today, modern navies and shipping lines provide essential services, even in an age of fast air transport.

Across the Oceans

Below: The earliest Egyptian boats were constructed from reeds of which there was a plentiful supply. The bow and stern were joined by a rope that could be tightened. By 3000 BC reed boats were journeying to Crete and Lebanon to bring back materials such as timber.

Man's natural element is the land. But even in the earliest days of civilization, curiosity probably led him to construct boats, and venture on to the water. The first boats were very simple. DUG-OUT CANOES of the type still used in some Pacific Islands were probably the earliest boats to appear. However, in order to travel long distances, more elaborate boats were needed. By

Reference

A **Aft** is a nautical term that means near or towards the rear of a ship.

Asdic is a form of SONAR in which pulses of ultrasound are beamed out horizontally beneath the ship and the reflected echoes are converted electronically into audible sound in the form of 'pings'. By aiming the sound beam the operator can find the range and bearing of a submarine from the echoes returned by its hull. The direction and speed of the submarine's movement are detected by the pitch, or note, of the returning echoes, using the DOPPLER EFFECT.

Astrolabe was an instrument used to measure the angle between a star and the horizontal. It was suspended by a ring from one hand and the star was viewed through two small holes at each end of a movable pointer. The angle could then be read off the circular scale.

B **Backstaff** was an instrument used to measure the angle between the Sun and the horizon at noon. A long stick, or staff, was aligned with the horizon using a slit at the end. A curved vane was then moved so that the top of its shadow aligned with the slit. The angle could then be read from the vane. By consulting his DECLINATION tables the navigator could quickly work out his latitude.

Battleship is the most heavily armed type of ship used in war and is designed to attack and defeat any other type of ship. In the 1600s to 1800s a battleship was called a SHIP-OF-THE-LINE. During World War I massive, heavily-armed and armoured ships were built, such as HMS *Dreadnought*, which carried 10 300-mm guns and had 275-mm thick armour.

Battleship HMS Dreadnought

2700 BC the Egyptians were building rafts of papyrus reeds, and their high-prowed wooden boats had begun sailing the waters of the River Nile and the Red Sea.

The early sailors did not venture far from land and so could easily find their position from landmarks along the coast. The only hazards to avoid were running aground or on to rocks. The Egyptian river boats had two men in the bows — one to test the depth of water with a long pole, the other to signal instructions to the helmsman.

Once out of sight of land, however, finding the boat's position was more difficult. With no landmarks to guide them the early seafarers had to discover new ways of finding out where they were. The Mediterranean Sea has been called 'the cradle of navigation', because it was there that sailors first ventured on to the open sea.

Below: The reed boat was basically just a raft, the natural buoyancy of the reeds keeping it afloat. To prevent the reeds from becoming waterlogged they were bound together in 2 bundles. Some drawings show reed boats with sails, which were probably made from rush matting.

The Mediterranean sailors

The Egyptians, Greeks and Phoenicians were the first to sail the Mediterranean and the earliest written record of navigation is provided by the Greek historian Herodotus. He tells us that when approaching land, sailors let down a weighted line with a lump of tallow on the end. The length of line gave the depth of water, and the gravel, sand or mud that stuck to the tallow told the experienced captain where he was. For example, yellow mud indicated that the ship was approaching the mouth of the River Nile.

Of all the Mediterranean sailors the Phoenicians were the most proficient. By about 1200 BC they had established trading posts all round the Mediterranean and had sailed into the Atlantic Ocean. They even reached the coast of Cornwall in England, where they traded with the tin

Biremes, triremes and quinquiremes were types of Greek and Roman galley. It is disputed whether these names meant that the vessels had 2, 3 or 5 banks of oars on each side or 2, 3 or 5 oarsmen to each oar. However, the latter seems likely, as in vessels with 3 or 5 banks of oars, the oarsmen on the higher levels would have had to pull impossibly long and heavy oars.

C

Caravel was a small, fast sailing ship, LATEEN-RIGGED except for the foremast, used mainly in the Mediterranean during the 1400s to 1600s. The *Niña* and *Pinta*, 2 of Columbus's ships, were caravels.

Carrack, or nao, was a small, broad, square-rigged merchant ship common in the 1500s. Columbus's ship *Santa Maria* was a carrack.

Carvel-built ships are made from long wooden planks with the edges of the planks butted against each other. The joints are sealed with pitch. Greek, Phoenician and Roman ships were carvel-built.

Clinker-built boats are made from long wooden planks and each plank overlaps the one below it. The Viking ships were built by this method. The technique persisted into the Middle Ages.

Cog boat, which appeared in Europe in about 1200, was the forerunner of the CARRACK and the GALLEON. It had a single square sail and a turret at each end.

Corvette is a small warship. In the 1700s a corvette was a sailing ship with a single, level deck and only 1 tier of guns. The modern corvette is an escort vessel. Corvettes used in World War II carried 2 100-mm guns, several anti-aircraft guns and anti-submarine armament.

French corvette, 1800s

Cross-staff was an instrument used to measure the angle between the Sun or a

Above: A Roman galley of about AD 50. In battle the main form of propulsion was by oars, but if there was a favourable wind the sails could be used on long trips. The wooden ram was bound with iron or bronze and used for sinking enemy ships.

miners. Their greatest achievement was to make the first known voyage round Africa. In about 600 BC, sailors set out from the Red Sea and worked their way round the coast. Each year they stopped, planted seed and then waited for the harvest before continuing their journey. Three years after starting out they had reached the Straits of Gibraltar.

The Phoenicians were superb navigators. They learned the importance of the stars, and Greek sailors still call the Pole Star the 'Phoenician Star'. The Greeks also learned to use the stars. In Homer's epic poem the *Odyssey,* written in about 850 BC, the hero Odysseus is described as having steered his boat by the Pleiades, Arcturus and the Great Bear.

The Greeks and Phoenicians also had formidable fighting ships. Each one was rowed by hundreds of slaves and manned with soldiers. At the bow was an underwater pointed ram for sinking enemy vessels. The Phoenicians pioneered this design of ship and the Greeks copied it, with success. At the battle of Salamis in 480 BC a mere 300 Greek galleys defeated about 1,000 Persian ships. Some 500 years later the Romans were still using a similar design of warship.

The northern sailors

When the Phoenicians visited Cornwall, they saw men who sailed to Ireland in boats made of skin. These were the Irish curraghs and Welsh coracles, which can still be seen today. But it was many centuries before the northern Europeans began to build boats to match those of the Phoenicians and Romans. When they did, some time between AD 500 and 700, they built long wooden boats, generally for warlike purposes. The European long boats eventually evolved into the Viking longships, which the fierce Viking warriors used for raiding and conquering large areas of Europe.

The navigators of the northern seas were faced with several problems not found in the Mediterranean. The weather could not be relied on to remain calm, and storms could cause waves up to about 9 metres high. The Sun and stars could not be used with such accuracy for navigation as the difference between the summer and winter risings and settings is much greater than in the Mediterranean. The greatest problem was the tide, particularly for sailors from the almost tideless Mediterranean.

The Vikings, however, overcame these problems and made long voyages across the Atlantic

Above: It is disputed whether the term bireme meant two banks of oars (**1**) or two men to each oar (**2**). The latter would seem to be more likely. In a trireme or quinqireme, if there had been three or five banks of oars, the top bank would have had oars that were too long and heavy for the oarsmen to manage.

star and the horizon. It consisted of a long stick, or staff, which was aligned with the horizon, and a small crosspiece which could be lined up with the Sun or star by sliding it along the staff. The required angle could be read from a scale along the staff.

Cruiser is a warship designed for speed. During World War I there were several classes of cruiser, which varied in displacement and armament according to the work they were required for.

D

Decca is a navigation system that uses radio signals transmitted by a master station and 3 slave stations. The navigator chooses the 2 most convenient slaves and the Decca receiver indicates the 2 time differences between the arrival of the master signal and the arrival of the 2 slave signals. These time differences can be plotted along curved position lines on a chart. In practice the lines are already plotted and where the 2 position lines cross is the ship's position. Decca only works within 160 km of the transmitting stations. *See also* DECTRA, LORAN.

Declination of the Sun or a star is its angular distance north or south of the equator. At the equinoxes (21 March and 21 September) the Sun is immediately above the equator and its declination is therefore zero. On any other day the Sun's declination can be found from the *Nautical Almanack.* To find the ship's latitude when it is on the opposite side of the equator to the Sun the declination has to be subtracted from the ZENITH distance (90° less the instrument-measured angular altitude of the Sun). When the ship and Sun are on the same side of the equator the declination is added to the zenith distance.

Dectra is a navigation system that uses radio sig-

Destroyer HMS Glasgow

Above: A portion of the Bayeux Tapestry, which tells the story of the Norman invasion of Britain in 1066. Here, the Norman fleet is setting sail from the French coast. They had square-rigged ships, each with a single sail. The steering oar on the starboard side of the ship is clearly shown.

Ocean. They used 'homing' winds to carry them where they wanted to go; in summer easterly winds carried them across to Iceland and Greenland; and by sailing south they could pick up prevailing westerly winds to carry them home. It is also generally believed that the Vikings reached America. Leif Eriksson, a Norwegian explorer, set out to find a land previously sighted by accident. He found it and after spending the winter there, returned home loaded with grapes. He therefore named the land Vinland, which is generally thought to have been what is now known as Newfoundland.

Sailors of the Pacific Ocean

As the Polynesians spread into the Pacific islands from Asia they were faced with the problem of sailing vast distances in their frail canoes. Even the larger double canoes, called *tonkiaka* or *kalia,* were vulnerable in really bad weather. As a result the Polynesians became sensitive to the weather and could forecast fairly accurately up to three days ahead.

They also developed navigation into a sophisticated art. In many of the islands, navigation schools were set up. There, young men were taught how to steer by the stars and how to recognize indicators of land, such as wave motions, cloud colours and birds. On the Isle of Arorae they were shown how to use DIRECTION STONES to determine the directions of nearby islands.

There were also a few navigational aids. In the Marshall Islands the natives had elaborate TWIG CHARTS to show the distance and bearings of other islands. In Hawaii there was an instrument called a SACRED CALABASH, which enabled the navigator to find out when his boat reached the same LATITUDE as the island.

Compasses and charts

Until AD 1000 navigation was mostly a matter of observation, memory, guesswork and luck. The scientific and mathematical skills of the Greeks had been denied to western Europe by the Church, which regarded them as black magic. However, they had been handed down to the Syrians and from them to the Arabs and by now were beginning to have more influence in Europe.

The compass, originally invented by the Chinese 1,000 years before, now reappeared in Europe. To begin with it took the form of a magnetic needle inside a straw, which was floated in a bowl of water. By 1180 sailors were using a needle on a pivot. This arrangement was later improved by mounting the needle on a card and pivoting them both together. Therefore, although it seemed like magic at the time, sailors could determine the approximate direction of the North Pole under any conditions.

The charts of the 1400s were far from accurate. Distance was very difficult to measure. Little was known of magnetic VARIATION or DEVIATION and

Right: A lateen-rigged Arab dhow. Such boats have been in use for over 1,500 years. In about AD 500 the Arabs were using seasonal monsoon winds to sail to India and back.

nals. It works in a similar way to DECCA, but is used for transatlantic crossings. There is a master and slave pair of transmitters in England and another pair in Newfoundland. The 2 pairs transmit signals alternately and by comparing their times of arrival and the time difference between the master and slave signals, the receiver indicates the ship's position. The Dectra system uses longer wavelengths (up to 1 km) than the Decca system and works over distances of up to 1,600 km.

Destroyer is an abbreviation for torpedo-boat destroyer. In World War I they were especially useful in combating U-boats (German submarines) after the development of the depth charge. Modern destroyers carry guided missiles.

Deviation is the amount by which a compass needle is deflected by the magnetic field caused by the metal and electrical parts of a ship.

Direction stones on the Isle of Arorae were used by the Polynesians to indicate the direction of nearby islands. Each stone had 2 poles set into it and a navigator setting off in the evening could take his course by aligning the poles. By the time they were out of sight it was dark and he could steer by the stars.

Doppler effect is used in ASDIC and in RADAR and satellite navigation. Sound or radio waves emitted by or reflected from a moving object change in frequency. If the source and receiver are moving towards each other, the frequency increases, and, in the case of asdic, a higher note is heard. If the source and receiver are moving away from each other the frequency decreases.

Dug-out canoes were the first form of boat. They were made from solid logs hollowed out with either tools or fire. Such boats appeared in many parts of the world and are still used in some places. Stability was achieved by placing stones

Peruvian Indians making dug-out canoe

Below: Ship development from the 1400s to 1600s.
1. Columbus's ship *Niña* was a caravel. When he sailed from Palos the *Niña* was lateen rigged, as shown here. However, she was rerigged with square sails in the Canary Islands.
2. A two-masted carrack of the mid-1400s. Columbus's ship *Santa Maria* was a similar vessel except that she had three masts.
3. A Spanish galeass of the mid-1500s, a type of ship obviously derived from the Roman galley.
4. *Ark Royal* was one of the English galleons that fought the Spanish Armada in 1588.
5. *Henry Grâce à Dieu*, a French galleon built in the 1540s.

taking correct bearings was therefore almost impossible. Charts were also very difficult to use because converting the sphere of the world on to a flat chart led to several distortions.

The problems were, to some extent, solved by Gerardus MERCATOR. He devised a new, flat projection of the world on which a straight line represented the true course between two points. Although the sizes of Greenland and Antarctica were greatly exaggerated on such MERCATOR PROJECTIONS, this did not concern the early navigators as they avoided these regions anyway. Unfortunately, Mercator's maps were not published until 1585, so they were not available to the great explorers.

The great age of exploration

In the early 1400s, the prospect of finding gold, spices, silks and new trade routes was a great incentive to the first explorers and the merchants who financed them. Spices were to be found in India, but the land routes had been closed by the expansion of the Ottoman Turk empire. Therefore men began to look for a sea route to India. Prince Henry of Portugal, also known as Henry the Navigator (1394–1460), set up a school at Sagres, and gathered together astronomers, mathematicians, navigators, masters, shipbuilders and instrument makers. By 1446 the Portuguese had reached the Azores and Madeira, and in 1461 they landed in Sierra Leone on the West African coast. In 1487, Bartholomew Diaz sailed round the southern tip of Africa, rediscovering the route sailed by the Phoenicians 2,000 years earlier. His crew forced him to turn back, but in 1498 Vasco da Gama reached India by this route.

By this time most European scholars believed that the Earth was round. However, 1° on the equator was thought to equal 90 kilometres. From this Christopher Columbus (1451–1506), an Italian explorer, estimated that the Earth's circumference was 32,830 kilometres and that Asia lay 4,800 kilometres away across the Atlantic Ocean. His plans for making the voyage were rejected by the Portuguese and English, but on 3 August 1492 he set sail with the *Santa Maria*, *Niña* and *Pinta*, with the blessing of the King and Queen of Spain. On 12 October, Columbus landed at San Salvador in the Bahamas, believing that he had reached Asia. From there he sailed on to Cuba, which he mistook for Japan. Columbus made three more voyages to the West Indies and was for a time Governor of Hispaniola, but when he died he was still unaware that he had reached America and not Asia.

The land that lay beyond the West Indies was

in the bottom or by lashing 2 canoes together.

F **Fore-and-aft** rigged ships are those with sails, usually triangular, set lengthwise to the ship. *See also* SQUARE-RIGGED.
Frigate is a medium-sized warship. In the 1700s it was a sailing ship that carried between 28 and 60 guns, but was not a SHIP-OF-THE-LINE. Today the term is loosely used to mean a CRUISER or large CORVETTE.

G **Galleon** was the large type of sailing ship that developed from the CARRACK towards the end of the 1500s. It was slimmer, lower and faster than the carrack and had 3 to 4 masts and several levels of upper deck. The galleon design was used for both merchant ships and warships for 250 years.

French frigate Hercule, *1700s*

Gyrocompass always points towards True North. This is because once the spinning gyroscope in its specially weighted frame has been aligned with the Earth's axis, it maintains this orientation despite changes in direction of the ship.

I **Inertial guidance system** uses gyroscopes to detect all the movements of a ship, i.e. speeding up, slowing down, starting, stopping and changing direction. The gyroscopes feed the information into a computer and, if the starting position of the ship is known, the computer can work out the ship's position at any time.

K **Kamal** was a simple Arab instrument for estimating LATITUDE. It consisted of a piece of card with a hole in the centre and a notch at the top. Attached to the middle of the card was a piece of string along which knots were tied at regular intervals. The navigator held the string in his mouth and sighted a known star through the notch. He then allowed the string to slip through his teeth until the hole in the centre of the card was aligned with the horizon. From the number of knots between his mouth and the card he could then work out his latitude.

L **Lateen-rigged** ships had one or more triangular sails, each of which was suspended from a long yard

first recognized as a new continent by the Italian Amerigo Vespucci (1451–1512), and eventually it was named after him rather than Columbus. Vespucci claimed to have made four voyages to the Americas, and in 1497 John Cabot, a Venetian, rediscovered Newfoundland. But it was left to Ferdinand Magellan, a Portuguese explorer, to find the sea route to Asia. In 1519 he sailed from Cadiz with five ships and eventually found his way through the narrow straits at the southern end of South America, now known as the Straits of Magellan. He was therefore the first to sail westwards into the Pacific Ocean. However, he never returned to Spain as he was killed by natives on the island of Mactan in the Philippines. In 1522 only one of Magellan's ships, the *Victoria*, returned to Spain, via the Moluccas and the Cape of Good Hope. This journey was not repeated for 50 years. In 1577 the English adventurer Sir Francis Drake in the *Golden Hind* set out on a voyage, the stated purpose of which was to open up the spice trade. His secret mission was to discover *Terra Australis Incognita* (now Australia) and to raid Spanish settlements and ships. He failed to find the southern continent, but in 1580 he returned to England laden with £10 million worth of plunder, having also completed the second voyage round the world.

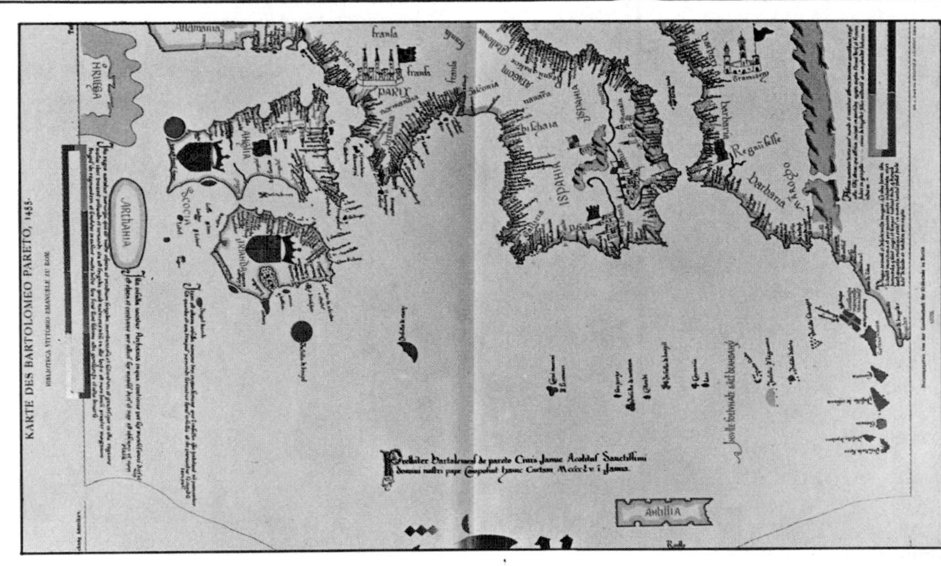

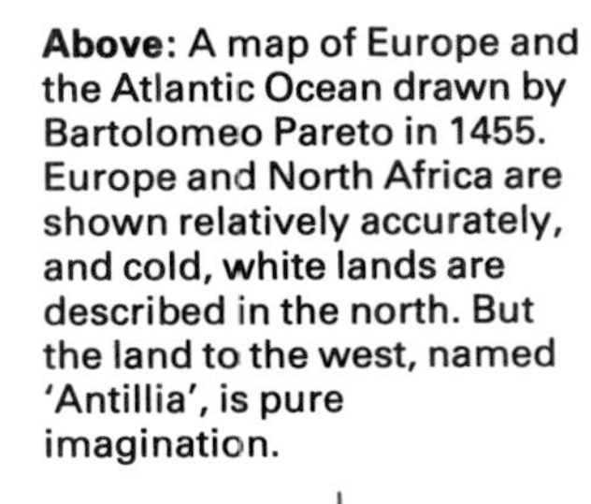
Above: A map of Europe and the Atlantic Ocean drawn by Bartolomeo Pareto in 1455. Europe and North Africa are shown relatively accurately, and cold, white lands are described in the north. But the land to the west, named 'Antillia', is pure imagination.

(wooden pole) attached to the mast at an angle of 45°. Lateen-rigged ships are still used, for example in Arab dhows, but in Europe they were replaced by SQUARE-RIGVED ships. However, lateen sails were the forerunner of the modern FORE-AND-AFT sails used in yachts.

Latitude of a point on the Earth's surface is its angular distance, north or south of the equator, measured from the centre of the Earth. Thus the equator is 0° latitude, the North Pole is 90°N, and the South Pole is 90°S. Any line on the Earth's surface parallel to the equator is a line of latitude and the circular lines of latitude decrease in circumference from the equator to the poles.

Log and line was an instrument used from the 1400s to measure the speed of a ship. It consisted of a weighted, triangular piece of wood, or log chip, on the end of a line. Along the line there were knots every 13 metres. The log chip was thrown over the stern of the ship and the line paid out freely for 30 seconds. Then it was hauled in and the number of knots that had passed over the ship's stern was the speed in NAUTICAL MILES per hour, or knots. The log and line was replaced by the PATENT LOG in the 1700s.

Longitude of a point on the Earth's surface is the angular distance east or west of the PRIME MERIDIAN measured from the centre of the Earth. The easiest way to measure longitude is by time; 1 hour of time equals 15° of longitude. Therefore if the time in London at 0° longitude is 12.00 noon and the time in New Orleans is 6.00 a.m. the difference is 6 hours and New Orleans is at 90°W.

Loran (*Lo*ng *ra*nge *n*avigation) is a navigation system that works on the same principle as DECCA and DECTRA. All three systems work by comparing the times of arrival of radio waves at a receiver. In the Loran system the waves are viewed on a screen, like that of a television. A dial is turned so that the waves coincide on the screen and the time difference is read from the dial. Like the Decca system, Loran uses long wavelengths and has a range of up to 1,280 km.

0° meridian marker

M

Mercator, Gerardus (1512–94) was a Flemish geographer, born Gerhard Kremer. His first

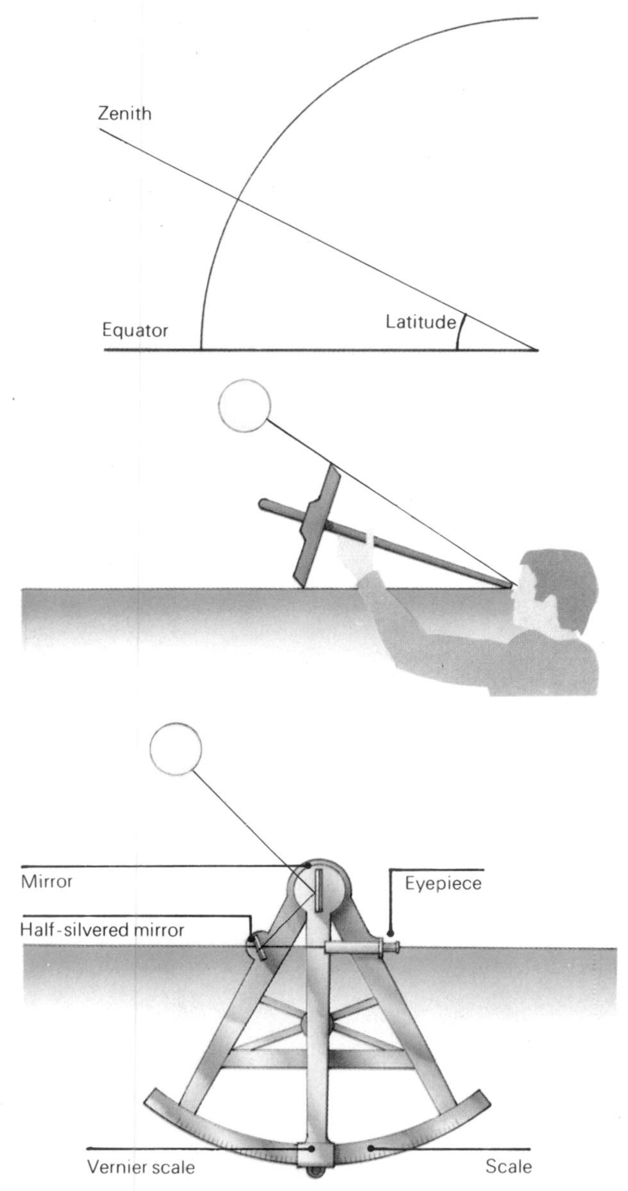

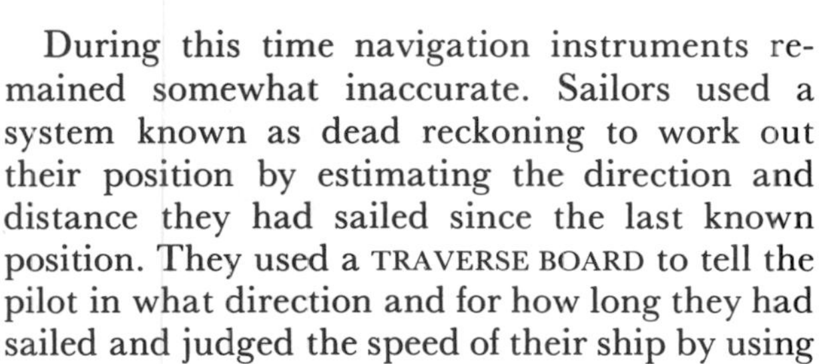

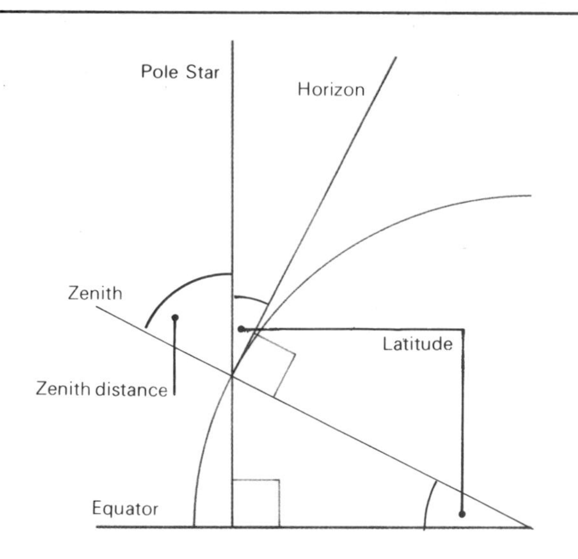

Far left: The latitude of any point on the Earth's surface is the angle at the centre of the Earth between a line drawn from that point and a line drawn from the equator.

Left: The same angle of latitude can be found by measuring the angle between the Pole Star and the horizon. During the day the angle between the Sun and the horizon can be measured. Latitude is found by adding or subtracting the declination of the Sun from the zenith distance (*see* DECLINATION).

Left: Using a cross-staff to measure the angle between the Sun and the horizon.

Left: A sextant. The navigator lines up the eyepiece and the half-silvered mirror with the horizon. Then by moving the arm he causes the image of the Sun produced by the index mirror and the half-silvered mirror to appear to rest on the horizon. The angle between the Sun and the horizon can be read from the scale at the bottom of the sextant.

During this time navigation instruments remained somewhat inaccurate. Sailors used a system known as dead reckoning to work out their position by estimating the direction and distance they had sailed since the last known position. They used a TRAVERSE BOARD to tell the pilot in what direction and for how long they had sailed and judged the speed of their ship by using a LOG AND LINE. Finding the ship's position was difficult. LATITUDE, or the ship's distance from the equator, was found by measuring the angle between the Sun or Pole Star and the horizon. For this there was the Arab KAMAL or the European CROSS-STAFF, BACKSTAFF, ASTROLABE and QUADRANT. If conditions were right for using such instruments, the navigator could estimate his latitude to within 32 kilometres.

Finding the ship's LONGITUDE was even more difficult. From the late 1400s it had been realized that knowing the correct time was important. Navigators compared the observed (ship's) time of certain astronomical events, such as eclipses of planets by the Moon or the position of the Sun at noon, with the times for such events recorded on land. However, the ship's clocks and hourglasses of the time were far too inaccurate. In 1714 the Board of Longitude was set up and a prize of £20,000 was offered to anyone who could devise an accurate method of finding longitude at sea. In 1729 John Harrison, a surveyor, set to work to produce a marine clock. Over the next 31 years he built three clocks, none of which completely satisfied the Board. But in 1761 he produced a watch-style chronometer that was taken on a trip to the West Indies and back. After the 147-day voyage the watch was only 1 minute 54 seconds in error. On a later trip it only lost 15 seconds.

Replicas of Harrison's chronometer were taken by Captain Cook on his voyages of exploration in *Resolution.* He could determine his position so

contribution to navigation was to make accurate globes with lines of latitude and MERIDIANS of longitude marked on them. These became available to navigators in 1541. Until this time men had been influenced by Greek geography, particularly Ptolemy's map of the world, which was somewhat inaccurate and included only Europe and Asia. With the discovery of new lands it became less easy to depict the world on a flat map until Mercator solved the problem with his MERCATOR PROJECTION.

Gerardus Mercator

Mercator projection is also known as a cylindrical projection. Imagine a cylinder of paper placed round a globe at the centre of which there is a light. The light casts a shadow of the Earth's surface on to the cylinder. When this is unwrapped it carries a map of the world on which the MERIDIANS of LONGITUDE are vertical and parallel and the lines of LATITUDE are horizontal and parallel. As one moves north or south away from the equator the lines of latitude spread out and Greenland and Antarctica are enlarged. A straight line on such a map does not necessarily represent the true distance between two points, but it does represent the true course – a much more important factor in navigation. On other projections the true course is represented by a curved line.
Meridian is a line of LONGITUDE that runs from the North to the South Pole.

N

Nautical mile is equal to 1.151 statute miles (6,076 ft or 1,852 metres). It is different from a statute mile because navigators have always calculated their position on the Earth's surface in degrees and it is therefore easier to measure the distance between two points in round numbers of degrees; 60 nautical miles equal 1°. However, until the 1600s, due to miscalculation of the Earth's circumference, the nautical mile was reckoned as 1,524 metres and the LOG AND LINE used the short knot of 13 metres. In 1635 the nautical mile was more accurately estimated as 1,853 metres and the knot as 15.5 metres. Even so the short knot was still being used in the 1800s. Modern instruments have allowed the nautical mile to be calculated to its present very high

Right: Brunel's steamship, the *Great Eastern,* which was completed in 1858. It was the largest steamship of its time and could carry 4,000 passengers or 10,000 troops. Altogether, 10,200 tonnes of iron and 5,430 square metres of canvas were used in its construction.

precisely that he was able to produce extremely accurate charts, some of which are still used today. The accuracy of Cook's navigation was further assisted by the work of Edmund Halley, generally better known for his discovery of Halley's Comet. He observed how his compass varied by different amounts from True North in different parts of the world. From this work came the first isogonic charts, which gave sailors the magnetic VARIATION in any part of the world. With such information they were then able to take accurate bearings.

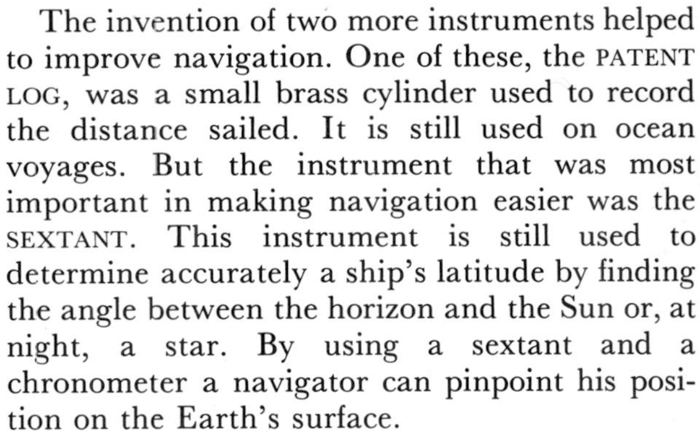

The invention of two more instruments helped to improve navigation. One of these, the PATENT LOG, was a small brass cylinder used to record the distance sailed. It is still used on ocean voyages. But the instrument that was most important in making navigation easier was the SEXTANT. This instrument is still used to determine accurately a ship's latitude by finding the angle between the horizon and the Sun or, at night, a star. By using a sextant and a chronometer a navigator can pinpoint his position on the Earth's surface.

Left: The *Charlotte Dundas* is known as the first successful steamboat. She was built in 1801 for Lord Dundas and named after his wife. In March 1803 she towed 2 loaded vessels along the Forth and Clyde Canal. However, further trips were prevented by the owners of the canal. They were afraid that the wash from the paddle wheel would damage the canal banks.

degree of accuracy.

Nocturnal was an instrument used in the 1500s and 1600s for finding local time at sea. It was a form of circular slide rule and had a fixed outer scale marked with dates and a movable inner scale marked with times and with two 'ears'. One ear was for use with the Great Bear (Plough) constellation, the other with the Little Bear. Pivoted in the middle was a long arm. The navigator set the Great Bear 'ear' to the correct date on the outer scale and sighted the Pole Star through the hole in the middle of the instrument. Then he moved the arm so that it lined up with the pointers of the Great Bear. The time was shown where the arm crossed the inner scale.

O

Octant was invented in 1731 by John Hadley. It was the forerunner of the SEXTANT, but the arm could move through an arc of only 45° and therefore the maximum angle that could be measured was 90°.

P

Patent log was a small brass cylinder with fins to make it rotate when towed in the water behind a ship. Each rotation caused dials on the side of the cylinder to record the distance the ship had travelled. Early patent logs were not very accurate and contrary currents and surging movements of the ship caused them to record greater distances than had actually been travelled. Later patent logs had a recording clock on board ship and a brass flywheel that evened out the surges of speed. Patent logs are still used on ocean voyages and the distance is recorded electronically.

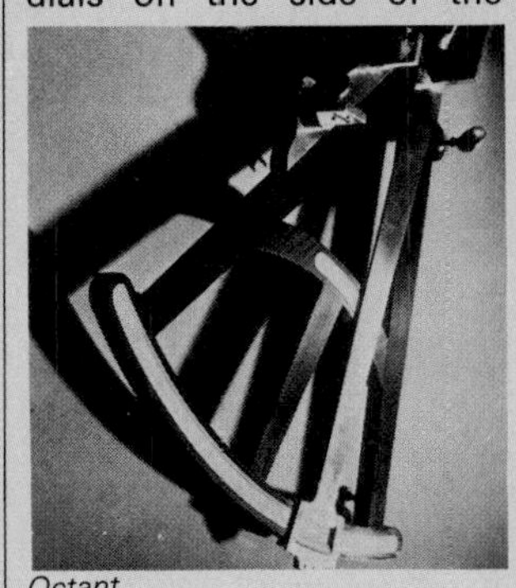

Octant

Port is the left-hand side of a ship when looking towards the bow. The origin of the term is uncertain. Formerly it was known as larboard, which may have been due to the fact that the ship was loaded, or laden, with cargo from the side nearest the port dock. *See also* STARBOARD.

Prime meridian, 0° LONGITUDE, is the MERIDIAN that passes through a point at the ROYAL OBSERVATORY, Greenwich, London. During the late 1700s and early 1800s each country recognized a different position for 0° longitude: the French, Paris; the Italians, Rome; and so on. But in 1844 a congress met

From sail to modern ships

The sailing ships used by the great explorers were developed from the European long boats. But for about 500 years the design of sailing ships changed very little. Wide, three- or four-masted ships with large, square sails were a familiar sight on the oceans of the world. The Spanish Armada (1588) and the fleets that fought in the battle of Trafalgar (1805) were composed of ships similar to those of Columbus. In the early 1800s, however, the arrival of steam propulsion caused a dramatic change in ship design. Ships needed to become more streamlined in order to cope with the increase in speed now provided by steam.

The early steamships were propelled by large paddle wheels, but by the mid-1800s propellers proved to be more efficient. However, steam engines were often unreliable and steam did not replace sail immediately. Many steamships were equipped with sails in case their engines broke down. The great clippers that sailed between America, China, Australia and Britain in the

Right: SMS *Rheinland* was a World War I German battleship. She had 29-cm-thick armour plating, 12 28-cm guns, 12 15-cm guns and 16 8·6-cm guns. Her top speed was 39 km/hour and her displacement 19,200 tonnes.

Wireless aerials
Lookouts
Searchlights
Navigating bridge
Gun turret
Crane for lifting out boats
28-cm gun
Anchor
Secondary armament, 15-cm guns
Secondary armament, 8.6-cm guns
Armour plating

in Washington and agreed that all the world's maps should be drawn with the prime meridian at Greenwich. The French and Brazilians did not immediately agree, only changing their maps in 1911.

Q **Quadrant** was an instrument used in the 1500s to 1700s to measure the angle between the Pole Star and the horizon. It consisted of a quarter circle of wood with a scale marked 0° to 90° along its curved edge. A weighted line hung from the right-angled corner. The Pole Star was sighted through 2 holes along 1 straight edge and the angle of the star was indicated by the line on the scale.

Quinquireme, *see* BIREME.

Radar scanner

R **Radar** (*Ra*dio *d*etection *a*nd *r*anging) uses radio waves to detect the position of objects. A rotating scanner emits a stream of radio-wave pulses. When these strike an object they are reflected back and picked up by the same scanner. Pulses from nearby objects return to the scanner more quickly than those from distant objects and the information is transferred electronically to a round screen, similar to that of a television. A dot in the middle of the screen indicates the ship's position and a bright radius sweeps round the screen at the same rate as the scanner. Objects appear as bright dots or areas on the screen, which is marked with a distance scale. Radar can also be used to detect the speed of objects using the DOPPLER EFFECT.

Royal Observatory at Greenwich, London, was founded by Charles II. He made the English astronomer John Flamsteed the first Astronomer Royal with the task of cataloguing the stars in order to help improve navigation. Thanks to the recent invention of the telescope Flamsteed was able to catalogue a large number of stars and compiled the first great star map of the age. He also observed the movements of the Moon and predicted the times of tides. The Royal Observatory is now in East Sussex.

S **Sacred calabash** was a navigation instrument

mid- to late 1800s were fast, sleek sailing ships that did not use steam at all.

As steam power became more reliable and coal and oil became cheaper, steamships replaced sailing ships for carrying cargo and passengers. Only on the really long-haul voyages were sailing ships still needed, as they did not have to stop at frequent intervals to take on coal. But by the early 1900s more efficient steam engines, particularly the steam turbine, ensured the supremacy of the steamship.

Wooden ships too became obsolete. Iron ships were more durable and in warships armour plate was needed against the new exploding shells that were being used. In the early 1900s there was a rapid growth in the construction of fighting ships as Europe prepared for World War I. Sleek, fast-moving battleships, cruisers and destroyers appeared. And the Germans developed the first practical submarines, which were powered by internal combustion engines *(see page 82)* on the surface and by electric batteries underwater.

Since then ships have changed little in basic shape. However, new materials, techniques and ideas have led to changes in the design of various parts, such as hulls and bows. Many ships are now powered by diesel engines and the installation of nuclear-powered engines in some ships has enabled them to avoid the need for refuelling. Ships have also taken on specialized roles. For example, car ferries, aircraft carriers, giant tankers, container ships and ice-breakers are all designed to do their particular jobs.

Below: 1. *Ark Royal*, the famous British aircraft carrier, was launched in 1950. She was refitted after 20 years and finally ended service in 1978.
2. The 2,330-tonne *Glückauf* was the world's first tanker. Ordered by Germany, she was built in Scotland in 1893.
3. A German IX-92 World War II U-boat, which carried 12 torpedoes and could travel at about 35 km/hour on the surface.
4. The 84,771-tonne *Normandie* was a French liner launched in 1932. She had turbo-electric power capable of producing 160,000 hp.

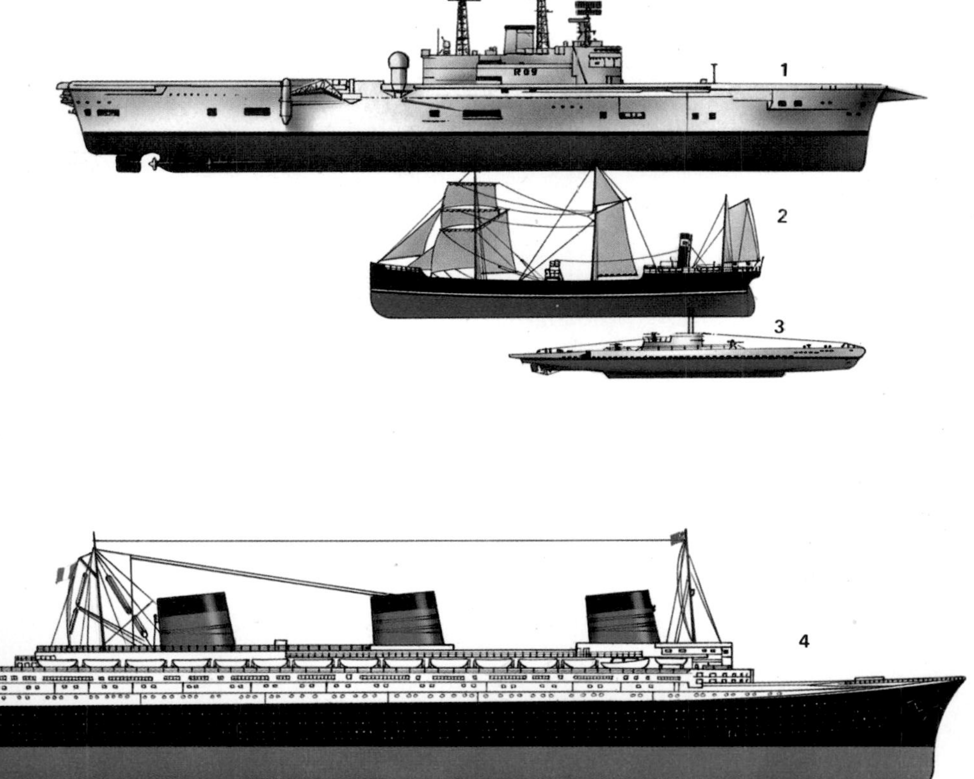

Modern navigation

Although aeroplanes have taken much cargo and many passengers from the sea, the increasing amount of trade and tourism round the world has led to an ever-growing volume of traffic across the oceans. As a result it is more important than ever for a navigator to know not only where he is, but also what hazards, such as rocks and other ships, are nearby.

The introduction of radio into ships has changed both communications and navigation. Instead of having to rely on flag and light signals, ships can now communicate with each other and with the shore without necessarily being in sight. In 1910 Ethel Le Neve and Harvey Crippen, who had poisoned his wife, were arrested on board ship in the Atlantic Ocean as a result of early radio communications.

The first radio direction-finding (RDF) system used in navigation was the D/F loop. This used a loop aerial which could be rotated until the signal from a shore-based transmitting station faded out completely. The loop was then at exactly 90° to the direction of the signal. By taking the bearings of two such signals the ship's position could be plotted on a chart. Today, RDF systems automatically indicate the ship's position on an instrument.

The modern DECCA, DECTRA and LORAN systems are considerably more accurate. They work on the principle that two signals transmitted simultaneously from stations at equal distances from a ship will arrive at the same moment. If the ship is nearer one station than the other, the signal from the first station will arrive a fraction of a second before the signal from the second station. The distances from the stations can then be calculated electronically and the ship's position can then be found.

Another instrument that has revolutionized

used by the natives of Hawaii. It consisted of a gourd hollowed-out into a bowl. A notch was cut in the rim and, opposite this at a certain distance below the rim, a sighting hole was drilled. Below this hole the gourd was drilled with a ring of holes. These ensured that the calabash was held level when filled with water. When the Pole Star could be sighted through the hole and the notch the navigator knew that he was on the same latitude as Hawaii.

Sextant is the modern instrument for finding LATITUDE by measuring the angle between the Sun or a star and the horizon. The arm of the instrument is moved so that the image of the star reflected by the index mirror and the half-silvered horizon mirror appears to rest on the horizon. The angle of the star can then be read from the scale. If it is turned on its side the instrument can also be used to measure angles between objects on the Earth's surface. The arm can be moved through an angle of 60° and this allows angles of up to 120° to be measured.

Ship-of-the-line was the name given to the largest type of sailing warship. The name is derived from the way in which naval battles were fought in the 1700s and 1800s. Two opposing fleets formed up in lines alongside each other and tried to pound each other to pieces with cannon.

Sonar *(So*und *na*vigation and *r*anging) is a method of detecting underwater objects using sound waves. A transducer emits pulses of ultrasonic (very high frequency) waves, which are reflected from the sea bed or

British ship-of-the-line (left) and frigate (right), 1800s

navigation is the gyroscope. A GYROCOMPASS always points towards True North and can assist the navigator greatly when he sets the ship's course. An INERTIAL GUIDANCE SYSTEM also uses gyroscopes to work out the ship's position. It works on the same dead reckoning system used by Columbus, except that it is far more accurate.

To avoid rocks, sandbanks and other vessels, ships can use two other electrical navigation aids: SONAR and RADAR. Sonar equipment emits pulses of ultrasound (very high frequency sound, inaudible to the human ear) and picks up the echoes reflected by underwater objects. In an echo-sounder a trace records the depth of water beneath the ship. In ASDIC, one of the earliest forms of sonar, the pulses and echoes are converted into audible sound in the form of 'pings' to detect the presence and movement of submarines. Radar uses radio waves to detect nearby ships and land in the same way as an echo-sounder uses sound waves. The radio waves reflected from an object are used to form a pattern on a screen, rather like a television screen, on which ships appear as bright dots.

Electrical aids alone, however, are not enough and navigators must learn many of the traditional methods of navigation. Knowledge of the rules of the sea, of how to take compass bearings and of how to use a chart are still important, particularly in crowded coastal waters. The chronometer and sextant are still essential instruments, especially in smaller boats. Lighthouses, first used in Egypt in 280 BC, and lightships are needed to mark dangerous rocks. Systems of buoys are required to mark safe channels, wrecks and sandbanks. Therefore the art of the navigator should not be allowed to die. The all too frequent collisions and wrecks that cause severe oil pollution of our seas and shores show what can result from lack of navigational skill.

Above: The supertanker *Universe Ireland* is over 300 metres long and 53 metres wide. Built in 1968, she has two steam turbines and can produce 37,400 hp.

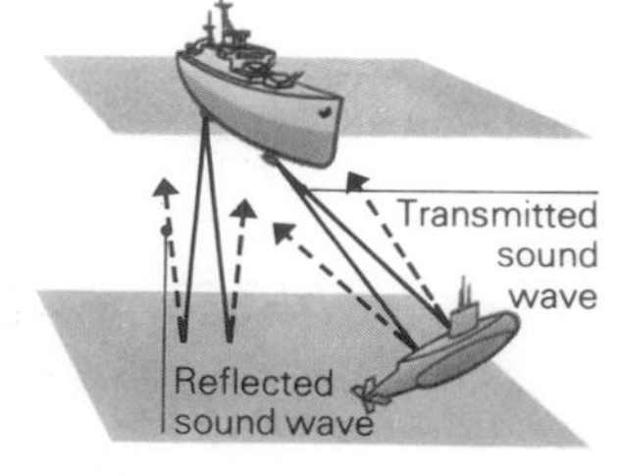

Below: Sonar equipment emits high-frequency sound impulses and from the reflected echoes can detect the depth of water underneath the ship, or the direction, range and speed of a submarine.

Left: A diesel engined patrol submarine which is capable of high underwater speeds and can maintain continuous submerged patrols. It is equipped to fire homing torpedoes.

underwater object and picked up by a receiver. *See also* ASDIC.

Square-rigged ships were those in which the principal sails were square and slung crosswise on horizontal booms attached to the masts. *See also* FORE-AND-AFT.

Square-rigged ship

Starboard is the right-hand side of a ship. The Viking ships had their steering oars on this side and the term literally means 'steering side' of the ship. *See also* PORT.

T **Traverse board** was a navigation aid used in the 1400s. Sailors could not read or write, so at half-hourly intervals they used the pegs on a traverse board to record the direction of the wind and the approximate speed of the ship. The pilot could later use this information to estimate the progress and position of the ship.

Traverse board

Trireme, *see* BIREME.

Twig charts were used by the Polynesians. The twigs were bound together with fibre and islands were marked with cowrie shells tied to the twigs. Long, straight twigs indicated the directions of islands; curved twigs showed how wave directions were altered by islands; and short, straight twigs indicated currents.

V **Variation** is the difference in degrees between True North and the direction in which a compass needle points. This difference is due to the fact that the Earth's magnetic poles are situated some distance from the geographical poles. Thus magnetic variation is different in different parts of the world. It also changes slightly from year to year, as the magnetic poles are gradually changing position.

Z **Zenith** is the point in the sky directly above an observer. The zenith distance is the angle between the Sun and the zenith. *See also* DECLINATION.

In developed countries, a car is regarded by many families not as a luxury but as a necessity. Yet the technology of mass production, which made cars cheap to buy, was introduced only in the first decade of this century.

Travelling by Road

As the villages, towns and cities of early civilizations grew, so the trading routes between them also grew in importance. Paths, along which men travelled on horseback, developed into roads where carts and wagons could travel. For over 5,000 years horse-drawn chariots, carriages and coaches remained the most efficient means of transport. However, in the 1700s and 1800s the Industrial Revolution gave rise to a series of inventions that led to mechanized transport and, gradually, horse-drawn carriages became obsolete.

The invention of steam power caught the imagination of many people. The first practical steam road vehicle was a gun-tractor built by the French engineer N. J. Cugnot in 1769 and used by the French army. Several inventors also attempted to produce commercially-successful steam cars. Richard TREVITHICK *(see page 111)*, for example, built a series of them between 1801 and 1811. His 'London Carriage' proved its worth by trundling slowly along Oxford Street in London in 1802. But Trevithick could not persuade anyone to take much interest in them so he turned his attention to the railways. By 1840 steam coaches had almost disappeared from the roads. There were probably two main reasons for this. First, they were frightening vehicles, inclined to belch smoke and sparks, and many people expected them to explode. Second, and probably more important, the rich and influential businessmen who owned the fleets of horsedrawn carriages, saw that steam coaches could upset their trade and so caused trouble for anyone operating them. A law was passed prohibiting the use of horseless vehicles unless a man walked in front carrying a red flag. The speed limit for such vehicles was set at 6 km/hour in the country and 3 km/hour in towns.

At the same time as people were trying out steam power, other experiments using mechanized transport were being devised. In 1826 Sam Brown, an English engineer, built a mechanical carriage that progressed with a series of loud bangs. Inside the CYLINDERS, which were similar to those of a steam engine, a series of gas explosions drove the PISTONS. After years of experimenting, a German engineer called Nikolaus Otto set up a factory to build gas engines. His production manager was Gottlieb Daimler (1834–1900). Otto devised a four-stroke cycle for his engines and they were a great success.

Meanwhile, an Austrian engineer, Siegfried Marcus, was working out ways to develop the use of petrol as a fuel. He used this portable, easily-vaporized liquid in several engines that he built between 1864 and 1874. But the first major advance was made by Daimler. In 1885 he devised the first high-speed, single-cylinder, petrol-driven engine. He tried this out first on a three-wheeled tricycle and afterwards on a four-wheeled car in 1886.

At the same time Karl Benz (1844–1929), another German engineer, was also producing successful engines, unknown to Daimler. While Daimler's main interest was in building engines, Benz saw the commercial possibilities in motor cars. In 1885 he built his first car, a three-

Below: Cugnot's steam tractor (1769) was the first self-propelled vehicle. It was intended for hauling field guns. However, it had two main disadvantages. The boiler was not strong enough to withstand high pressures and so tended to run out of steam very quickly. Also, the heavy boiler was supported in the front axle and this made steering difficult.

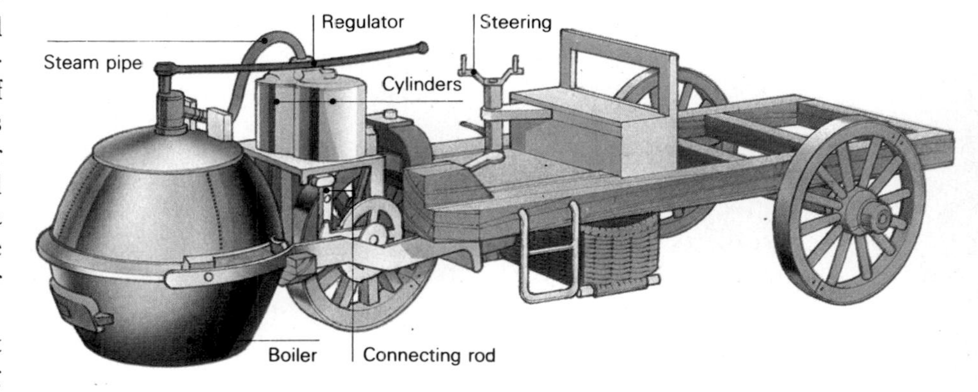

Reference

A Accelerator is the pedal used to control the speed of a car. It is linked to the throttle in the CARBURETTOR by a cable, rod or air tube.

Air-cooled engines are cooled by air which may be blown through ducts by a fan. Such engines have fins on the CYLINDER block to help remove the heat.

Alternator, *see* GENERATOR.

Antifreeze is a mixture of chemicals added to the water in the cooling system to stop it freezing. The main constituent is ethylene glycol, an alcohol, which reduces the freezing point of water. Anti-corrosion agents are also added to the mixture.

Austin 7 was also known as the 'Baby Austin'. Over 250,000 were sold between 1922 and 1938. 4-cylinder, 747-cc, side-valve engine; top speed 72 km/hour.

Automatic transmission uses an automatic CLUTCH and an EPICYCLIC GEAR SYSTEM to keep the engine speed (roughly) constant.

B Battery (more correctly called an accumulator) contains a series of electrical cells, each of which has an output of 2 volts. Each cell consists of 2 lead-compound plates immersed in sulphuric acid. Charging the battery causes chemical changes to take place in the

Austin 7

plates. The chemical changes are reversed when the battery discharges.

Bayonne Bridge has the second longest steel arch span in the world (503.5 metres). It bridges the Kill Van Kull between Staten Island, New York, and Bayonne, New Jersey.

Bearing is a device that supports a fixed part on a rotating wheel or shaft and reduces the friction between them. Ball bearings use steel balls between an inner race (fixed to the stationary part) and an outer race (fixed to the rotating part). Roller

Bayonne Bridge

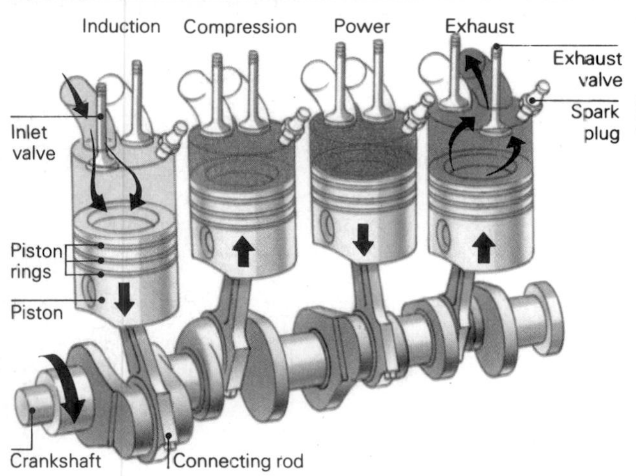

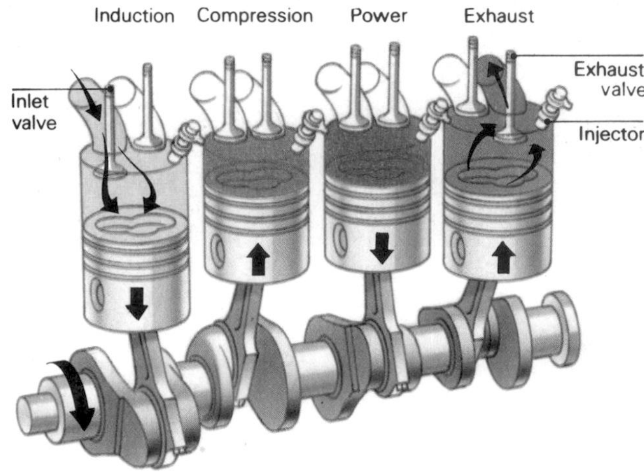

Far left: A four-cylinder petrol engine.

Left: The same four-stroke cycle occurs in a diesel engine. However, only air is drawn in during the induction stroke. During the compression stroke the air becomes hot and when fuel is injected into the combustion chamber it ignites spontaneously, resulting in the power stroke.

wheeler with a tubular steel chassis. It had a 0.5 horsepower engine and could reach about 15 km/hour. Three years later Benz became the first man to offer cars for sale to the public.

Internal combustion engines

Nikolaus Otto's four-stroke engine cycle was originally designed for use in a gas engine. However, Otto's work led the way towards petrol engines and today most cars, motor bikes, buses and lorries are propelled by engines that have four-stroke cycles. This cycle is so called because only one stroke in four produces power. The other three strokes are concerned with drawing in the fuel (induction), compressing it (compression) and letting out the exhaust gases (exhaust).

In a petrol engine the induction stroke begins as the piston inside the cylinder moves down. This occurs as a result of the turning movement of the CRANKSHAFT. At the same time the inlet VALVE at the top of the cylinder opens and a mixture of air and petrol vapour is drawn in. During the compression stroke the piston moves upwards again, the inlet valve closes and the air and petrol are compressed into a small space. Near the end of the compression stroke the gases are ignited by a spark from the SPARK PLUG. They burn explosively and the gases that are produced as they burn, expand rapidly. Thus the piston is forced downwards. This power stroke causes the crankshaft to continue turning. When the power stroke has ended, the piston once again moves upwards. At the same time the exhaust valve opens and the waste gases are expelled. At the end of the exhaust stroke the piston is ready to begin the next induction stroke.

In practice an engine usually has more than one cylinder and each cylinder fires in turn. Thus in a four-cylinder engine, the type commonly used in cars, when the first piston is performing its induction stroke, the second piston is on its compression stroke, the third is on its power stroke and the last is on its exhaust stroke. For smooth running of the engine the cylinders do not fire in the order one, two, three, four (one is at the front of the engine). In most British cars the firing order is one, three, four, two.

In 1890, a few years after Benz and Daimler had begun producing their petrol engines, Herbert Stuart, a British engineer, patented the first compression-ignition engine. In 1893 a German engineer, Rudolf Diesel (1858–1913), produced a better version which he perfected in 1897. This type of engine has since been known as the diesel engine. It has a four-stroke cycle and works in much the same way as the petrol engine. However, the fuel is ignited in a different way. During the induction stroke only air is drawn into the cylinder. During the compression stroke the compression ratio (the amount by which the air is compressed) is high, much higher than in a petrol engine, and as a result the air gets very hot. At the top of the compression stroke both valves are closed and fuel is injected into the cylinder. It immediately ignites in the hot air, causing the power stroke, which is followed by the exhaust stroke.

The advantage of a diesel engine is that it can run on cheaper fuel than petrol. Diesel fuel has a higher flash point (the temperature at which it

Below: A Bardon motor car in about 1900. Bardons were built from 1899 to 1903. The first model had a 2,000 cc, 4–5 hp engine mounted under the floor. The engine had a single cylinder that contained 2 opposed pistons. The final drive consisted of a double chain. The 1901 model had 2 cylinders mounted under the bonnet. The 1902 model had a shaft drive instead of the chain drive.

bearings use steel rollers instead of balls. A bush bearing consists of a replaceable soft metal alloy sleeve in between the moving parts. A shell bearing is a sleeve bearing composed of 2 halves (*see also* BIG END).

Benz Velo was the first car to go into quantity production, as a 3-wheeler in 1886 and as a 4-wheeler from 1892 to 1901. Single-cylinder (2-cylinder from 1897), horizontally-mounted, 2,540-cc engine; top speed 25 km/hour.

Big end is the larger end of the piston rod, a connecting rod, joined to the crankshaft by a shell bearing. Knocking of the big end is due to failure of the bearing, caused by lack of oil.

Brakes are the means by which a moving vehicle can be stopped. Usually, drum brakes or disc brakes, both operated by a hydraulic system, are used and act on all 4 wheels.

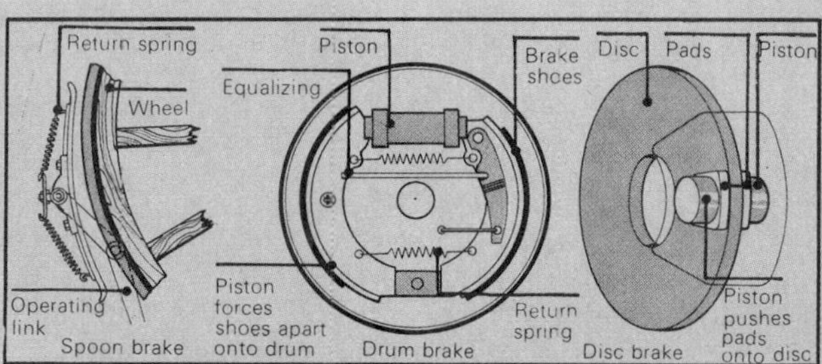

Types of brakes

C **Camber** of a road is the slightly rounded shape of the surface. It allows water to drain away.

Camshaft is a shaft on which cams (unevenly distributed projections) are carried. Thus, as the shaft rotates, a rod touching a cam is made to move up and down. The camshaft operates the VALVES, DISTRIBUTOR and oil pump. In many engines it is situated within the cylinder block, but some engines have one or two overhead camshafts situated under the rocker cover at the top of the engine.

Capacity is the total volume of fuel mixture drawn into an engine during a complete cycle as the pistons each move from top to bottom of their cylinders, i.e. the total volume of an engine's cylinders excluding the combustion chambers. It is measured in litres (l) or cubic centimetres (cc).

Carburettor is the device used to supply the engine with the correct mixture of air and vaporized petrol. It operates by engine suction

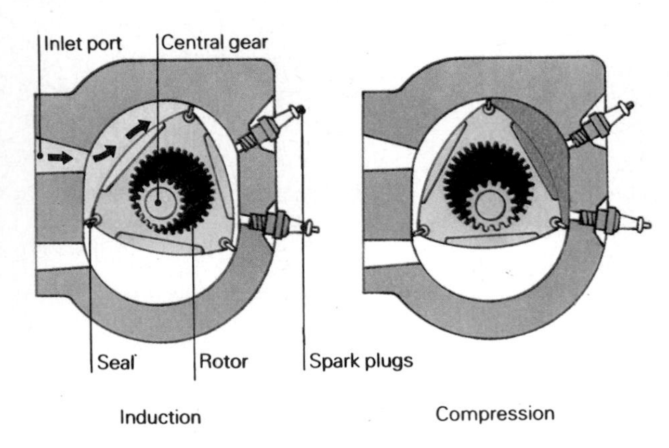

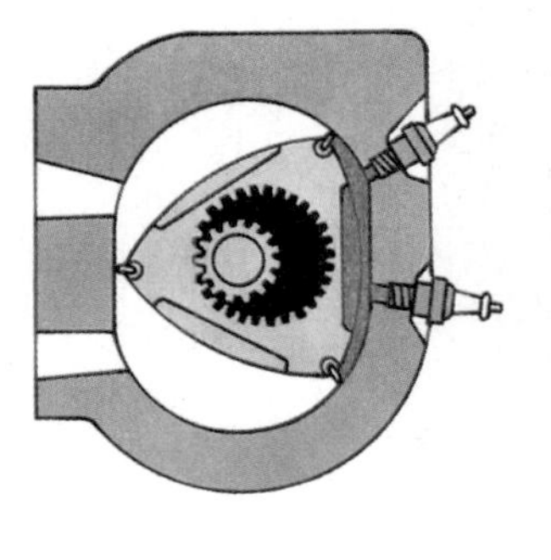

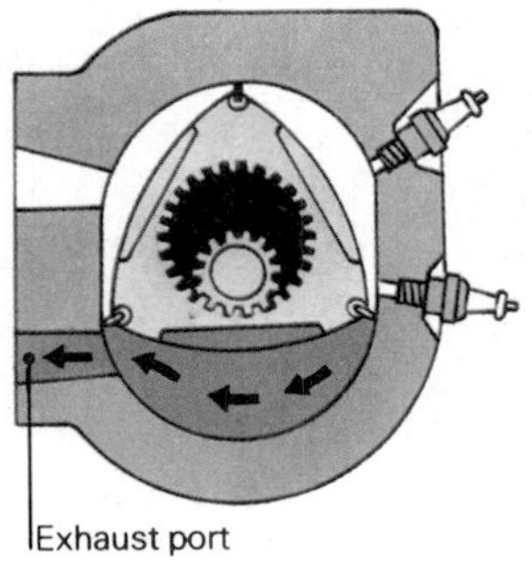

Left: In a Wankel engine the rotor takes over the function of the pistons. As it moves round, fuel and air are drawn in through the inlet port. When the trailing seal has passed the inlet port, compression begins. When fully compressed, the mixture is ignited by spark plugs. The resulting power 'stroke' forces the rotor to continue turning. Exhaust gases leave through the exhaust port.

and the amount of fuel entering the engine is controlled by a butterfly valve called the throttle. Petrol from the petrol tank is pumped into a reservoir called the float chamber, which maintains a constant pressure supply to the carburettor. Inside a carburettor there are 1 or more jets, small holes through which the petrol is vaporized. A fixed jet carburettor has a main jet, used for constant high-speed running, and 2 or more other jets to allow the engine to idle or accelerate rapidly. A variable jet carburettor has only 1 jet. When the throttle is opened the increased suction from the engine causes a needle to be pulled out of the jet, allowing more petrol through. *See also* CHOKE.

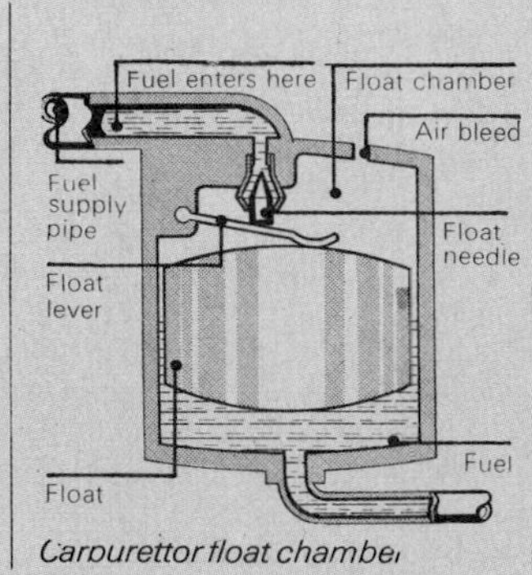

Carburettor float chamber

Choke is a device in a CARBURETTOR for increasing the ratio of petrol to air in the mixture that enters the engine, either by reducing the amount of air or by increasing the amount of petrol. Choke, or venturi, is also the name given to the narrowing of the air passage in a carburettor where the air passes over the jet. This narrowing causes the speed of the air to increase, which results in a lowering of the air pressure and petrol is sucked into the air-stream.

Clutch is a device for controlling the engagement of the engine drive with the gearbox. In a single-plate clutch the friction plate is forced against the engine FLYWHEEL by several springs. In a diaphragm clutch the springs are replaced by a single diaphragm spring. There are also several kinds of automatic clutch. The type commonly used in AUTOMATIC TRANSMISSION systems is a fluid-coupling called a torque converter.

Coil is the electrical device in a car used to create high voltage (high tension) current. Low voltage current flows from the battery to a primary coil. The current passes through 2 contact breaker points in the DISTRIBUTOR. As the rotor arm of the distributor rotates the points open and close to interrupt the flow of current. The build up and fall of low voltage current in the primary coil induces pulses of high voltage current in an inner secondary coil.

Compression ratio is the ratio of the volume of fuel mixture when the piston is at the end of its induction

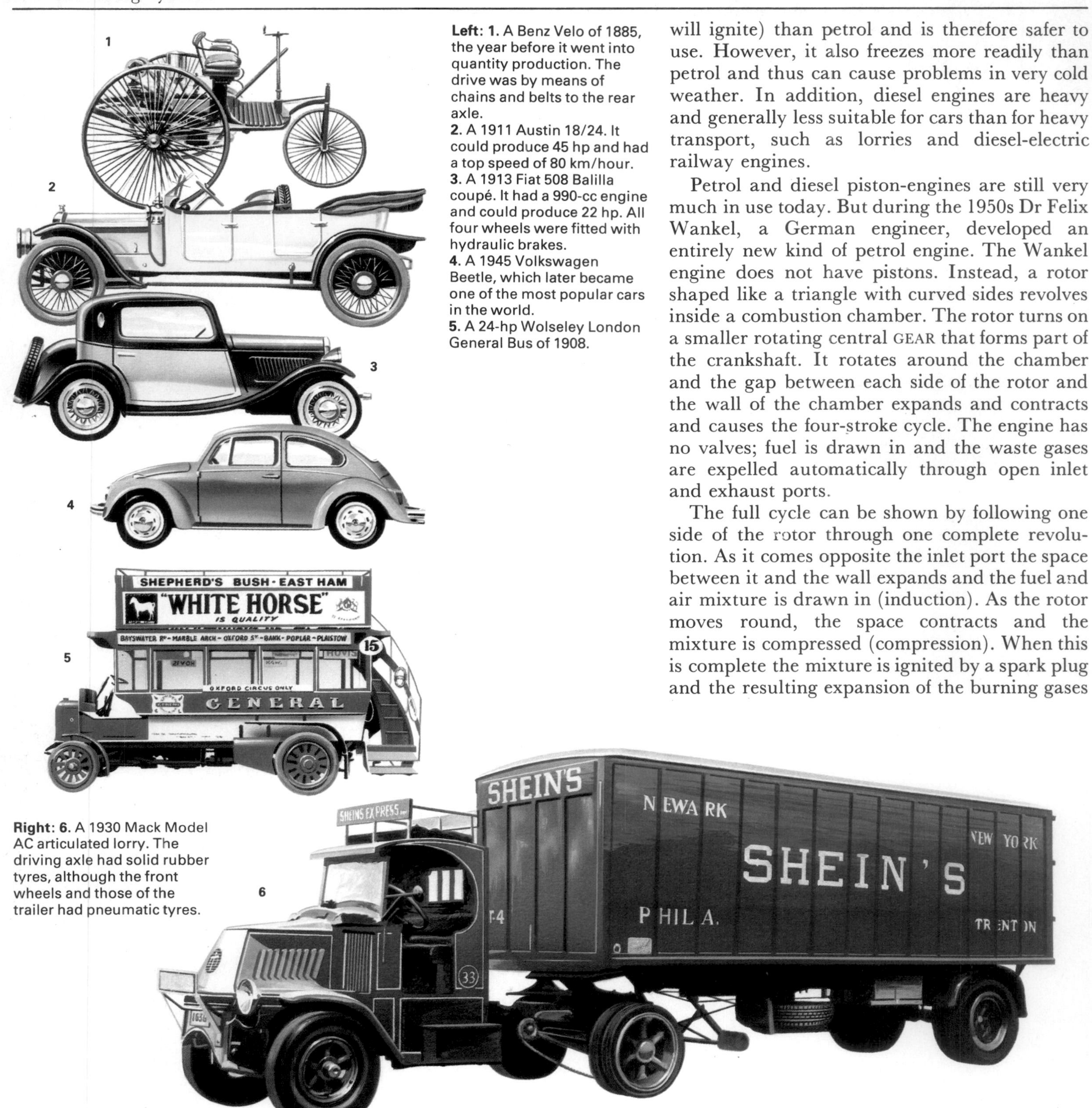

Left: 1. A Benz Velo of 1885, the year before it went into quantity production. The drive was by means of chains and belts to the rear axle.
2. A 1911 Austin 18/24. It could produce 45 hp and had a top speed of 80 km/hour.
3. A 1913 Fiat 508 Balilla coupé. It had a 990-cc engine and could produce 22 hp. All four wheels were fitted with hydraulic brakes.
4. A 1945 Volkswagen Beetle, which later became one of the most popular cars in the world.
5. A 24-hp Wolseley London General Bus of 1908.

Right: 6. A 1930 Mack Model AC articulated lorry. The driving axle had solid rubber tyres, although the front wheels and those of the trailer had pneumatic tyres.

will ignite) than petrol and is therefore safer to use. However, it also freezes more readily than petrol and thus can cause problems in very cold weather. In addition, diesel engines are heavy and generally less suitable for cars than for heavy transport, such as lorries and diesel-electric railway engines.

Petrol and diesel piston-engines are still very much in use today. But during the 1950s Dr Felix Wankel, a German engineer, developed an entirely new kind of petrol engine. The Wankel engine does not have pistons. Instead, a rotor shaped like a triangle with curved sides revolves inside a combustion chamber. The rotor turns on a smaller rotating central GEAR that forms part of the crankshaft. It rotates around the chamber and the gap between each side of the rotor and the wall of the chamber expands and contracts and causes the four-stroke cycle. The engine has no valves; fuel is drawn in and the waste gases are expelled automatically through open inlet and exhaust ports.

The full cycle can be shown by following one side of the rotor through one complete revolution. As it comes opposite the inlet port the space between it and the wall expands and the fuel and air mixture is drawn in (induction). As the rotor moves round, the space contracts and the mixture is compressed (compression). When this is complete the mixture is ignited by a spark plug and the resulting expansion of the burning gases

stroke to the volume when the piston is at the end of its compression stroke. In modern petrol-engined cars the ratio is about 8:1 or 9:1. Diesel engines have compression ratios of between 12:1 and 25:1.
Crankshaft is the shaft that, through connecting rods, converts the up-and-down motion of the pistons into a rotary action.
Crown wheel and pinion are two bevel GEARS which transmit the rotation of the propeller shaft to the DIFFERENTIAL. The pinion is the driving gear at the end of the propeller shaft and the crown wheel is fixed to the cage of the differential.
Cylinder is a tubular space in the engine block in which gases from the carburettor are ignited, so forcing the piston down to drive the engine. Most cars have 4 cylinders and their total CAPACITY is given as the engine size.

D **Derailleur gear** on a modern bicycle consists of several different sizes of SPROCKET on which the chain can run. The mechanism lifts the chain from one sprocket to another and a tensioning device takes up slack as the chain moves from a larger to a smaller sprocket.

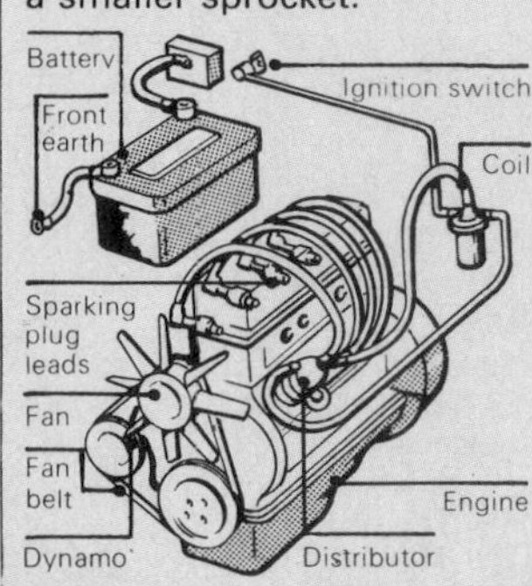

Distributor in ignition system

Differential is an assembly of 4 bevel GEARS that allows the driven wheels of a car to rotate at different speeds when the car turns a corner.
Distributor is the device, operated by a gear on the camshaft, that distributes the high voltage current from the COIL to the SPARK PLUGS. The rotor arm is permanently connected to the high tension lead from the coil. As it rotates, it passes on the pulses of high voltage current to each distributor terminal in succession. This causes each spark plug to produce a spark. On the distributor shaft there are raised cams (1 for each spark plug) which push against a contact breaker arm, forcing the contact breaker points apart. After each cam has passed a spring presses the points together.
Draisienne was a 2-wheeled hobby horse made in 1818. Its front wheel was steerable and it could be made to reach 16 km/hour.
Dynamo, *see* GENERATOR.

E **Epicyclic gear systems** are used in bicycles and in cars with automatic transmission. This

keeps the rotor turning (power). When the mixture is fully burned, the space begins to decrease and the waste gases are expelled (exhaust).

The motor car through the years

Although Daimler and Benz did not invent the motor car, they did make it into a viable commercial proposition. Two brilliant engineers, they brought together a number of inventions and improved on them so that the modern car evolved from their efforts. After Benz, many other manufacturers opened up, including such famous names as Panhard and Levassor, Renault, Peugeot, Austin, Morris and Ford.

Rene Panhard (1841–1908) and Emile Levassor were building cars with engines underneath in 1889. But by 1891 they were producing cars with engines at the front, which kept them clear of road dirt and helped improve the cars' balance. They also used sliding pinion gears for changing speed. These were a great improvement on the belts and pulleys used by Benz as more power could be transferred to the wheels.

Independent suspension was already in use, having been introduced in 1873 on a steam carriage. The CARBURETTOR was invented in 1884 and the modern float-type carburettor was first used in 1892. The steering wheel, which replaced the steering lever, or tiller, was first introduced in 1894 (*see* STEERING SYSTEM). In 1895 the Michelin brothers in France began producing pneumatic (air-filled) TYRES, which had been invented in 1888 by a Scottish vet called John Dunlop (1840–1921).

Below: The Ford Model T was popular in America and all over Europe. This 1909 version comes from a museum in Portugal. Henry Ford (1863–1947) built his first motor car in 1893 and in 1908 he thought up the idea of manufacturing cars on an assembly line. Using this system, together with interchangeable parts, he set out to build a car that was cheap and reliable. The success of the Model T paved the way for all later mass-produced cars.

The repeal of the 'red flag' law and the raising of the speed limit in 1896 to 22.5 km/hour, provided further incentive for improving parts of the car. By 1899 multi-cylinder engines had appeared, together with the honeycomb RADIATOR and SHAFT DRIVE, which gradually replaced the chain drive used on most early cars. Daimler had introduced the floor-mounted ACCELERATOR and Renault had adopted the UNIVERSAL JOINT used on modern cars.

As the speed of cars increased, so braking efficiency also had to improve. The first cars used wooden blocks applied to the rear wheels. In 1902 Renault fitted drum BRAKES to his cars and in the same year Frederick Lanchester (1868–1946), a British car manufacturer, invented the disc brake. However, this did not become widely accepted until the late 1950s.

During the early 1900s the motor car industry was expanding very rapidly, particularly in America. Many parts of the modern motor car date from this period, including AUTOMATIC TRANSMISSION, the DISTRIBUTOR, electric starting and lighting, the DYNAMO and hydraulic braking.

Until about 1915 comfortable cars were an expensive luxury. In 1903, however, the Cadillac and Oldsmobile companies in America introduced a new principle — cars with interchangeable parts. This was the first step towards mass-production. Henry Ford took up this idea and in 1908 began manufacturing the now famous FORD MODEL T. By 1927, when the factory making this model closed down, 15 million of these cars had been produced. This was the world's first mass-produced basic car. Because it was cheap and easy to run and maintain, it proved to be an immensely popular car.

After World War I manufacturers found themselves with markets for both luxury and mass-produced cars. Those who could afford them wanted the best models and many high-quality cars were built to a fine standard of craftsmanship. Such cars as the Bugatti-Royale Type 41 and Hispano-Suiza H6B were probably among the finest cars ever built. Unfortunately, the combination of the American Depression in the 1930s and competition with mass-produced

---

type of system uses 2 or 4 small pinions (planetary GEARS) mounted on a carrier. These revolve round a central sun gear and mesh with an outer ring called the annulus. A car that has AUTOMATIC TRANSMISSION uses 2 or more sets of epicyclic gears. The input drive is transmitted to the first annulus and the output is driven by the last annulus. Changing gear is achieved by using band brakes and clutches automatically applied to certain parts of the system at appropriate speeds. A bicycle has 2 sets of 4 planets, 2 sun gears and a single annulus. Changing gear is achieved by moving the sun gears horizontally so that they mesh with different parts of the system.

Expressway, *see* MOTORWAY.

F

Flywheel is a heavy wheel fixed to the rear end of the CRANKSHAFT. When the engine is running the momentum of the flywheel keeps the crankshaft turning smoothly in between individual power strokes. The rim of the flywheel has gear teeth which mesh with the STARTER MOTOR pinion when the engine is being started.

Ford Model T, also known as the 'Tin Lizzie', was mass-produced by Henry Ford between 1908 and 1927. 4-cylinder, 2,898-cc engine; 2-speed (forward) EPICYCLIC gearbox (and a reverse gear), operated by pedals; top speed 64 km/hour.

Forth Road Bridge was the first major suspension bridge built in Britain (1958–64). It has a main span of 1,006 metres, the longest in Britain, and side spans of 408 metres. It is the northernmost (latitude 56°N) major bridge in the world and high winds caused great problems during its erection.

Forth Road Bridge

Fuel injection is used on some cars instead of a CARBURETTOR. A measured amount of fuel is pumped through a fine nozzle into the inlet port of each cylinder at the beginning of the induction stroke.

G

Gears are toothed wheels mostly used to transmit movement from one rotating shaft to another. Simple spur gears

cars, caused many of the custom-built car firms to go out of business. Only a few manufacturers of high-quality cars, such as Lancia and Rolls Royce, survive from this era.

Many firms that produced cheaper cars also went out of business as a result of the intense competition. Two of the most successful cars in Europe in the 1920s were the AUSTIN 7, or Baby Austin, and the MORRIS COWLEY. One advantage was their low horsepower. Thus in 1921 when car tax was raised to £1 per horsepower they became cheaper to run than the Ford Model T.

Over the years car design continued to improve and many innovations were added as manufacturers vied with each other for a share of the ever-increasing market. One of the great testing-grounds for cars was the race track. Many standard items on modern cars began life on a racing car, including detachable wheels (which allowed spare wheels to be carried) and radial tyres. Streamlining, too, was first introduced into racing cars, but found its way into the popular car market as the family cars of the 1930s reached higher speeds. Improvements in fuel systems and engine technology led to lighter, more powerful engines and the resulting increase in power-to-weight ratio caused improvements to acceleration, speed, road-holding and braking.

In addition to cars other vehicles were now being constructed. The first lorry was built by Daimler in 1896 and by 1910 there were vans, trucks, lorries and fire-engines on the road. Until

Right: 1. A Draisienne hobby horse of 1819. It had a steerable front wheel, an arm rest and a saddle. This type of hobby horse became very popular.
2. A Matchless Ordinary penny farthing bicycle of 1883.
3. A Starley Rover of 1888. It had solid rubber tyres and no brakes.
4. An 1894 Hildebrand and Wolfmüller motor cycle. Its two horizontal cylinders drove the rear wheels directly and it had a top speed of about 45 km/hour — faster than any car of the time.
5. A modern Lambretta, one of the most popular makes of motor scooter.

Right: 6. A 1925 Harley-Davidson 7/9 hp motor cycle with a sidecar attached.

have teeth set at the same angle as the shaft; in helical gears the teeth are set at an angle. If 2 different sized gears mesh together, the larger is called the gear and it rotates more slowly than the smaller, which is called the pinion. A bevel gear is shaped like the bottom half of a cone. Bevel gears are used to change not only the speed but also the axis of rotation.

Generator, or dynamo, is an engine-driven device that produces an electric current. The term dynamo is sometimes used to mean the D.C. generator of a car, which produces direct current (current flowing in one direction only). An A.C. generator, or alternator, produces an alternating current. It is simpler and more reliable than a D.C. generator and can charge the battery when the engine is idling.

Golden Gate Bridge

Gladesville Bridge, across the Parramatta River, Sydney, Australia, is the world's longest-span reinforced concrete arch bridge. Completed in 1964, its main span is 304.8 metres long.

Golden Gate Suspension Bridge, built between 1933 and 1937 had, until 1964 (*see* VERRAZANO NARROWS BRIDGE), the longest span in the world (1,280 metres). It links the 2 sides of the entrance to San Francisco Bay Harbour.

H **Handbrake** is a device used to operate the rear brakes by means of cables and/or rods.

Horsepower (hp) is a unit of power used to describe the power output of an engine (1 hp = 550 ft-lb per sec; 1 ft-lb is the amount of work done by lifting a 1 lb weight through a vertical distance of 1 ft). This can be calculated from the work done by the pistons. However, some of this power is used up in overcoming friction in the engine. The proportion of power that remains can be measured by applying a brake (in an instrument called a dynamometer) to the drive shaft and is known as the brake horsepower (bhp).

Humber bicycle, first made in 1890, was the first to have the true diamond-shaped frame that, until the cross-frame became popular in the 1960s, was standard on all bicycles.

the 1920s they ran on petrol and had solid rubber tyres. But diesel engines were more economical and pneumatic (air-filled) tyres were becoming more reliable and by the early 1930s most lorries were similar to those of today.

Horse-drawn buses were replaced in the mid-1800s by horse-drawn trams, which ran on rails set into the road. Steam trams were tried in the late 1800s, but were not a great success. Two other methods of replacing the horse were found. One was the cable car, which was a tram hauled along by a cable set into a groove in the road. San Francisco's cable cars, first devised in 1873, are still in use. The other method was the use of electricity. Batteries could not produce sufficient power for trams, so various forms of overhead cable were tried. In 1888 Frank Sprague, an American, devised a swivel trolley pole that was attached to the roof of the tramcar and ran underneath the cable. This system was later used in tramways all over the world.

Electric cars are generally thought of as a modern idea, but the first electric carriage was in fact built in 1837 by Robert Davidson, a Scottish engineer. Electric buses appeared in Paris (1881) and London (1888) and in 1897 London's first taxi cabs were powered by batteries. They could travel at 14 km/hour and had a range of 48 kilometres. In 1899 Camille Jenatzy, a Belgian, achieved 106 km/hour, breaking the land speed record of the time, in his bullet-shaped electric car called *La Jamais Contente.*

Today, electric cars do not have the speed or range needed by most car drivers. Their use is limited to milk-floats and similar vehicles that do a lot of stopping and starting but do not travel great distances. However, more advanced electric cars and vans are being developed and some can reach 80 km/hour. The cleanness of such cars has great appeal in a world which is now conscious of how much pollution is caused by exhaust fumes from internal combustion engines.

Bicycles and motorcycles

While some people were trying to produce steam carriages another form of mechanical transport was catching on. In about 1800 there was a craze for two-wheeled 'hobby horses', which were built of wood and iron. They consisted of two wheels, one behind the other, joined by a bar which carried a seat. The rider straddled the bar and propelled the machine by 'walking' it along the road.

The craze soon died down, but a Scottish blacksmith, Kirkpatrick Macmillan (1813–78), followed up the idea and invented the first bicycle in 1839. It had a steerable front wheel carried on a forked rod. It was driven by pedals at the front which worked backwards and forwards driving rods and cranks attached to the rear wheel. In 1861 two brothers who owned a Paris pram factory produced a bicycle called the Michaux Velocipede. This had pedals that turned the front wheel, and as a result the wheel could only turn as fast as the rider pedalled. The immediate solution to the problem of increasing the speed of the bicycle was to increase the size of the front wheel. However, wooden wheels, with their heavy spokes and bone-shaking iron rims, were too heavy, so experiments began to develop SUSPENSION WHEELS. James Starley (1831–81), the manager of an English sewing machine factory, used suspension wheels on his famous

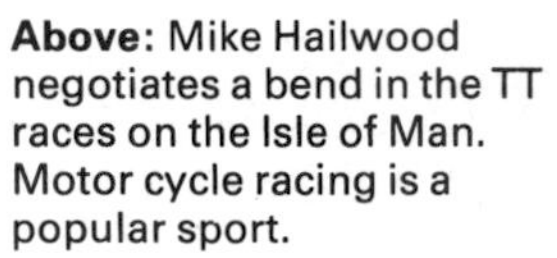
Above: Mike Hailwood negotiates a bend in the TT races on the Isle of Man. Motor cycle racing is a popular sport.

Left: Bicycles are used as a cheap and healthy form of transport by people all over the world, such as these Peking schoolchildren. The tricycle cart in the background is heavily laden.

Humber Estuary Suspension Bridge, started in 1972 and at present scheduled to be opened in mid-1979, will have the longest span in the world (1,410 metres).

K **Knocking**, or pinking, is a hammering sound heard in the engine. It is due to premature firing of the fuel and can be caused by using unsuitable fuel (*see* OCTANE RATING).

M **Manifold** is a pipe or chamber that has several openings. A multicylinder engine has two manifolds attached to the cylinder head. The inlet manifold consists of pipes or channels that carry fuel from the carburettor to the inlet ports. The exhaust manifold carries waste gases from the exhaust ports to the exhaust pipe.

Morris Cowley

Minor roads connect hamlets and villages with each other and with PRIMARY and SECONDARY ROADS. In some places minor roads may be little more than tracks with only thin base courses (road metal). A road without a base course is described as unmetalled.

Mont Blanc Tunnel, completed in 1962, is the longest road tunnel in the world (11.6 km). It is 9 metres high and has a single, 2-lane carriageway.

Morris Cowley, the most famous being the 'Bullnose Morris', was produced in thousands between the 2 World Wars. 4-cylinder, side-valve, 11.9-hp, 1,550-cc Hotchkiss engine; 3-speed gearbox; top speed 85 km/hour.

Motorway, or expressway, is a multiple-carriage TRUNK ROAD with 2 or more lanes in each carriageway. Vehicles can only gain access to a motorway via slip roads and there are no junctions except at the ends of the motorway.

Moulton bicycle was introduced in 1962 and its design has been copied, with modifications, many times since. Moultons have wheels of between 35 and 50 cm diameter, an F-shaped cross-frame with no high crossbar, and suspension — a coil spring at the front and a rubber shock-absorber at the rear. The height of the saddle and handlebars can be adjusted freely.

Moulton bicycle

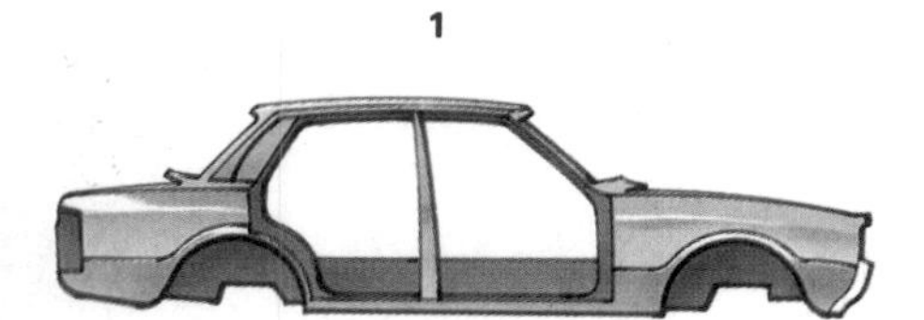

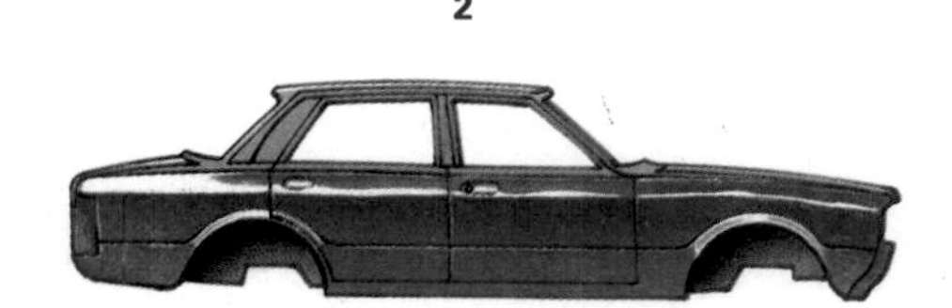

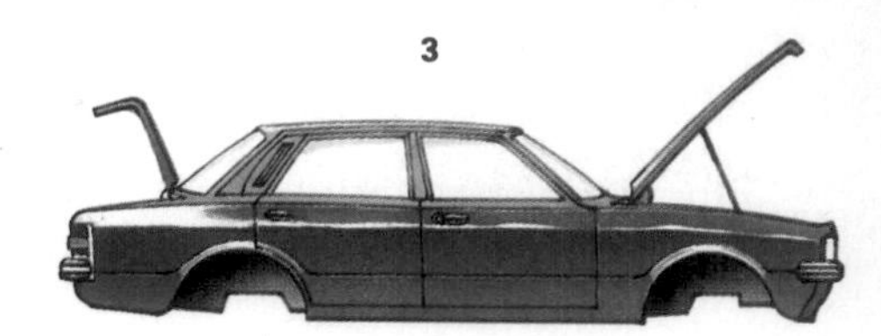

Above: Stages on a car assembly line.
1. The underbody and superstructure are welded together and the doors, boot and bonnet are fitted.
2. The body is thoroughly washed and sanded and then sprayed with two coats of primer paint and three coats of top colour.
3. The trimmings and fittings, such as windows, bumpers, steering, wiring and lights, are added.
4. The body is lowered on to the engine and transmission.
5. The wheels are fitted, the suspension, steering and all other components are examined and checked and the car is test-driven on rollers.
6. The finished car is tested for water leaks. Finally it is given a short road test and is then ready to be sold.

PENNY FARTHING BICYCLE, with its huge front wheel and tiny rear wheel. In early models of this bicycle the wheels were driven by pedals directly attached to the hub, but by 1885 there were versions with a form of chain drive and brakes on both wheels. Such bicycles were difficult and dangerous to ride so Starley also designed cumbersome three- and four-wheeled versions for the less adventurous riders.

Starley invented the type of suspension wheel in use today, with criss-crossing wires. In 1885 his nephew, J. K. Starley, designed the first safety bicycle, the Starley Rover. It resembled modern bicycles and its rear wheel was driven by a chain. However, its solid rubber tyres made it something of a bone-shaker.

Dunlop's pneumatic tyres made cycling more comfortable after 1888, but at first they had to be stuck and bandaged to the wheel rims. Removable tyres were invented in 1890. Later safety bicycles incorporated springing to help reduce the bumpiness of the ride still further. Today, bicycles have lightweight frames, gears and freewheeling devices.

Right: The Audi NSU Ro80 was one of the first cars to use a Wankel engine. Its 2-rotor, 995-cc engine gives it a maximum speed of 180 km/hour. The main advantages of a Wankel engine are its small size and smooth running. However, its high fuel consumption is a problem that has not yet been solved.

The first motorcycle was a Michaux Velocipede fitted with a steam engine in 1869 and over the next 15 years various steam bicycles were produced, some of which could reach 96 km/hour. They were probably very uncomfortable and hot to ride.

During the late 1800s and early 1900s, however, petrol-engined pedal cycles (now called mopeds) appeared. Some of these were converted into two-seater tricars by fitting a chair in front or behind. The side-car appeared in 1904 and is still sometimes used to provide extra seating capacity.

By the beginning of World War I motorcycle design had progressed considerably, largely due to the popularity of racing. The frames had become heavier, the seats lower and the front forks were sprung to reduce road-shocks. Some machines even had gearboxes with two or three gears. Today motorcycles are built in various sizes, ranging from small 250-cc (*see* CAPACITY) models to powerful 1,000-cc racers.

Building cars

Before producing a new car a manufacturer spends a great deal of time and money on working out the design. A prototype is built and thoroughly tested. When everything is seen to work satisfactorily the car can be mass-produced.

Identical parts, such as body-panels, engines and wheels, are made and taken to a point on the assembly line. Each mechanic stays in one place and as the line moves forward he fits a particular part to every car as it passes him. Parts are added in a carefully devised sequence, beginning with the body and finishing with the wheels. At the end of the line each identical car is tested before being sent out to the customer.

N **New River Gorge Bridge**, West Virginia, USA, is the world's longest steel arch bridge. Completed in 1976, it has a main span of 518.2 metres.

O **Octane rating** of a petrol is a measure of its anti-knock properties. A sample of the petrol is tested in a special engine and compared with a mixture of iso-octane and heptane. The percentage of octane in the mixture comparable to the petrol gives the octane number. Petrol grades are sold according to minimum octane ratings. In Britain there are 4 grades: 2-star (minimum 90 octane), 3-star (minimum 94.96 octane), 4-star (minimum 97.99 octane) and 5-star (minimum 100 octane). In other parts of Europe only 2 grades are generally available; premium, which varies between 92 and 100 octane, and regular, which varies between 78 and 95 octane.

P **Pavement** of a road consists of the various layers laid down on the SUBGRADE. Flexible pavements have surfaces of fine stones bound together with asphalt or tar. Below this there is a base course of gravel, well compacted down and often bound together with similar bituminous material to that used in the surface. Below the base course a sub-base course of crushed rock rests on the subgrade. Rigid pavements have surfaces of portland cement, usually reinforced with steel bars, laid on to the sub-base course.

Paving with granite blocks, c.1840

Penny farthing bicycles had front wheels up to 127 cm in diameter and rear wheels of about 60 cm diameter. The frame consisted of a single curved tube from the top of the front wheel-fork to the top of the rear wheel-fork. The saddle was mounted on the frame near the handlebars.

Piston is a circular plunger that fits inside a CYLINDER. A modern engine piston is made of a light aluminium alloy. It has 3 grooves around the upper part. Two of these are for compression rings, which prevent gases

4

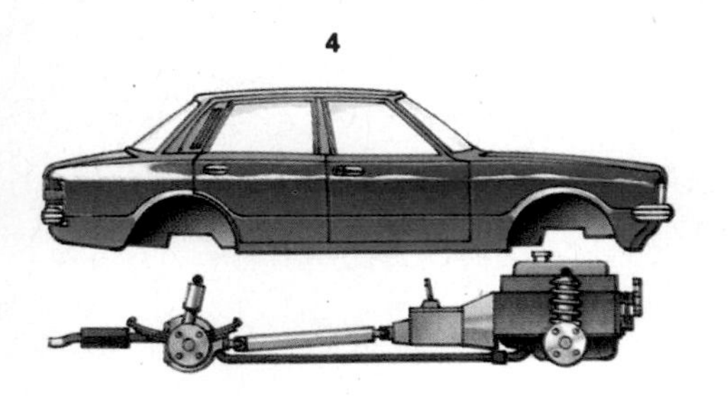

5

6

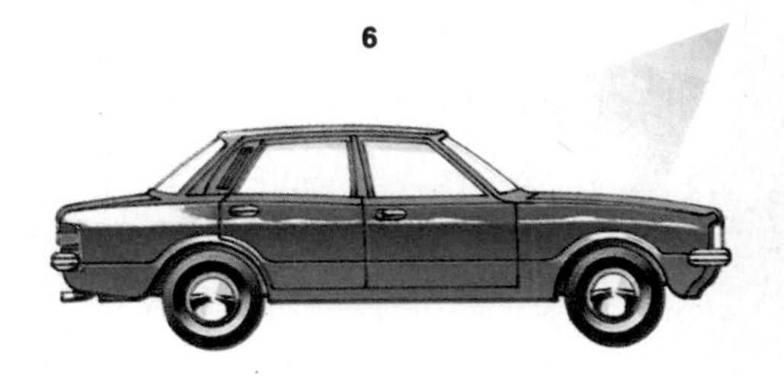

Left: Mass-produced cars are put together on an assembly line. Here, as the line moves forward, each Audi is having its body trim added before the engine, transmission and wheels are fitted.

Below: Some of the most expensive makes of modern car are still hand built without an assembly line. Here skilled mechanics are assembling Rolls Royces.

The moving assembly line was first used by Henry Ford to build the Model T and today nearly all cars are built by this mass-production method. It is the fastest and most economical method of building cars. The main problems are keeping each mechanic supplied with all the parts he needs, finding the optimum speed of the line and ensuring that sufficient time is given for each stage to be completed.

How a car works

The main systems of a motor car are the engine, cooling system, CLUTCH and GEARS, TRANSMISSION, suspension, STEERING, BRAKES, fuel, exhaust and electrical systems.

The modern car engine, although it works on the relatively simple four-stroke cycle, is a complex piece of machinery. The CYLINDERS in which the pistons move up and down form part of the cylinder block. The PISTONS are connected by piston rods to the CRANKSHAFT, which lies in the lower part of the cylinder block, known as the crankcase. The up-and-down movement of the piston rods causes the crankshaft to turn.

The cylinder head bolts on to the top of the cylinder block. It contains the combustion chambers, SPARK PLUGS and VALVES. The valves fit into the inlet and exhaust openings at the top of each cylinder. They are operated by a rotating CAMSHAFT, which is driven by a belt or chain connected to the crankshaft.

The engine needs to be kept lubricated with oil, otherwise the friction between the moving parts would cause them to overheat and seize up. Below the crankcase is a reservoir of oil called the sump. An oil pump, operated by the camshaft, pumps oil through narrow passages in the cylinder block, which feeds oil to where it is needed.

Many engines are cooled by water which is kept circulating through channels in the cylinder

---

from escaping from the combustion chamber, and 1 is for a scraper ring, which prevents oil from entering the chamber. The piston is connected to the small end of a connecting rod by a gudgeon pin, which lies horizontally inside the piston. The gudgeon pin may be free or fixed to the connecting rod. Other kinds of piston are used in the hydraulic systems of a car, such as the BRAKES.

Pont de Quebec, across the St Lawrence River in Canada, has the longest cantilever span in the world (549 metres). Work on this bridge began in 1899. After a major structural failure, a major accident and the loss of 87 lives the bridge was finally opened in 1917. Intended originally as a railway bridge, it now has a railway track and 2 carriageways running along its 987-metre length.

Primary roads connect towns and cities with each other and with TRUNK ROADS.

R

Radiator is used to cool the water of an engine's cooling system. It is composed of very many thin channels and so has a large surface area from which heat is lost. Cooling is aided by a fan.

Pont de Quebec

S

Sault Sainte Marie Bridge, Michigan, USA, is the world's largest bascule bridge (drawbridge). Built in 1941, its opening span is 102 metres.

Secondary roads lead from PRIMARY ROADS and link small towns and villages to each other.

Servo-assisted brakes use the vacuum created in the inlet manifold by suction from the engine. When the brake pedal is depressed, the small movement of fluid in the master cylinder operates an air valve. This allows air to pass to one side of a

block by an engine driven pump. Hoses connect these channels to a radiator fanned with air where the water is cooled before passing back into the system.

The rotating movement of the crankshaft is transferred to the wheels via the clutch, gearbox and transmission. The clutch is a device for connecting the engine to the gearbox. Most cars have a single-plate clutch. When the driver lifts his foot off the clutch pedal, a friction plate, which is connected to the gearbox, is forced by springs against the engine flywheel and both turn together.

The gearbox contains a number of toothed wheels, or gears, to transmit the engine power to the wheels. The engine produces the most power when running at high speed and the gears are used to slow down the speed of rotation so that the wheels need not turn so fast, particularly when the car is starting from rest. In a simple gearbox the gear lever is used to move two or more gears so that they interlock (mesh) and turn together. Changing gear with the early gearboxes could be a noisy and tiring business, but with the invention of synchromesh, gear-changing has become easier. Synchromesh is a device that synchronizes the speeds of two gearwheels, making sure that when they actually mesh they are rotating at the same speed.

In modern gearboxes all the gears are constantly in mesh. Gear-changing is achieved by moving two selector rings that revolve with the transmission shaft. Sliding one of these rings backwards or forwards causes it to engage a small dog gear attached to the required main gear. Today some cars have AUTOMATIC TRANSMISSION, a system which uses an automatic clutch and an arrangement of special gears to select the most appropriate gear ratio for the speed of the car.

The transmission system conveys the power to the wheels. In many rear-wheel drive cars the transmission shaft of the gearbox is connected by a UNIVERSAL JOINT to the propeller shaft. This type of joint allows the rotation to be transmitted while at the same time allowing the propeller shaft to move up and down. The rear end of the propeller shaft is connected to the final drive unit by another universal joint. The final drive unit

Brake drum
Slave cylinder
Brake shoe
Brake lining

Above: In a drum brake an increase in brake fluid pressure in the slave cylinder forces the two brake shoes apart and the linings make contact with the inside of the drum.

Below right: A 'ghost' drawing of a Ford Fiesta Ghia 1100. Fiestas are a range of useful hatchback cars of which the Ghia is the most luxurious. The Ghia 1300 is the fastest, having a top speed of 158 km/hour. The Ghia 1100 has a top speed of 145 km/hour and its actual engine capacity is 1117 cc. Like the other Fiestas it has a 4-cylinder transverse engine and front wheel drive. Its braking system is servo-assisted and there are disc brakes on the front wheels and drum brakes on the rear wheels. The steering is rack and pinion and both front and rear suspensions include coil springs and telescopic dampers. On the Ghia 1300 there is an additional anti-roll bar at the rear.

Left: Safety is an increasingly important factor in modern car design and manufacturers use crash tests to find out how each part of a newly-designed car will behave in a collision. The front end of the car is designed to crumple on impact, so that some of the shock is absorbed. Other safety features that can be tried out in such tests include collapsible steering columns, safety windscreens, head rests and impact-absorbing bumpers.

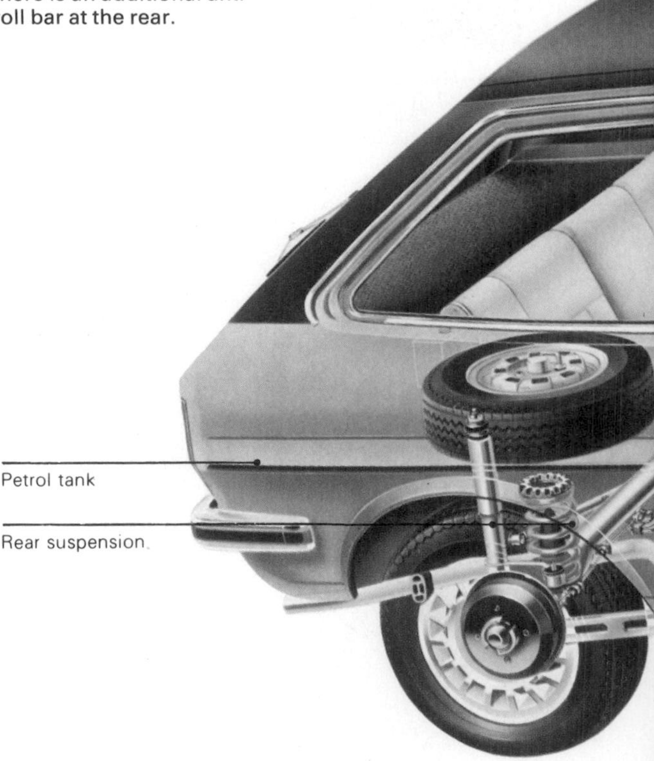

diaphragm while at the same time allowing suction from the inlet manifold to create a partial vacuum on the other side. The diaphragm therefore moves and a rod attached to the diaphragm presses against the piston in the master cylinder, thus amplifying the effort used in depressing the brake pedal. Some brake servo systems use the vacuum to assist the depression of the pedal directly, rather than the master cylinder piston.

Shaft drive is the system of transmitting the power of the engine using gears, UNIVERSAL JOINTS and rotating solid metal shafts.

Shoulder of the road is a continuation of the base course (*see* PAVEMENT) on either side of the riding surface.

Silencer is a chamber in the exhaust pipe in which the waste gases are allowed to expand and the noise released during expansion is reduced. In a baffle silencer the plates, or baffles, increase the distance travelled by the gases. A straight-through silencer, which causes less reduction in engine power, contains a perforated tube surrounded by a sound-absorbent material.

Solenoid switch is an electro-magnetic device used in starting. When the ignition switch is turned, current from the battery flows through the coil of the solenoid. This causes a metal plunger to make contact with the two terminals connected to the BATTERY and STARTER MOTOR. Thus the large current needed to turn the motor does not have to flow down the long cables to the ignition switch.

Spark plug contains a central core of metal, or inner electrode, connected to a terminal at the top. This electrode is insulated by a ceramic material from the metal plug casing, or outer electrode. At the base the 2 electrodes are separated only by a small gap — the spark gap. The terminal is connected to 1 of the leads from the DISTRIBUTOR and the metal casing screws into the cylinder head, which is connected through the car body to the earth terminal of the BATTERY. A pulse of high voltage current from the distributor discharges, causing a spark to pass across the spark gap.

Speedometer

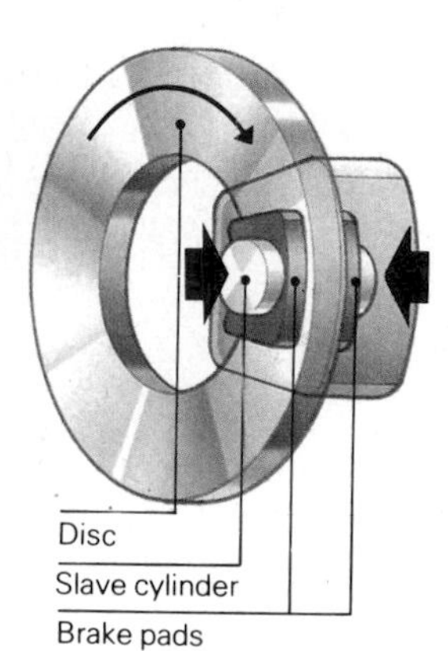

includes a set of DIFFERENTIAL gears. This allows the wheels at the ends of the two drive shafts to revolve at different speeds when the car is turning a corner.

In some front-wheel drive cars the engine and gearbox are mounted transversely (across the car) and power is transmitted from the gearbox directly to a differential gear system. From this a drive shaft, connected at both ends by universal joints, leads to each wheel.

The suspension system is needed to provide a comfortable ride and to keep the wheels on the road. The earliest form of suspension, used on carriages in the 1800s, was the leaf spring. This is still used in cars, particularly on the rear axles. Other forms of suspension include rubber, air, fluid-gas and coiled metal springs and torsion (twisting) bars, generally in conjunction with hydraulic shock absorbers to damp down the movement of the springs.

Several different types of steering mechanism are used but they all work on a principle first devised in the 1700s. Before the 1700s the axle pivoted in the centre, but in the new system the turning action of the steering wheel caused the axle to slide to the right or left and so change the direction of the wheels which were pivoted at each end of the axle. Power steering, used on many modern cars, uses oil under pressure to assist the movement of the steering mechanism.

For braking, cars use drum brakes, disc brakes or both. A common system uses drum brakes on

Above: In a disc brake an increase in brake fluid pressure in the two slave cylinders forces the brake pads inwards and they make contact with the disc.

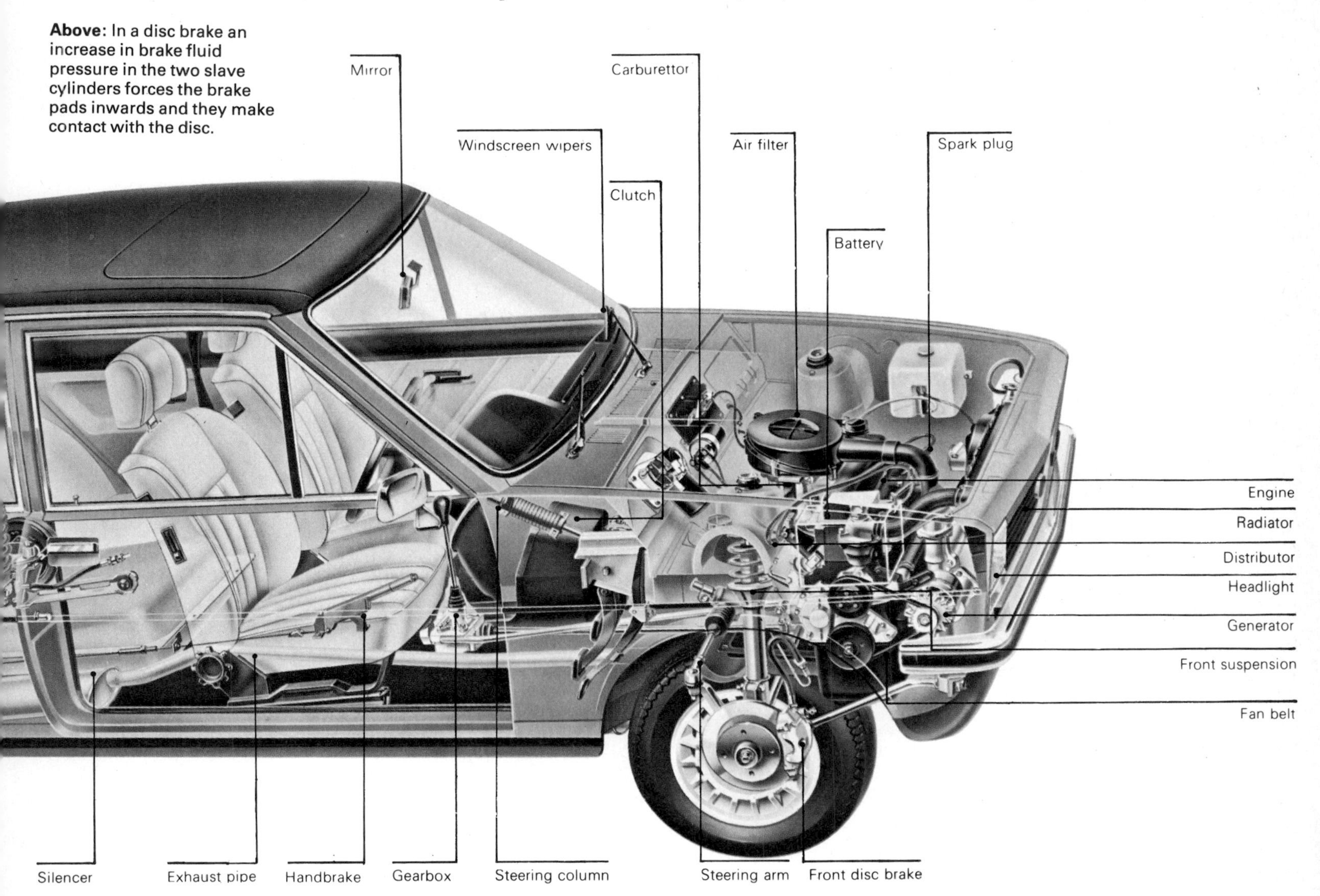

Speedometer is driven from a point on the transmission that rotates at a speed directly proportional to the speed of the wheels. A flexible drive cable rotates a magnet inside a metal drum. The drum, which is connected to the speedometer indicator needle, also tries to turn in the magnetic field, but is prevented from turning very far by a coiled spring. However, the faster the magnet rotates the further the drum and needle move round. A speedometer generally also incorporates an odometer, which measures the distance travelled.

Sprocket is a small, toothed wheel, which can be turned by a roller chain; the rollers fit in between the teeth.

Steering wheel and column

Starter motor is the electric motor used to turn the CRANKSHAFT during the process of starting the engine. When the motor turns, a pinion moves along a screw-threaded shaft and engages the teeth on the FLYWHEEL. When the engine fires, the pinion is driven back along the shaft and disengages.

Steering system consists of a steering wheel and column linked to the wheels. In a rack and pinion steering system a pinion (*see* GEARS) on the end of the column moves the rack, which is a flat, toothed rod linked to the steering rods, from side to side. In a worm and peg system a worm gear, turned by the column, moves a peg backwards or forwards. The peg is linked by arms to the steering rods.

Subgrade of a road is the soil or rock on which the PAVEMENT is laid.

Supercharger is a device for increasing the amount and pressure of air and fuel taken into the cylinders so that more fuel is burned. Some types are driven by the engine via belts and pulleys; turbochargers use a turbine which is driven by the exhaust gases.

Suspension wheel works on the principle that a heavy weight will hang safely from a thin, light wire. Thus, in a bicycle the weight of the rider and bicycle is made to hang from the top of the wheels rather than stand on thick spokes.

T

Throttle, *see* CARBURETTOR.

Transmission is the means by which engine power is transferred to the wheels.

Left: The Dover end of the Channel Tunnel, which was intended to link England and France. The project was abandoned in 1975, owing to rising costs. However, if it is ever built, the chalk rock beneath the English Channel will be easy to dig through and will require very little support.

Below right: A bridge has to withstand its own weight and that of any traffic it might carry. The longer the bridge's span, the greater the load it must bear.
1. A beam bridge needs several supports to carry the load.
2. In an arch bridge the load is directed outwards along the arch.
3. The load of a suspension bridge is held by the two main supports and cables attached to the banks.
4. In a cantilever bridge the main load is carried by the supports. The load in the middle of the long span is held by the arms attached to the bank.

the rear wheels and disc brakes on the front. The brakes are operated hydraulically. Pressing down on the foot pedal increases pressure in a fluid-filled master cylinder. This is transmitted through pipes to the smaller slave cylinder in each brake. In a drum brake two brake shoes are pushed outwards to make contact with the inside of a brake drum. In a disc brake the disc is gripped by two brake pads. In some cars the brakes are SERVO-ASSISTED by a system that uses the vacuum created in the engine's inlet MANIFOLD. Most cars have a secondary system called a HANDBRAKE.

The fuel system consists of a petrol tank, fuel pump and CARBURETTOR. The function of the carburettor is to vaporize the petrol, mix the petrol vapour with the correct proportion of air and to supply the engine with the required amount of this mixture.

The exhaust system carries away the waste gases from the engine and consists of a long pipe connected to the exhaust manifold. Some way along the pipe a SILENCER is fitted and this reduces the noise of the expanding gases.

The electrical system consists of a number of components. The BATTERY supplies power for the lights, wipers and other equipment and also for starting. When the starter switch is operated the electric STARTER MOTOR turns the engine, drawing in fuel from the carburettor. At the same time the coil produces a high tension (high voltage) current which flows to the DISTRIBUTOR. The rotor arm of the distributor is turned by a mechanical connection to the camshaft of the engine and feeds the current to each spark plug via four leads. When the engine is idling, the battery provides all the power. At higher speeds an engine-driven GENERATOR provides the necessary current and recharges the battery.

Roads, bridges and tunnels

The first roads developed from the tracks and paths trodden by Stone Age man. The increase in trade between settlements that resulted from the arrival of the wheel meant that such tracks had to be improved. Roadways were levelled and ditches were dug at the sides to provide drainage.

Stone roads appeared in China in about 3500 BC and in the Mediterranean area in about 1500 BC while at the same time in northern Europe log roads were in use. Many of these roads formed long trade routes. But the Romans were the first to build a system of scientifically constructed roads. They built more than 85,000 kilometres of road, linking the various parts of their Empire. These long, straight roads were constructed in several layers, topped with paving stones set in mortar.

After the decline of the Roman Empire, the roadways of Europe fell into disrepair and only began to improve again in the late 1600s. The STAGECOACHES *(see page 69)* ran on stone-surfaced toll roads built on much the same principle as the massive Roman roads. The French roads of the late 1700s were the best in Europe. In 1775 Pierre Tresuaget devised the modern method of road construction in which the load is carried by the soil underneath the road instead of by the surface of the road itself.

Three British engineers also made major contributions to road building. John Metcalf, who built many roads in the mid-1700s, stressed the importance of good drainage. Thomas

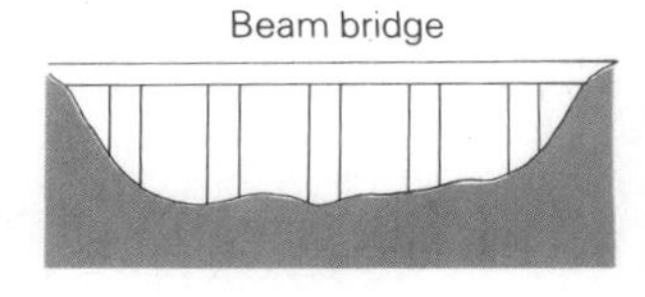

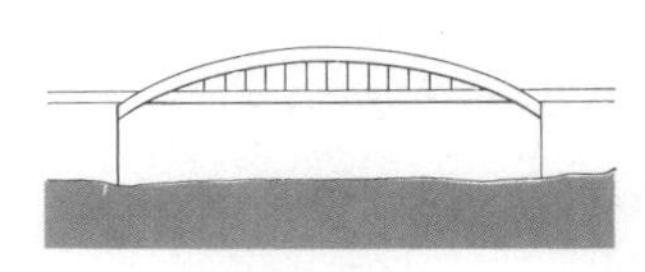

Suspension bridge

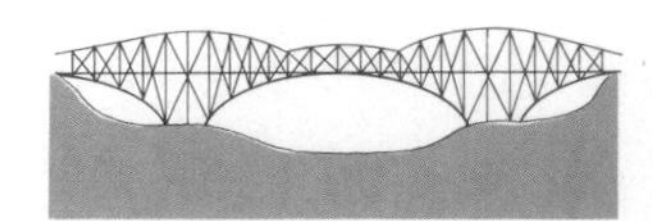

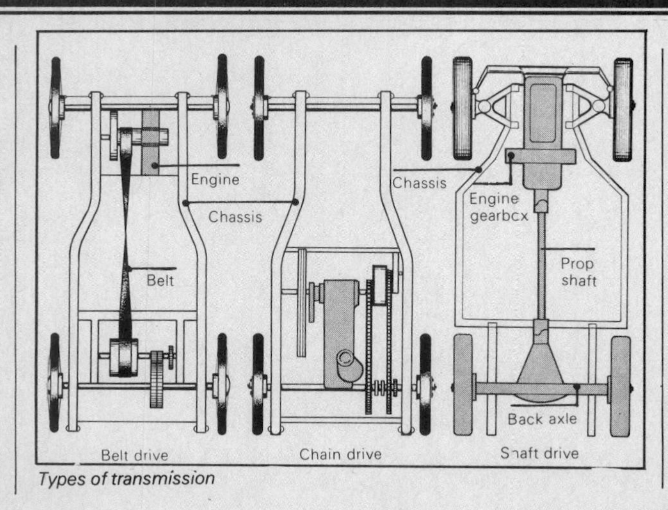

Types of transmission

See AUTOMATIC TRANSMISSION and SHAFT DRIVE.

Trunk roads, or arterial roads, carry long distance traffic, avoiding city centres.

Two-stroke cycle is a system of engine operation in which every second stroke of the piston is the power stroke. Each stroke combines 2 of the strokes of a 4-stroke cycle. On the upstroke fuel and air mixture is drawn into the crankcase through the inlet port. Towards the end of the upstroke the mixture in the cylinder is compressed. Combustion of the fuel begins the downstroke, or power stroke, and towards the end of the downstroke fuel mixture passes into the cylinder through a transfer passage, forcing the waste gases out through the exhaust port.

Tyres are inflatable rubber rings round the wheels of a car. They have flexible sidewalls and thick treads, which are grooved for draining away water. Some tyres have an inner tube, but today most tyres are tubeless, having a soft rubber lining which forms a seal between the wheel rim and the tyre. Inside the outer rubber layer of a tyre there is a casing made of layers (plies) of fabric. A cross-ply tyre is so-called because the cords (threads) used to run across the tyre (at right angles to the direction of rotation). Today, however, cross-ply cords run at an angle of 40° or less. In a radial-ply tyre the cords run across the tyre, and there is a layer of breaker plies, with diagonally-running cords, underneath the tread.

U

Universal joint, or Hooke joint, consists of 2 yokes (forks on the ends of

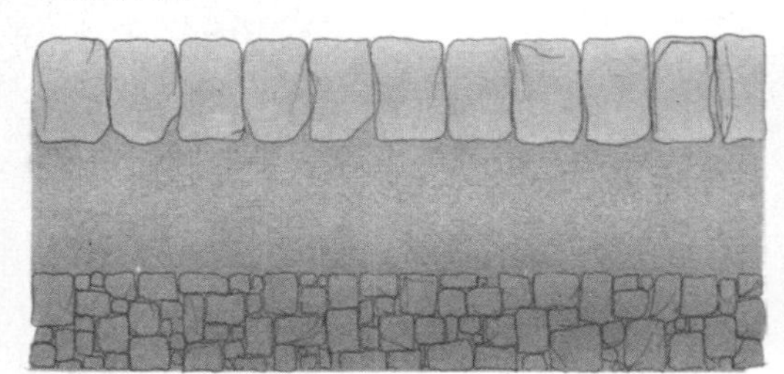

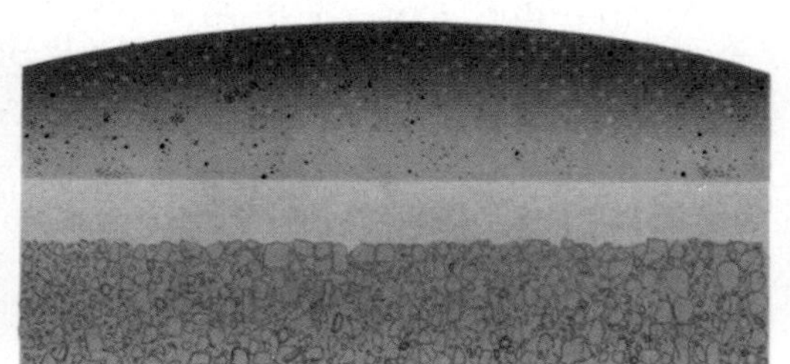

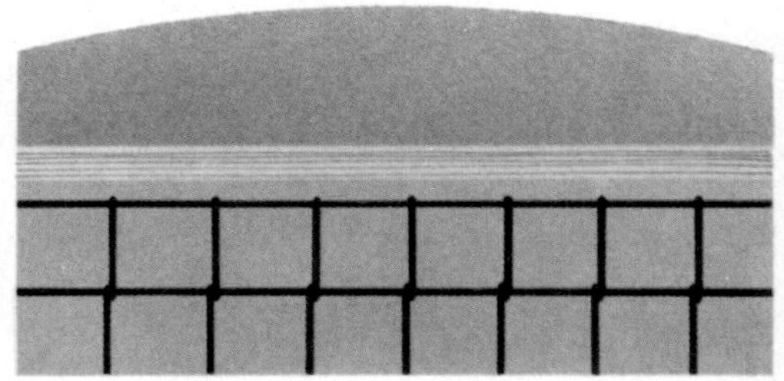

Telford in the late 1700s emphasized the need for solid foundations. In 1816 John McAdam devised a method of surface construction that used small stones compacted together. Roads that use this type of hard surface are still referred to as macadamized roads.

Above: A Roman road had a foundation of stone blocks set in mortar. Above this was a layer of smaller stones bound together with cement. The surface was made up of close-fitting stone blocks. A modern road used by light traffic has a sub-base course of large stones. Above this is a base course of ash or gravel, well rolled down. The surface is tarmacadam (small stones bound together with tar or asphalt). A modern road used by heavy traffic, such as a motorway, may have a base course of reinforced concrete. Above this is a layer of waterproof paper. The surface is another layer of concrete, perhaps topped with asphalt.

Left: From above, the Harbor Freeway in Los Angeles, USA, can be seen as a dual carriageway with single carriageway roads crossing over it. The minor roads are connected to the freeway by curving link roads.

Modern roads, built for cars and heavy lorries, have hard, cambered (rounded) surfaces and solid, load-bearing foundations composed of several layers. Each layer spreads the load over a wider area, so that the underlying soil, or subgrade, does not subside under the weight of the traffic above. The surface of the road may be asphalt, a mixture of bitumen and fine stone chips, or concrete reinforced with steel mesh.

Bridges for spanning rivers and similar obstacles have been in use for thousands of years. The simplest beam bridge consists of a plank across a small stream, and suspension bridges in the form of log footways suspended over ravines are still in use in some parts of the world. Stone and wooden arch bridges were first used by the Romans. During the Industrial Revolution bridges that could carry enormous weights became necessary. The large iron bridges that were built were tremendous feats of engineering.

Modern bridges are built from steel and reinforced concrete, materials which allow them to have both wide spans and elegant constructions. Large beam bridges use several beams supported by a number of piers. A cantilever bridge is a type of beam bridge in which the number of piers is reduced. The beams project out from the tops of the piers and are fixed at each end, either to the bank or to the ends of other cantilevered beams. In a suspension bridge the roadway is suspended by steel ropes from cables which pass over towers and are then anchored to the ground.

Tunnels are bored where hills and mountains are too steep to pass over. The Romans were the first to build tunnels, using picks and shovels. Tunnelling became easier with the discovery of explosives, but today it is mostly done with machinery. Tunnels are more common on railways than on roads, as trains have more difficulty than cars in negotiating steep inclines.

shafts) pivoted on a central, cross-shaped spider. Both shafts can rotate even if there is a small angle between them. However, if the angle becomes too great, the velocity of one shaft becomes greater than the velocity of the other. On modern front-wheel drive cars this problem is solved by using special kinds of universal joints called constant velocity joints.

V **Valves** are mushroom-shaped pieces of metal used to open and close the inlet and exhaust ports of a 4-stroke engine. They are operated by the CAMSHAFT. In an engine with overhead valves and a camshaft in the cylinder block the movement of a cam is transferred to a tappet, which rests on the cam. In the tappet there is a push rod, which is linked to one end of a rocker arm above the cylinder head. The other end of the rocker arm, which pivots on the rocker shaft, touches the top of a valve stem. Thus when the tappet and push rod are moved upwards by the cam the valve is lowered and the port is opened. The port is closed by a spring fixed to the valve stem. In side-valve and overhead camshaft engines the tappet operates the valve stem directly.

Verrazano Narrows Bridge, completed in 1964, is the longest suspension bridge in the world. Its main span is 1,298 metres long and it bridges the entrance to New York harbour. It has 2 decks and carries 12 lanes of road traffic. Its 4 main cables are each 90 cm in diameter.

Y **Yerbo Bueno Island Tunnel**, San Francisco, has the largest diameter of any tunnel in the world. It is 17 metres high, 23 metres wide and 165 metres long. It carries 2 decks of traffic.

Verrazano Narrows Bridge

Most canal construction ended when the golden age of the railways began. But some canals, such as the Panama Canal, the St Lawrence Seaway and the Suez Canal, are still vital arteries of world trade.

Canals

Canals are man-made waterways used for water transport and irrigation or drainage. No-one knows how canals first evolved, but in hot countries large irrigation ditches could have carried small boats and eventually people probably realized the advantages of transporting heavy loads by water. The earliest known canals were built in Mesopotamia in about 5000 BC.

The early canals could only pass over flat areas of land. If the countryside became hilly the canals could not be made to rise with it. However, when locks were introduced, it became possible for canals to cross several different levels of countryside. The Chinese used a simple kind of lock in about AD 100. This consisted of a weir — a dam over which water falls — in which there was a removable gate, called a flash lock. A boat moving from the higher level to the lower would hurtle down with the rush of water when the gate was opened. A boat moving in the other direction had to be winched up against the flow.

In the late 900s the Chinese were using two flash locks together, thus inventing the modern pound, or chamber, lock. This consists of two gates in between which is a chamber of water. The level of the water in the chamber can be altered to either of the two levels at each end of the lock. The early Chinese chamber lock used guillotine gates. These could be raised to allow water to flow in or out of the lock. Similar locks were in use in Europe by the late 1300s.

Most modern locks have mitre gates, a system invented by Leonardo da Vinci in 1487 for a canal in Milan, Italy. At each end of the lock

Right: The entrance to the canal tunnel at Worsley, the start of the original Bridgewater canal. Inspired by a visit to the Canal du Midi in France, Francis Egerton, Duke of Bridgewater, commissioned James Brindley to build a canal to carry coal to Manchester from his estate at Worsley. In this scene an early form of crane is being used to load a barge. The canal was later extended to the River Mersey at Runcorn.

Reference

B **Briare Canal**, built by Hugues Cosnier in 1642, was the first true summit-level canal. From Briare it rises 39 metres to a summit plateau 6.2 km long. From there it drops 81 metres to join the River Loing at Rogny. It has 40 locks and at one point between the summit and Rogny 6 of these form a staircase that drops 20 metres.

Bridgewater Canal was the first of many canals built in Britain in the 1700s. These included the Grand Trunk Canal, which linked the Mersey to the Trent, the Grand Union Canal, which linked the Grand Trunk Canal with the Thames, and the Severn Waterway and Gloucester and Sharpness Canal, which provided the final link from the Thames to the Bristol Channel.

C **Caledonian Ship Canal** was built by the Scottish engineer Thomas Telford (1757–1834) between 1803 and 1822. It linked Loch Lochy and Loch Ness along the line of the Great Glen of Scotland with Loch Linnhe (sea) in the west and the Moray Firth in the east. Built so that sailing ships could avoid the long and hazardous journey round Cape Wrath, it soon became obsolete when steam power arrived and ships increased in size.

Canal du Midi, also known as the Languedoc Canal, was built by Paul Riquet between 1666 and 1685. It links the River Garonne, which flows into the sea at Bordeaux, with the Mediterranean Sea via the Bassin de Thau at

Corinth Canal

Sète. From Toulouse it uses 26 locks to rise 61 metres in 54 km to the summit. The descent to the sea is a drop of 185 metres over 190 km, and 74 locks are used for this. Riquet went to great lengths to ensure that the summit had sufficient water and built a dam and over 60 km of feeder channels.

Corinth Canal in Greece is a deep sea-to-sea ship canal that connects the Ionian Sea via the Gulf of Corinth to the Aegean Sea, saving the long journey round the Peloponnisos. The canal was first attempted in about AD 60 by

there is a pair of wooden gates hinged to the sides of the chamber. When they close they form a V-shape. The name 'mitre', a carpenter's term, refers to the join between the two gates when they are closed. The V-shape points upstream so that the water pressure helps to seal the gates more effectively. The greatest pressure on the gates, and hence the most effective seal, occurs when there is a difference in water level on either side of the gates.

Many of the early canals follow a wandering course. The reason for this is that the engineering techniques needed to cross high ground were not available. However, since the 1600s several methods have been used to overcome such problems. Tunnels were bored through hills and tall viaducts crossed over valleys. In high-rise canals, water was pumped to the top in order to maintain the water level of the higher locks. Step locks consist of a series of locks. The lower gate of each lock acts as the upper gate of the next. Boat lifts are used to haul boats up slopes, either in tanks of water or on trolleys, from one level to another. A modern lift at Montech in France pushes boats up in a pool of water sealed by a movable dam. In some places boats are lifted down from one canal to another canal or river.

Above: A horse-drawn barge on the Grand Western Canal, Devon, where it crosses (by means of an aqueduct) the disused Tiverton branch railway line.

Below: A chamber, or pound, lock. First the upper gates and sluices close. The lower sluices open and water falls to the lower level. Then the lower gates open and a boat enters the chamber. The lower gates close and upper sluices open to let water rise to the upper level. The upper gates open and the boat leaves.

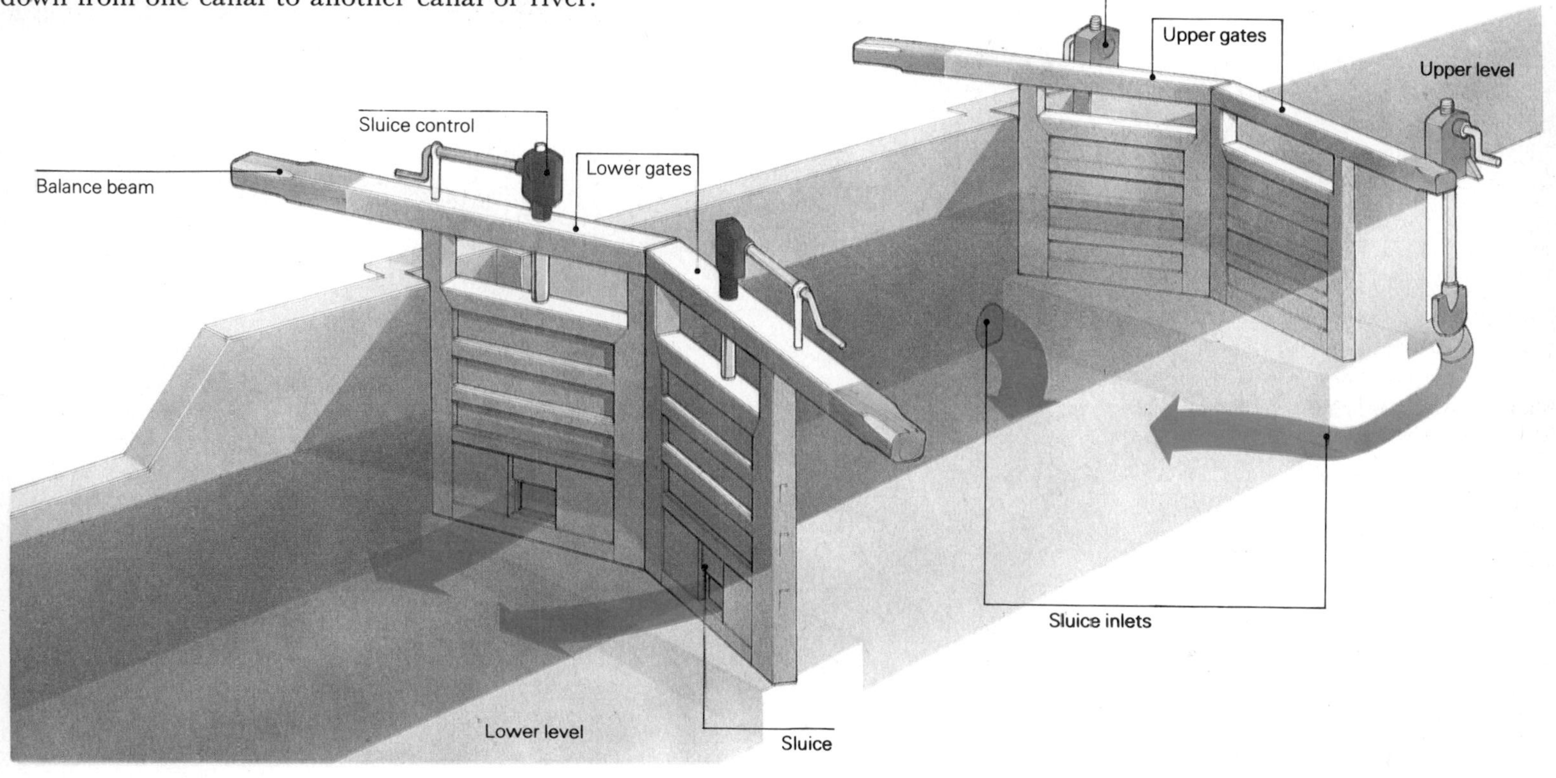

the Roman Emperor Nero. In the late 1800s the shafts he had begun were reopened and continued. The 6.3-km canal is today an awesome sight. Its almost vertical, sheer cliffs, cut through solid rock, rise 86 metres above the water. Being only 8 metres deep and 25 metres wide the canal can only take vessels of up to 10,000 tonnes and the largest ships have to be towed through with care.

D **Dry dock** is used when the hull of a ship needs to be inspected or repaired. As the dock is pumped dry, the ship comes to rest on keel blocks arranged to fit the shape of the hull. Modern dry docks often have guidance systems so that the ship is accurately positioned on the blocks. Most dry docks take one ship at a time, but the largest dry dock in the world, the Lisnave dock, Lisbon, Portugal, can take two or more ships. It is 518 metres long, 97 metres wide and 16.15 metres deep.

E **Erie Canal** is 580 km long and has 82 locks. Built between 1817 and 1825, it links Buffalo on Lake Erie with Albany on the River Hudson. This canal eased the flow of produce from the mid-west prairies to New York and, when it was opened, was the longest canal in the world (580 km).

F **Floating dock** is a form of dry dock. The dock is first submerged and the ship positioned above it. Water is then pumped out of ballast tanks and the dock rises, lifting the ship with it. Some floating docks can lift ships of up to 80,000 tonnes.

P **Panama Canal** was first thought of by Ferdinand de Lesseps after his success with the SUEZ CANAL. He proposed a sea-level canal via Lake Nicaragua, but in attempting this project between 1881 and 1889 he was beaten by the terrain and the climate. He then began a

The de Lesseps attempt to dig the Panama Canal

Inland canals became popular in Europe between the 1600s and 1800s. Two French canals were among the greatest engineering achievements of their time. The BRIARE CANAL, completed in 1642, linked the River Loire and the River Seine. The CANAL DU MIDI, completed in 1681, linked the River Garonne with the Mediterranean. This canal inspired the building of the BRIDGEWATER CANAL between Worsley and Manchester in England in 1761, which began a 70-year era of canal construction in Britain during the Industrial Revolution. This was brought to an end by the coming of the railways. Only the Manchester Ship Canal (1894) was built after this period.

The success of the European canals showed the Americans the possibilities of inland water transport. In 1825 the ERIE CANAL, between Buffalo and Albany, was built, and in 1829 the WELLAND CANAL by-passed the Niagara Falls between Lake Ontario and Lake Erie. A series of later canals linked the Great Lakes with the Mississippi, Ohio and Susquehanna Rivers and, finally, in 1959 the whole inland waterway system was linked to the sea via the ST LAWRENCE SEAWAY.

In addition to inland waterways, several sea-to-sea ship canals have been built. The most famous of these is the SUEZ CANAL, built by the French engineer Ferdinand de Lesseps (1805–94) and opened in 1869. Other major ship canals include the CORINTH CANAL (1893) and the PANAMA CANAL (1914). However, most of the early ship canals are now too narrow to take modern ships. The CALEDONIAN SHIP CANAL (1822), for example, is obsolete. Only the Suez Canal, which has no locks and crosses desert, has been relatively easy to enlarge.

The tremendous increase in the size of ships has created similar problems in docks. The early docks, built to take the steamships of the 1800s, are not large enough for modern container ships. Thus new docks have had to be built.

A dock is an area in which ships can be moored and loaded or unloaded, generally alongside a quay. The quay is often equipped with cranes and rails for railway wagons. Many modern docks have facilities for handling containerized goods and wheeled freight or passenger transport that can roll off or on a ship via an opening in the bow or stern.

Right: A tanker on the Welland Canal passing under a vertical lift bridge. The tanker is on its way from the Great Lakes to the sea via the St Lawrence Seaway.

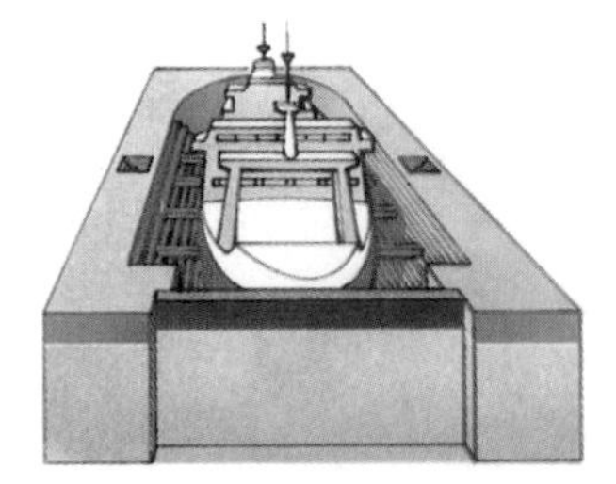

Above: A dry dock. The ship is floated into the dock and positioned over the hull supports. The steel gate is then floated into place and sunk and the water is pumped out of the dock.

In many places the tides do not affect the working of the dock. For example, in Melbourne, Australia, the water rises only 1 metre during a spring tide (the most extreme form of tide). However, in places where the difference between high and low tides is much greater, docks have to be enclosed by an entrance lock. An extreme example of tidal variation occurs at Bristol in England, where the difference between high and low water can be as much as 15 metres. However, inside the enclosed docks ships remain in deep water and can leave or enter at any time via the locks. A DRY DOCK is simply a basin that can be pumped dry.

Early docks all had mitre gates similar to those used on canals. Modern lock gates, which have to be deeper than those of the 1800s, are of various kinds. Some are mitre gates; others are flap gates, hinged at the lower end; others are traversing caissons, which move sideways on rails; and others are caissons, which are floated into place and then filled with water to sink them.

canal across the narrowest point of the isthmus at Panama, but was again forced to abandon it owing to problems with the rock strata and the malaria and yellow fever contracted by his workmen. The American government remained interested in such a canal, but political problems, concerned with who had the right to build, maintain and regulate it, delayed the project for several years. Work began in 1904 and the canal was finally completed in 1914. It is 85 km long, 91.5 metres wide and 8.2 metres deep. Between Libon Bay in the Caribbean Sea and Balboa on the Pacific coast it rises 22 metres to the Gatun Lake, which is the main source of the canal's water supply.

S St Lawrence Seaway was built to open up 3,830 km of navigable waterway between the head of Lake Superior and the Atlantic Ocean. Work on the seaway began in 1954 and it was opened in 1959. From Montreal harbour at the head of the St Lawrence River it rises 52 metres to Lake Ontario, bypassing the Lachine rapids and the International rapids. From Lake Ontario, the WELLAND CANAL leads to Lake Erie, 45 km away. From there, 2 canals with locks form the final links with Lake Huron and Lake Superior.

Suez Canal is a sea-to-sea ship canal that provides a link between the Mediterranean and the Red Sea. The canal was first dug during the early Egyptian civilization, but fell into disuse and had become blocked by about AD 700. In 1854 the Frenchman Ferdinand de Lesseps was granted a concession to build a new canal. Work began in 1859 and was completed in 10 years. It is a sea-level canal 169 km long and has no locks. Originally 60 metres wide and 7.5 metres deep, it was enlarged in 1954 to 150 metres wide with a main channel 13 metres deep in order to accommodate the growing number of larger ships.

Suez Canal, 1869

W Welland Canal, which links Lake Erie and Lake Ontario, was originally built with 40 locks. In 1912 it was improved as an early part of the St Lawrence Seaway project and the 98-metre rise was covered by 8 locks, each 236 metres long, 24 metres wide and 8 metres deep.

In 1829 George Stephenson's locomotive, the *Rocket*, reached the then unheard of speed of 58 km/ hour. Today Japanese 'lightning trains' reach 210 km/ hour, providing the world's fastest regular service.

The Railway Revolution

The various components of railways, such as tracks, locomotives and wagons, were all invented at different times for different purposes. They were brought together at the beginning of the 1800s, when England was in the middle of the Industrial Revolution, to satisfy the urgent need for an efficient method of transporting large quantities of goods.

The result revolutionized everyone's way of life. Because goods could be moved cheaply and easily, industry flourished and new towns grew up. The railways offered speed and ease of transport and many more people could now afford the time and money to travel. Gradually, people realized that they could live in the country and work in the town. Thus our modern way of life – a way of life that thousands of commuters, businessmen and holidaymakers now take for granted – has been shaped for us by the railways.

Tracks, horses and steam

The idea of using tracks originated in mines during the 1500s. Hauling a wagon along the uneven floor of a mine was nearly impossible and so two parallel rows of planks were laid down for the wheels to run on. To keep the wagon on the planks a vertical bar ran in the central gap. Wooden rails were soon invented and some mines had wagons with flanged wheels, like those of modern railcars, to keep them on the rails.

When iron wheels came into use, in the early 1700s, it was found that they wore down the wooden rails, and iron rails were introduced. At the same time, mine railways were being extended to carry coal to nearby wharves on rivers and, later, canals. Where the railway was on a suitable slope the loaded wagons were allowed to roll down to the wharf. In other places horses were used to pull trains of wagons.

All these early railways were private works or mine lines. In 1803 the first public line, the Surrey Iron Railway, was opened. It carried goods between Wandsworth and Croydon. The first horse-drawn passenger line, the Oystermouth Railway, was opened four years later and carried passengers between Oystermouth and Swansea. Horse-drawn trains continued to operate for a number of years, particularly in France, where steam power did not arrive until the 1830s, and in America, where they survived on short lines long after the arrival of steam-powered locomotives.

Steam power was first discovered by the Greek engineer HERO in about AD 50. His 'steam engine' worked on the same principle as a modern lawn sprinkler. Jets of steam from two bent tubes attached to a hollow sphere filled with boiling water caused the sphere to revolve. However, slave labour was cheap and the idea of using steam power for doing useful work did not occur to anyone until the late 1600s. The French

Below: As long ago as the 1500s wooden trucks in mines had flanged wheels that ran on wooden rails. Flanged wheels are still used today to keep rolling stock on the rails.

Reference

A **Abt,** Roman (1850–1933) was a Swiss locomotive engineer. Having trained under RIGGENBACH he became chief engineer of the International Company for Mountain Railways in 1875. In 1882 he devised his rack rail system. This was a great improvement on previous systems as the rack consisted of a single upright plate into which the rack slots were machined. By 1929 there were 72 railways using this system.

Arthur Kill Bridge is a vertical lift bridge that links Staten Island, New York, with New Jersey. The 170-metre lifting span remains horizontal as it is raised by counterweights at each end.

Articulated steam locomotives have 2 sets of cylinders each driving an independent set of wheels. The frames on which the wheel axles revolve are hinged together and the assembly supports the boiler.

B **Best Friend of Charleston** began the first steam passenger service in America on Christmas Day 1830 on the South Carolina Railroad. This was an American-built locomotive with a vertical boiler and it hauled regular traffic from January to June 1831, when the boiler exploded. Parts of the locomotive were used in another locomotive, the *Phoenix.*

Blucher, built in 1814, was George STEPHENSON'S first locomotive. Its vertical cylin-

Locomotive, The Best Friend of Charleston, *December, 1830*

physicist Denis PAPIN discovered that expanding steam could be used to force a piston up a tube. In 1698 he built a model steam engine in which the piston was forced up by steam and pushed back by the air pressure on the other side when the cylinder was cooled.

One of the problems in early mines was flooding and pumping water out by hand was hard work and very slow. In the same year as Papin built his model engine, Thomas SAVERY, an English engineer, built a machine that drew water up by creating a vacuum in a chamber at the top of a pipe. The vacuum was caused by filling the chamber with steam and then cooling it, thus condensing the steam to a few drops of water. The vacuum sucked the water some way up the pipe and then steam pressure was used to blow the water out completely. There were no moving parts and the movements of water and steam were controlled by valves. Savery's steam pump was used in mines, but it employed steam under high pressure and was liable to explode.

In 1712 another English engineer, Thomas NEWCOMEN, invented a better steam pump. It worked on the same principle as Savery's, but the steam pressure and vacuum were used to make a piston move up and down, as in Papin's engine. The piston was attached to one end of a rocking beam, the other end of which was fitted to a water pump. Newcomen's pump was thus the first engine to convert heat energy into useful mechanical energy. The steam was under fairly low pressure and so the machine was not as dangerous as Savery's. Newcomen's steam pumps were used in mines for many years. The last was dismantled in 1934.

However, the most important advance in the development of steam engines was made by the Scottish engineer James WATT. In 1764, after repairing a Newcomen engine, he realized how inefficient it was. The steam in the cylinder had to be cooled down on each stroke and enormous amounts of steam were needed to heat it up again. Watt built an improved steam engine with a separate condenser. Thus the cylinder always remained hot. By 1769 Watt had developed a much more efficient engine than Newcomen's. As a further improvement, instead of using a vacuum he devised a system of moving the piston by introducing steam first to one side and then the other. In 1774 he began manufacturing his machines and by 1800 he had sold about 500. In 1781 he added a system of mechanical cranks and levers that converted the back-and-forth

Right: A pithead scene in about 1820 showing the Newcomen engine used to pump water out of the mine. A coal fire was used to boil the water in the large round boiler. Steam was fed to the cylinder, causing the piston to rise and thus pushing one end of the wooden beam upwards. When the piston reached the top of the cylinder, the steam condensed, leaving a partial vacuum. Atmospheric pressure pushed the piston down again and the other end of the beam rose. This end was attached to the water pump.

ders were linked by connecting rods and cranks to gears, which turned the wheels. At one time it had a chain drive to the front wheels of the tender.

Blue Train, inaugurated in 1939, runs between Cape Town and Pretoria in South Africa and is the most luxurious train in the world. Passengers travel in private rooms for the whole of the 26-hour journey and the carriages run silently with no vibration. The service is like that of a first class hotel.

Brakes, in the early days of railways, were hand-operated levers on each wagon. In the late 1800s, two automatic brake systems came into use, the American Westinghouse air brake and the British vacuum brake. They are operated by the driver in the locomotive and tread or disc brakes act on all the wheels of the coaches. If a coach is disconnected from the locomotive its brakes are automatically applied.

Britannia Bridge, the first iron box-girder bridge, was built across the Menai Straits in North Wales between 1845 and 1850 by Robert STEPHENSON. It was originally designed as a suspension bridge, but Stephenson used plate girders to build two long tubes and during construction he found that the supporting cables were unnecessary. Trains ran inside the tubes across four spans. The two longest spans over the water measured 140 metres. In 1970 the bridge was destroyed by fire and replaced with a steel arch bridge.

Section of Britannia Bridge

Brunel, Isambard Kingdom (1806–59) was the son of Sir Marc Isambard BRUNEL. He trained under his father and assisted in building the Rotherhithe Tunnel in London. Between 1829 and 1831 he designed the Clifton Suspension Bridge over the Avon Gorge in Bristol. Lack of funds delayed the work and others completed the construction. Between 1833 and 1841 he laid out the 188-km Great Western Railway line from London to Bristol, building a magnificent series of bridges and tunnels. He also designed a few locomotives, which were not successful. But despite his railway achievements Brunel was not primarily a railway engineer. His motive for building the Great Western Railway was to provide a railway link as part of the journey from London to New York. To provide the sea link

Above: Before the arrival of steam power, trains were drawn by horses. This print shows the railway from St Etienne to Lyons in 1829. Later, the line was the first in France to use steam locomotives, although at first only for goods trains.

Left: Trevithick's circular track in north London in 1808. He charged passengers a shilling a ride in a coach pulled by *Catch-me-who-can,* his fourth locomotive. This was 22 years before the first permanent steam passenger railway was built.

movement of the piston into the rotary movement of a wheel. Watt's steam engines, developed just when they were needed in the middle of the Industrial Revolution, came to be used in factories all over England.

With Watt's invention steam transport became possible. Steam road carriages, such as those built by the English inventor Richard TREVITHICK and others, appeared on the roads *(see page 81)*. But they had only limited success and it was at this point that steam engines and railway tracks were brought together. At first, stationary steam engines were used to haul wagons up slopes by cable. Then in 1803 Trevithick developed the first steam locomotive. Engineering technology had improved since the days of Savery, and Trevithick found that he could use steam under high pressure and do away with the condenser used by Watt, thus making the whole engine lighter. In 1804 Trevithick's second locomotive, PENYDARRAN, successfully hauled a loaded train both ways along 14.4 kilometres of railway in South Wales.

Trevithick demonstrated his fourth locomotive, *Catch-me-who-can*, on a circular track in London in 1808. He advertized it as 'Trevithick's travelling machine without horses, impelled by steam' and it was the first occasion on which a train pulled by a steam locomotive carried fare-paying passengers, albeit round in circles! Unfortunately, Trevithick, already plagued by bad luck in his numerous experiments, could not interest anyone with sufficient money to back him and he gave up.

By this time, however, others had caught on to the idea of steam traction on railways. In 1812, John Blenkinsop (1783–1831) had four locomotives that successfully ran on a rack railway at the Middleton colliery near Leeds. These locomotives were the forerunners of those that run on the rack railways in mountain areas today. In 1814, William Hedley (1779–1843) built the PUFFING BILLY at the Wylam colliery on the River Tyne. This locomotive, and the two others that were built soon after it, did not use the rack system, relying on the natural grip between the rails and the wheels.

The success of these locomotives prompted other mine owners to build steam railways and the English inventor George STEPHENSON was commissioned to build a railway at Killingworth colliery. His first locomotive, the BLUCHER, was a success and he went on to build several more locomotives for use in collieries.

Stephenson was not only an excellent engineer. He also had the vision to see the prospect of a Britain criss-crossed with railway lines. His enthusiasm and shrewd tactics aroused the interest of others, and in 1821 he was appointed chief engineer of the Stockton and Darlington Railway. This, the first public steam railway, was built to carry coal from Durham to the coast. There was no canal and so there was no competition. It was opened on September 27 1825 with a train of passenger-carrying wagons pulled by Stephenson's latest locomotive, LOCOMOTION, built for the occasion.

Stephenson's most famous locomotive is the ROCKET. The design of this locomotive added

---

he built 3 great steamships. The *Great Western,* built between 1837 and 1838, operated the first scheduled transatlantic service. The *Great Britain,* built between 1843 and 1845, was the first iron steamship and the first fitted with a screw propeller. The 12,000-tonne *Great Eastern* was not a commercial success, but in 1866 was used to lay the first transatlantic cable.

Brunel, Sir Marc Isambard (1769–1849) was a civil engineer. Born in France, he settled in England in 1799. He devised an early form of mass-production for making naval pulley blocks. In 1818 he devised the tunnelling shield for boring through soft ground, and between 1825 and 1843 he used it to construct the Rotherhithe Tunnel under the River Thames.

Thames Tunnel being built

C

Canadian, inaugurated in 1955, is one of the great trains of the world. It runs on the Canadian Pacific Railway between Vancouver, Montreal and Toronto. Diesel-hauled throughout the journey of 69 hours 10 minutes, it passes through magnificent scenery and the coaches include an observation lounge with bedrooms.

Cascade Tunnel, on the Great Northern Railway, USA, is the longest tunnel in North America. Opened in 1929, it is 12.542 km long.

Centennial-type locomotives are the most powerful diesel-electric locomotives (6600 hp). From 1969 General Motors Corporation built 47 of these.

Central Railway of Peru reaches the highest altitude of any railway in the world. On a branch line at La Cima it reaches 4,818 metres. On the main line, which reaches a height of 4,781 metres, there are 21 zig-zag reversing stations and the gradient is 1 in 23.

Classification of locomotives is according to the number of wheels or axles. In the Whyte system, steam locomotives are given 3-figure numbers that indicate, in order, the number of front-carrying wheels, the number of driving wheels and the number of rear-carrying wheels. Thus, the 6-wheeled NORTH STAR is classified as 2-2-2 and the 1950s American design, which had a 4-wheeled carrying bogie at the front, four coupled driving wheels and no rear-carrying wheels, is classified as 4-4-0. In the classification

several new ideas and its basic features have been used in all later locomotives. In its original form it could travel at 58 kilometres per hour and it proved its worth by winning the Rainhill trials in 1829. These trials were held by the Liverpool and Manchester Company who were looking for the best locomotive to use on their new railway. In 1830, the Liverpool and Manchester Railway (L&MR) was opened and continued to use locomotives of the *Rocket* type for several years.

The opening of the Liverpool and Manchester Railway is generally thought of as the beginning of the railway age. It was built to compete with existing canals and roads and its success amazed even its builders. The profitability and usefulness of railways being obvious, during the next 90 years over 1,100,000 kilometres of rail were built around the world.

The growth of railways

By 1830 both France and America were using horse-drawn trains and they began experimenting with steam traction by importing a few English-built locomotives. Within a few years, however, both countries were building their own locomotives and developing their own railway systems. After France, other European countries opened steam railways and by 1860 Europe was covered by a network of tracks linking up towns and cities.

In America, the railway era began in 1830 on the Baltimore–Ohio line with the maiden trip of the first American-built locomotive, *Tom Thumb*. As in Britain and Europe tremendous expansion of the railways followed, but in America this had the effect of opening vast new areas of territory. The first trans-continental railway was com-

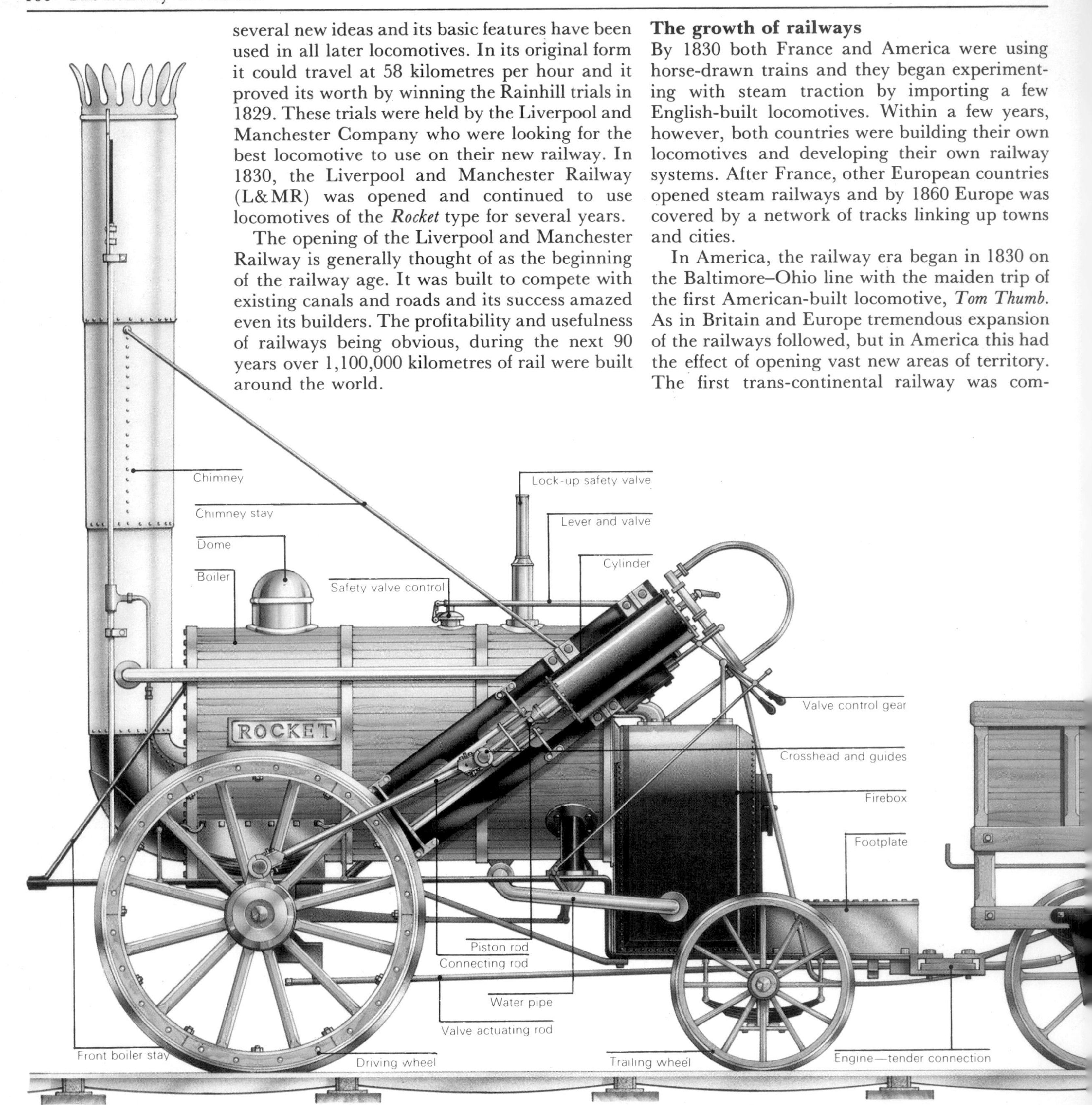

2-6-4T the T indicates a TANK LOCOMOTIVE. In the continental system of classification, which counts axles instead of wheels, these 3 classifications become 1-1-1, 2-2-0 and 1-3-2T. Modern locomotives, particularly diesel and electric, are also classified according to the number of axles, but a system of letters is used to describe the driving axles (A=1, B=2, C=3, D=4, E=5). Thus, a modern 2-D-1 locomotive is equivalent to a 4-8-2 steam locomotive. If the suffix o is used, this indicates that each driving axle is individually driven. A common type is Co-Co.

Couplings, in the early days of railways, were of the chain and hook type. Screw couplings were first used on passenger coaches by the Liverpool and Manchester Railway to eliminate jolting. However, in Britain they were never used on goods wagons. In America the chain and hook couplings were replaced by a rigid link fixed wih a pin. During the 1890s this was superseded by the automatic coupling – a hook that engages a similar hook on the next coach – which later came to be used in Europe as well. In modern coaches automatic couplings now incorporate electrical connections.

Link and pin coupling

D **Dead man's handle** is a lever in the driving cab of a modern locomotive that the driver has to hold down to keep the train moving. If he becomes incapacitated and releases the handle, the train automatically comes to a halt.

F **Festiniog Railway** in North Wales was the world's first public narrow-gauge railway (597 mm). It was built to carry wagons loaded with slate from Blaenau Ffestiniog 20.8 km down to Porthmadog. Thus the gradient (averaging 1 in 90) is all 1-way and increases slightly on the curves to prevent heavy trains from sticking. It closed in 1946 and has since been modified

Festiniog Railway

Above: An early railway print showing, at the top, Stephenson's *Rocket* winning the Rainhill trials in 1829. By 1833 first class passengers either travelled in comfortable, stagecoach-style carriages or in their own carriages on top of flat wagons (middle). Second class passengers had to stand in open-topped wagons (bottom).

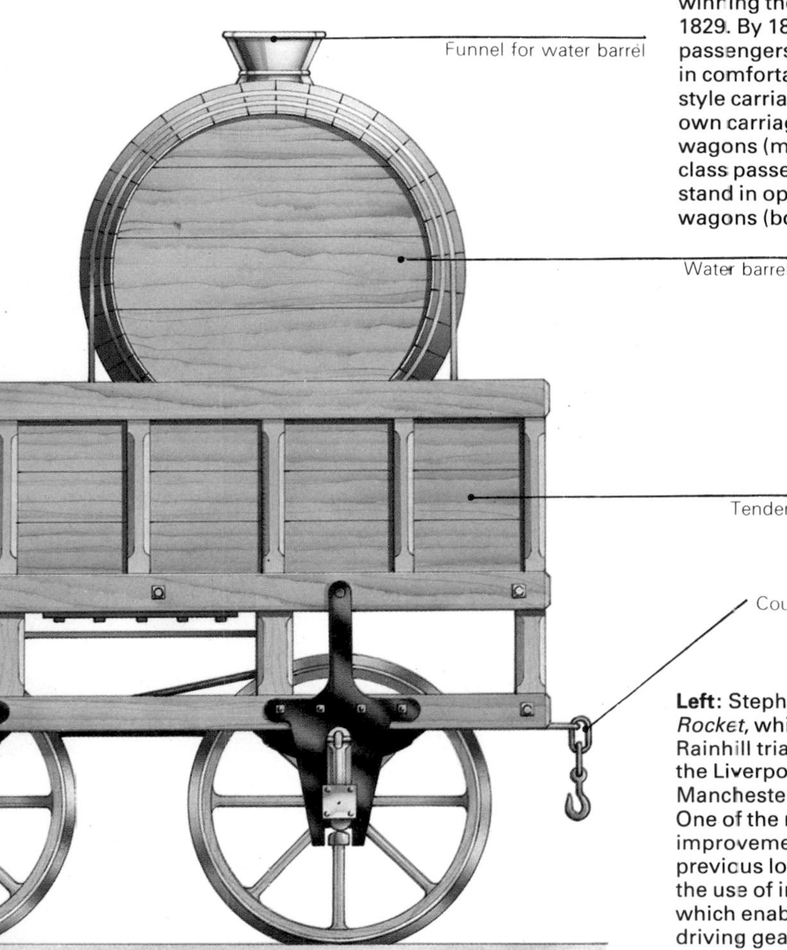

Left: Stephenson's 1829 *Rocket*, which won the Rainhill trials and opened the Liverpool and Manchester Railway in 1830. One of the main improvements over previous locomotives was the use of inclined cylinders, which enabled a simpler driving gear to be used.

pleted in 1869 with the meeting at Promontory in Utah of the Central Pacific and Grand Union Pacific lines. By 1920, America had 407,000 kilometres of track, Britain had 32,700 kilometres, Germany 11,500 kilometres and France 9,400 kilometres. After 1920, various factors, particularly competition with the increasing amount of road traffic, caused railways to decline, with the result that many lines have since been closed.

In the early days of railways there were many small companies. One of the earliest problems that resulted from this was the use of different GAUGES (the distance between the inner faces of the rails). George Stephenson selected, for no particular reason, a gauge of 1,435 millimetres. However, he had the vision to foresee that all the railways in Britain would eventually link up and saw to it that all the northern lines used this gauge. The great civil engineer Isambard Kingdom BRUNEL, however, was an individualist. He built the Great Western Railway with a gauge of 2,140 millimetres. The advantage of such a broad gauge is that it can carry much heavier loads. On the other hand, narrow-gauge railways are cheaper to build.

When the Great Western and Midland railways met at Gloucester, the 'battle of the gauges' began. Transferring passengers and cargo from one railway to the other was expensive and caused great inconvenience. The government held trials and, although the broad gauge was proved to be the best, there were many more kilometres of the narrower track and in 1846 the Gauge Act standardized Stephenson's 1,435-millimetre gauge.

The Great Western Railway gradually changed its track and by 1892 all of Britain's main railways were using the standard gauge. In France and America it had been adopted from the beginning and so there was no problem. Gauges do vary round the world, however, and there are at present six different standard gauges. The largest of these is 1,676 millimetres and the smallest is 1,000 millimetres.

Wagons and carriages

The original intention of Stephenson and the other railway pioneers was to speed up the transport of coal and other goods. Freight transport is still an important part of railway

to carry tourists uphill as well as down.

Flying Hamburger was the first German high-speed diesel train. From 1933 this twin-unit locomotive hauled trains between Berlin and Hamburg at an average speed of 124 km/hour.

Flying Scotsman is Britain's prestige train. At present it runs between London (King's Cross) and Edinburgh in 5 hours 35 minutes. Originally called the *Special Scottish Express*, it was nicknamed the *Flying Scotsman* during the 1860s. It received the name officially in 1923 and was given Gresley Pacific (4-6-2) locomotives. These were replaced in the 1930s by the streamlined A4 class Pacifics. The train is now hauled by high-powered *Deltic* diesel-electric locomotives.

Forth Railway Bridge, built between 1882 and 1890, is the oldest and second largest cantilever bridge in the world. Its 2 main spans are 520 metres long. With the TAY BRIDGE disaster in mind, the designer spent 2 years conducting wind-pressure tests before it was built. The main struts are tubular.

G

Garratt locomotives are ARTICULATED STEAM LOCOMOTIVES. The boiler is carried on a cradle slung between engine units and there are no wheels underneath it. This allows the boiler to be large in diameter and shorter than in the MALLET arrangement. The first Garratts, weighing 32 tonnes, were built in 1908. The largest were the 4-8-4 + 4-8-4 Beyer Garratts, weighing 256 tonnes built for New South Wales Railways in 1952.

Gauge is the distance between the inside faces of the rails. The 6 principal standard gauges in use around the world are 1,676 mm, 1,600 mm, 1,524 mm, 1,435 mm, 1,067 mm and 1,000 mm. Today the 1,435-mm gauge is used in many countries including Britain, North America, China, Japan, Australia, Mexico and most of Europe. Some countries have more than one gauge. For example, Switzerland has 3,518 km of 1,435-mm track and 1,183 km of 1,000-mm track; and Argentina has 23,235 km of 1,676-mm track, 3,086 km of 1,435-mm track and 13,461 km of 1,000-mm track. In addition there are various narrow-gauge railways around the world, ranging from 597-mm to 914-mm gauges.

Mixed-gauge points system

Boiler
Regulator valve
Chimney
Dome
Regulator valve lever
Safety valve
Firebox
Piston
Blast pipe
Boiler flues
Smoke stack
Main steam pipe

Above: A steam engine. Hot air from the firebox passes down the flues and steam is raised in the boiler. The steam passes via the regulator valve to the cylinders. Valves allow the steam to pass first to one side of the piston and then the other.

Below: *Der Adler* (The Eagle) was built by Robert Stephenson. On December 7 1835 it opened the first German railway, from Nüremberg to Fürth.

operation, but there has been considerable technological improvement. Due to improved suspension systems wagons can now travel much faster and there are many wagons specially designed to carry particular kinds of freight.

The Liverpool and Manchester Railway was the first steam railway to begin a regular passenger service and the first carriages were simply stagecoaches on flanged wheels. Soon, larger carriages were being constructed by placing two or three stagecoach bodies on a single chassis. In Europe, open-topped wagons were introduced as a form of cheap-fare passenger carriage.

The modern style of coach was developed in America, where the early railways often had sharper curves than in Europe. Instead of a short coach set on a single chassis they began to use a long coach on independent trucks, or bogies. The coach could move on central pivots on the bogies and so negotiate sharp corners more easily.

The interiors of coaches have either followed the American open-plan arrangement with a central gangway or the European type with separate, isolated compartments. The first luxury PULLMAN parlour and sleeping cars were introduced in America in the 1860s. But in Europe, and particularly in Britain, it was some time before passenger comfort, below the level of first class, was given much attention. By the 1870s, public pressure had forced the companies to give up some of their precious passenger space (and hence revenue) for lavatories and, later, corridors so that everyone had access to them.

The age of steam locomotives

The principle on which a steam locomotive works is very simple. Burning fuel in the firebox heats water in the boiler and produces steam. This passes to each cylinder and VALVES operate to pass steam first to one side of the piston and then the other. From the earliest days, however, modifications have been made to this basic system to improve the performance of locomotives.

One of the most important discoveries was made by Trevithick. In his locomotive *Penydarran* the exhaust steam was pushed out via the chimney. Each blast of the exhaust created a partial vacuum in the chimney and helped to draw the fire through the boiler flue. All subsequent steam locomotives used this effect

H

Hell Gate Bridge, which carries the Pennsylvania Railroad to Long Island, New York, is the largest steel arch bridge in the USA. Its main span is 297.8 metres long and carries four tracks.

Hero (born *c.* AD 20) was a Greek engineer who lived during the last years of the age of Greek science. In addition to inventing his 'steam engine' he extended the work of Archimedes on levers and constructed gears.

Huey P. Long Bridge across the Mississippi River near New Orleans is the longest railway bridge in the world. Opened in 1935 it is 7.009 km long and has 8 river spans, the longest of which is 241 metres. The bridge also carries road traffic.

The Indian Pacific *passenger express*

I

Indian Pacific is Australia's great train that runs from Perth on the shore of the Indian Ocean to Sydney on the shore of the Pacific Ocean. Diesel-hauled for most of the time, the train travels at a moderate speed and the journey takes 65 hours 45 minutes. On the Nullarbor Plain in Western Australia the Indian Pacific runs on the world's longest pieces of continuous straight track (478 km).

K

Kanmon Tunnel between Honshu and Kyushu on the extension of the SHINKANSEN line in Japan is the longest underwater tunnel in the world. It was completed in 1974 and is 18.7 km long.

Key West Extension, completed in 1912, was built to extend the Florida East Coast Railway 182 km into the sea, with the object of shortening the sea passage to Cuba. Most of the track (135.5 km) is laid on coral reefs called keys. The remainder is built on 27.5 km of bridging and 32 km of embankment. Damaged by a hurricane in 1936, it is now only open to road traffic.

and, later, a steam blower was placed in the chimney to keep the fire drawing when the locomotive was stationary.

In *Locomotion,* Stephenson used a single, straight-through flue. In 1827, Henry Booth, Secretary of the Liverpool and Manchester Company, patented the multi-tubular boiler, which Stephenson incorporated into the *Rocket.* Instead of a single flue running through the boiler, locomotives now had a number of small tubes. This provided a much greater steam-raising capacity. Much later, in the early 1900s, a system of SUPERHEATING was developed. In this system, the steam is passed through small pipes in superheater flues in the upper part of the boiler. It is then delivered to the cylinders at a much higher temperature and has a much greater power of expansion.

All the early locomotives were built with vertical cylinders and the driving wheels were turned by gears. Stephenson's *Rocket,* however, used direct drive from the pistons on to crank-pins on the wheels and the cylinders were at an angle of about 45°. By 1850, the design of locomotives had changed considerably. Locomotives like the 2-2-2 NORTH STAR, built by George Stephenson's son Robert, had a single pair of

Above right: Tank engine No. 1247 near New Bridge on the North York Moors Railway in England. Built in 1899, this 0-6-2T locomotive was the first British Rail steam engine to be preserved.

Right: A typical American train of the mid-1800s. The 4-4-0 engine was woodburning, hence the enormous, spark-arresting chimney. The carriages were of the open-plan type with central corridors and they were carried on 2 4-wheel bogies.

Kicking Horse Pass on the Canadian Pacific Railway originally had a gradient of 1 in 23. In 1920 the track was relaid and 2 spiral tunnels were constructed. The gradient was thus reduced to 1 in 45 and more heavily loaded trains could be hauled over the pass at less cost.

King George V, the Great Western Railway locomotive No. 6000, was built by G. J. Churchward and made its first public appearance in 1927 at the Iron Horse Centennial in America. This type of locomotive was one of the few 4-6-0 designs that was a noticeable improvement on the 4-4-0s. After steam had been phased out from British railways, *King George V* was loaned to H. P. Bulmer, the cider makers of Hereford, to haul an exhibition train in 1968 and is now preserved at their railway centre. On March 2 1979 this now-famous locomotive hauled a train from Paddington to Didcot to celebrate the 125th anniversary of Paddington Station. Unfortunately, an overheated axle-box prevented it from making the return journey.

L

Linear electric motor is one that produces motion in a straight line instead of a rotary motion. Only prototype linear motors have yet been built and they are mostly linear induction motors. Such a motor consists of a moving body containing coils and a stationary track formed from a metal strip. Electric current passed through the coils creates a magnetic field that travels in a straight line and the effect of this is to propel the moving body along the track.

Locomotion was the locomotive built by George STEPHENSON for the opening of the Stockton and Darlington Railway. It was similar to earlier locomotives in that it still had vertical cylinders which were set into the top of the boiler. However, it was one of the first to have outside coupling rods on the wheels.

The Locomotion, *c.1890*

Lower Zambezi Bridge in Mozambique is the second longest railway bridge in the world. A steel truss bridge, it is 3.677 km long and has 33 main spans of 7.9 metres each.

M

Magnetic levitation (MAGLEV) is one of the ways in which a hovertrain

driving wheels and four carrying wheels. The American 4-4-0 design (*see* CLASSIFICATION) had a four-wheeled bogie at the front and four driving wheels. This long wheel-base design rode better on the American tracks, which were rougher than British tracks. In Britain, the long boiler design began to develop. The smokeboxes of short-boilered locomotives tended to overheat, whereas in long-boilered locomotives more heat was absorbed along the fire tubes.

The early American locomotives burned wood and their vast funnel-shaped chimneys were designed to prevent sparks from flying out. Other well-known features of such locomotives included the large lamp for lighting up the track at night and the cowcatcher, which was necessary for sweeping away branches and other obstacles.

In Britain, the first locomotives burned coal. However, they emitted large quantities of dirty black smoke and for several years coke was used instead. In about 1860 the idea of a brick arch in the firebox was developed. This improved the combustion of the fuel, and coal, much cheaper than coke, could once again be used.

At about the same time steel came into use. Boilers built of steel could withstand higher steam pressures and locomotives became larger and more powerful. During the late 1800s the compound engine was introduced. In this type of engine, steam is used first in a high-pressure cylinder and then allowed to expand in a larger low-pressure cylinder. Compound engines use less fuel and water and were popular in Europe.

In the late 1800s and early 1900s even larger locomotives were built. In Britain the outstanding 4-4-0s appeared, followed by the 4-4-2s (Atlantics) which were a great success on long-distance express routes. In America the 4-4-2s were followed by the great 4-6-2s (Pacifics), 4-6-4s (Hudsons) and 4-8-4s. The largest locomotives of all were the articulated MALLETS and GARRATTS, built for hauling heavy trains up steep slopes. The largest locomotive ever built was the UNION PACIFIC BIG BOY 4-8-8-4.

Over the years a vast number of different types of locomotive were built for different tasks and engineers were constantly trying to build better locomotives. However, by the 1940s the end of the age of steam was in sight. The 4-6-0 MALLARD still holds the world speed record for a steam locomotive (202 kilometres per hour). However, many steam locomotives are in use in various parts of the world and even where diesel and electric traction have superseded steam many locomotives have been preserved by enthusiasts.

Modern locomotives

The first diesel-electric locomotives appeared in 1924 and today most so-called diesel locomotives are of this type. Their engines power generators which provide electricity for the electric traction motors that turn the wheels. A few diesel-mechanical locomotives exist, together with a few diesel-hydraulic locomotives which use a system similar to that of a motor car with AUTOMATIC TRANSMISSION *(see page 81).*

Unlike steam locomotives, the basic design of diesel-electric locomotives is much the same all over the world. After a few initial experiments it was found that the best arrangement was to mount the body on two independent bogies and to power the driving axles separately.

Diesel-electric locomotives are more efficient than steam locomotives and can go further without refuelling or servicing. In addition, they are easy to use as multiple units and this is particularly useful in America, where as many as

Below: A cutaway of a Do-Do General Electric 5,000 hp diesel-electric locomotive. The diesel engines are used to turn the traction generator which keeps the batteries charged. Electric current from the batteries powers the traction motors, which are linked to the axles by gears.

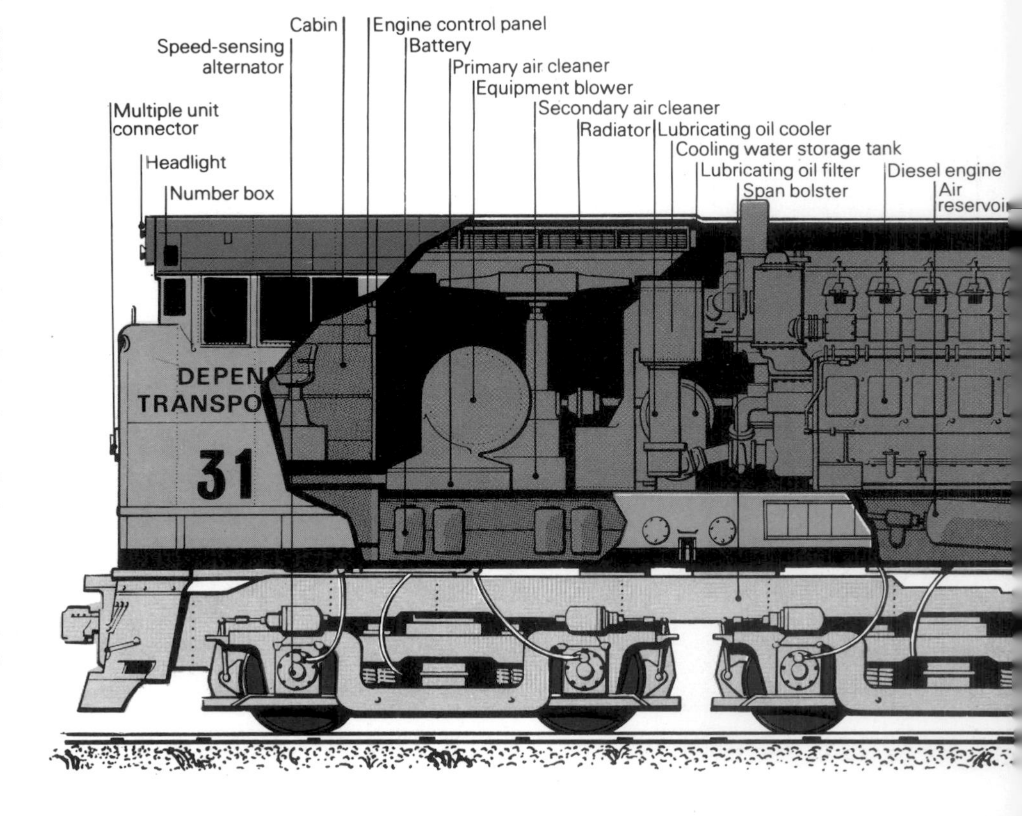

might be raised off its track. The principle used is the repulsion effect between 2 strong magnetic fields created by electromagnets in the train and on the track.

Mallard is the streamlined A4 class Gresley Pacific locomotive No. 4468 that holds the recognized world speed record for a steam locomotive. On July 3 1938 it achieved a speed of 202 km/hour down Stoke Bank near Peterborough in England. The engine was severely damaged.

Mallet locomotives were ARTICULATED STEAM LOCOMOTIVES. The earliest type, which appeared in 1889, was a compound locomotive, with the rear end of the boiler fixed to the rear chassis. The largest compound Mallets were the 2-10-10-2s built in America in the 1920s. Later Mallets were simple-expansion locomotives. They had very long, narrow boilers. The largest type was UNION PACIFIC BIG BOY 4-8-8-4.

Marsh, Sylvester (1803–84), an American entrepreneur, built the first mountain rack railway, the Mount Washington Cog Railway, in 1868. He used a ladder rack system with round bars.

Mistral is the train that runs from Paris to Nice on the French Riviera. It earned the name because the 4-4-0 locomotives that pulled the train during the late 1880s had an early form of streamlining to combat the terrific Mistral wind that sweeps down from the Maritime Alps. The Mistral has always been a high-speed train and today the electric locomotives average 121.5 km/hour and in places reach 160 km/hour. The journey takes 6 hours 51 minutes.

Mont Cenis Tunnel (Fréjus Tunnel) was the first of the major tunnels bored through the Alps. Work began in 1857 and it took 14 years to build. The 14-km tunnel was bored from both ends and ventilation problems were solved by using water-powered fans and a horizontal diaphragm that divided off the upper part of the tunnel as an exhaust duct.

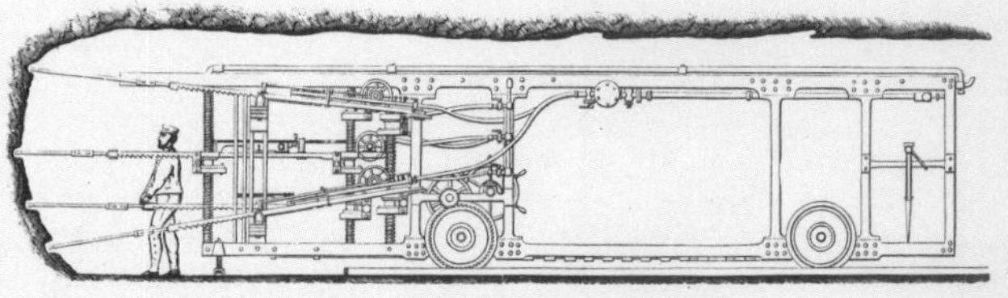

Sommellier drilling machine, Mont Cenis Tunnel, 1871

N **Newcomen**, Thomas (1663–1729) was an English engineer. In 1698 he went into partnership with

Above: A massive 2-10-2 freight locomotive in a coal yard in China.

Right: The A3 class Gresley Pacific (4-6-2) No. 4472 was the first to be named *Flying Scotsman.* Built in the 1920s, it has since been restored.

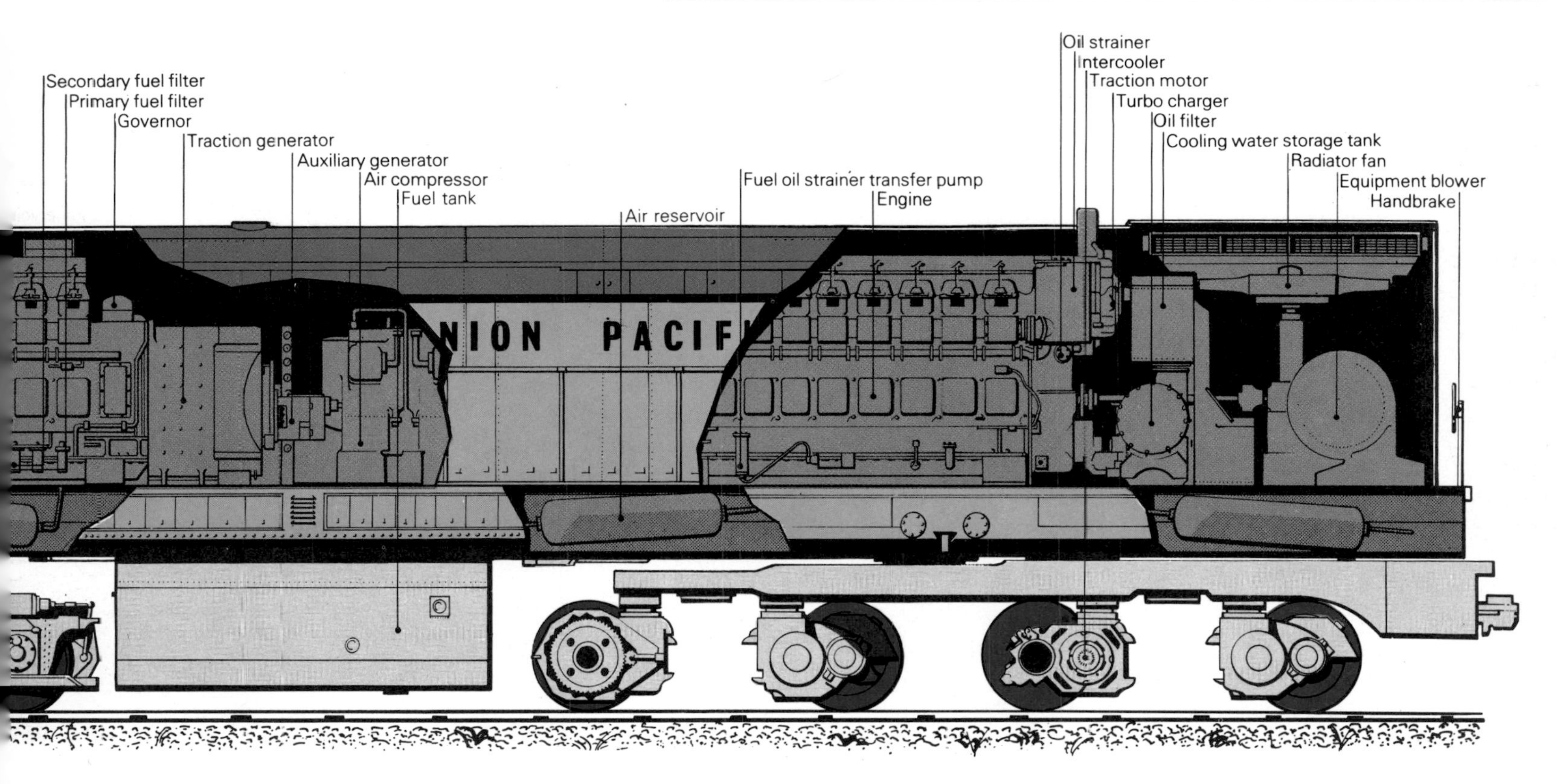

SAVERY. He built an improved steam engine and to do this he had to construct highly polished cylinders into which pistons would fit to form an air-tight seal.

Northern Line on London's underground railway is the longest continuous tunnel in the world (27.8 km).

North Star was the 2-2-2 locomotive built in 1837 by Robert STEPHENSON for the broad-gauge Great Western Railway. It hauled the first train on this line in 1838 and remained Great Western Railway's best locomotive for several years.

O

Orient Express is the famous train that ran from Paris to Constantinople (now Istanbul). The service from Paris to Vienna, Budapest, Bucharest and Varna on the Black Sea was inaugurated in 1883 and the link to Constantinople was completed by boat. In 1889, the route from Budapest was changed and the whole journey, now via Belgrade, was completed by train in 67 hours 35 minutes. When the SIMPLON TUNNEL was opened in 1906 a rival service was introduced. Today, the *Orient Express* runs from Paris to Vienna and Budapest, with through carriages to Bucharest four days a week. The *Simplon Orient,* now called the *Direct Orient,* runs from Paris to Belgrade via Lausanne, Milan, Venice and Trieste; through services are provided to the cities of Istanbul and Athens by the *Marmara* and the *Athens Express.*

P

Papin, Denis (1647–1712) was a French physicist. In 1769 he developed a steam digester – the forerunner of the modern pressure cooker – and went on to experiment with steam engines.

Penydarran was Richard TREVITHICK's second locomotive, built in 1804. It could haul about 25 tonnes and climb a 1 in 36 gradient. Its horizontal boiler had a U-shaped flue and a single cylinder drove the wheels on one side by means of a system of gears.

Pilatus Railway in Switzerland is the steepest rack railway in the world – in places 1 in 2. The carriages are stepped and the railway uses the Locher rack system, in which a pair of horizontal

Mount Pilatus Railway

Left: An electric train on the Rhaetian Railway in Switzerland showing the pantograph that collects current from overhead wires. The pantograph is spring-loaded, so that it always follows the contours of the wire, and it can be retracted.

Right: The cab of a modern diesel-electric locomotive. The controls are easier to handle than those of a steam locomotive, but the driver still needs a high degree of concentration.

Above: Electric trains often receive their current from a third rail, which runs either in between the two carrying rails (top) or to one side (bottom).

five diesel-electric units may be used to pull the heaviest trains.

The first electric locomotives were built in the 1830s, but until the 1880s they were powered by batteries, which were heavy and needed recharging at frequent intervals. In 1883, the first true electric railway was built by Magnus Volk in Brighton and this railway still operates as a tourist attraction along the sea front. Electric current is collected from a third rail.

One of the most important applications of electric traction is in underground railways. In England, the first section of the Metropolitan railway had been opened in 1863, but steam trains were used, with the result that the tunnels filled with smoke. After 1890 electric locomotives were brought into use on the line between the City of London and Stockwell.

Today, many electric trains are multiple units in which a number of individually-propelled coaches can be controlled by a single driver from either end of the train. The system was invented by the American Frank Sprague in 1897. Sprague had also invented the swivel trolley pole in 1888 for the Union Passenger Railway in Virginia and this led to the modern pantograph which collects current from overhead wires.

Electric trains have several advantages. They are clean, accelerate rapidly (especially multiple-unit trains) and can climb gradients more easily than steam or diesel locomotives. However, they are not cheap to run and the cost of electrification is enormous.

Monorails

In 1824, a wooden monorail was built in England between the Royal Victualling Yard and the River Thames to carry horse-drawn trains. During the late 1800s several trestle railways were built, the last of which was the Listowel to Ballybunion line in Ireland, which ran from 1887 to 1924. The train was supported on a central rail about a metre from the ground. On either side of the trestles which supported the central rail were guide rails. The locomotive and carriages were divided into two, with the wheels in the middle.

Between 1903 and 1910 a series of monorails were built on which the trains were guided and kept upright by gyroscopes. A more successful system is that on the Wuppertal Railway in Germany, which has been in use since 1901. There the cars are suspended from a single line of wheels running on an overhead rail. Another successful monorail was invented by the Swedish industrialist Dr Axel L. Wenner-Gren. In this Alweg system, the cars straddle a central beam on which they are driven by rubber-tyred wheels. There are four rows of wheels (two on each side)

pinions grip a central rail that has teeth on both sides. The Pilatus Railway was opened in 1889 and electrified in 1937.

Puffing Billy was built in 1814 by William Hedley (1779–1843), who also built *Wylam Billy* and *Lady Mary*. These 3 locomotives had boilers like those of Trevithick locomotives and their 2 cylinders drove the wheels through a system of rods, cranks and gears. They travelled at walking pace and hauled loads of about 50 tonnes. All 3 locomotives had long lives. *Puffing Billy* was still in use in 1859 and is now preserved in the Science Museum in London.

Pullman, George (1831–97) was the founder of the Pullman Car Company. In Chicago in 1858 he rebuilt 2 coaches into sleeping cars and in 1865 he built the first Pullman Car. He founded his company in 1867 and introduced dining cars (1868), chaircars (1875) and vestibule cars (1887).

R

Re 6/6 electric Bo-Bo-Bo locomotives, built in Switzerland from 1972, are the world's most powerful electric locomotives. Weighing 122 tonnes, they produce 10,450 hp and can travel at 140 km/hour.

Rhaetian Railway in Switzerland uses 4 spiral tunnels to reach the mountain valley tourist resort of St Moritz. At one point 1 of the tunnels is vertically above another This railway also uses snow galleries to prevent the line being covered by avalanches, and magnificent masonry viaducts. The Landwasser viaduct carries the line from a tunnel entrance in a vertical cliff face. It has a gradient of 1 in 29 and is built in a series of angles so that it appears to curve.

Riggenbach, Nikolaus (1817–95) was experimenting with rack railways at the same time as MARSH. He patented his ladder-rack in 1863 and used it on the Kahlenberg Railway near Vienna, which opened in 1874. It is generally held that the bars of Riggenbach's ladder were better shaped to mesh with the pinion than those of Marsh's ladder.

Rocket was the STEPHENSON locomotive that won the Rainhill trials in 1829 and opened the Liverpool and

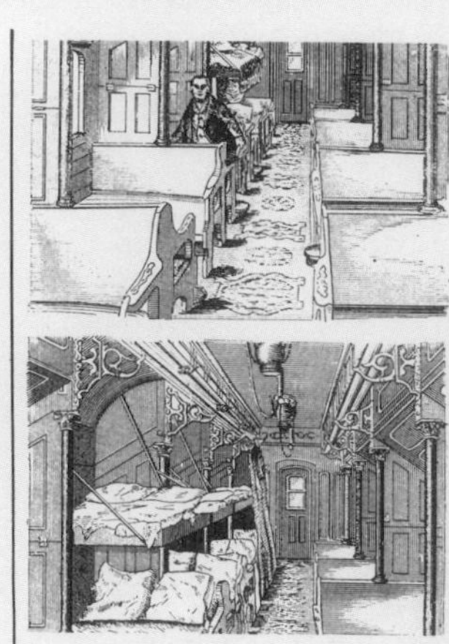

Pullman sleeping car, 1880

that support and guide the cars. The Tokyo–Haneda Airport railway is an Alweg monorail that has run successfully since 1964.

Above: The Snowdon Mountain Railway, opened in 1896, is a tourist line that uses the double Abt rack system. It is Britain's highest railway and climbs to a height of 1,064 metres.

Below: The electric Wuppertal monorail that runs along the valley between Barmen and Elberfeld in West Germany. It is 13 km long and for most of the way runs over the River Wupper.

Mountain railways

In the early 1800s trains were hauled up steep inclines by stationary engines. From this developed the idea of self-acting cableways on which loaded descending wagons hauled up empty wagons. Today, much the same principle is used on funicular railways. A descending coach attached to one end of a cable is balanced by an ascending coach attached to the other end. Such railways are generally single-track, with a short double-track section where the coaches pass each other.

However, engineers wanted a system by which a locomotive could move up a steep slope under its own power. The Fell system, first used across the Mont Cenis pass in 1868, had two extra horizontal wheels that gripped a central rail. However, rack railways have proved to be the best for negotiating steep slopes.

The first rack railway was built in 1812 by John Blenkinsop. Alongside one of the rails he laid an additional toothed rail, or rack, which meshed with a cog, or pinion, on the locomotive. In 1874, Nikolaus RIGGENBACH built a mountain railway in Austria. His rack was a ladder of bars

---

Manchester Railway in 1830. It had the first multitubular boiler and an external firebox with a water-jacketed top and sides. The 2 upturned exhaust steam nozzles in the smokestack had constrictions at their ends to intensify the action of the exhaust blast in drawing the fire. Its 2 outside cylinders were inclined and there was direct drive on to the wheels. In its original form it had a top speed of 58 km/hour; when the cylinders were lowered, the top speed was increased to 80 km/hour.

Royal George, built by Timothy Hackworth in 1827, was the first 6-coupled (0-6-0) locomotive and the first to have direct drive on to the wheels.

S **St Gotthard Tunnel** is one of the major tunnels through the Alps. It is 14.99 km long and was bored between 1872 and 1882. The builders had great problems with weak rock and water inflows.

Savery, Thomas (1650–1715) was a prolific English inventor. His most important invention was the Miner's Friend, a form of steam pump for removing water from coal mines.

Severn Tunnel is the longest tunnel in Britain (7.011 km) and when it was completed in 1886 it was the longest

St Gotthard Tunnel

underwater tunnel in the world. It was built by Sir John Hawkshaw for the Great Western Railway extension to South Wales. During construction a tremendous inflow of water from the Great Spring was encountered and a massive steam pump was used to remove 7 million litres of water an hour.

Shinkansen Hikari are the Japanese 'lightning trains' that run every 15 minutes between Tokyo and Osaka. A new line was specially built for the service, which was inaugurated in 1964, and the bullet-nosed, multiple-unit electric trains travel at about 210 km/hour nearly all the way. In 1975 the line was extended to the island of Kyushu in the south and the new track was built to take trains travelling at 260 km/hour. The extension included 222 km of tunnels, one of which was the KANMON TUNNEL, and 117 km of bridges and viaducts.

Siemens, Dr Ernst Werner von (1816–92) built the first electric railway at the Berlin Trades Exhibition in 1879. The 550-metre line was straight and level and the

Above: The fine vaulted roof over the platforms of St Pancras Station in London. The station is a classic example of Victorian Gothic architecture. It was designed in 1865 by Sir George Scott and was opened in 1866.

Middle: Massive marble pillars support the roof of this station on the Moscow Underground system.

Bottom: A wooden trestle bridge built in 1900 on the Northern Pacific Railway, between Cul-de-Sac and Grangeville, Idaho, in America. The photograph was taken in 1910 and the bridge has since been replaced with a steel structure.

shaped to mesh with the pinion. The rack invented by Roman ABT is today the most popular. It consists of a single-toothed rail that meshes closely with the pinion. First used in 1886 on the Blankenburg Railway in Brunswick, it is now used in many parts of the world. Many railways, such as the Snowdon Mountain Railway in Wales, use a double Abt rack system. This has two racks and the teeth of one are staggered in relation to the teeth of the other. A special rack system was devised by Colonel Locher-Freuler for the very steep railway (1 in 2, in places) at PILATUS in Switzerland. The rack rail has teeth on each side that mesh with a pair of pinions.

In the early days of rack railways, engineers realized that the boiler of a conventional steam locomotive would dry out at one end when the train was on a gradient. At first, vertical boilers were used to overcome this problem, but engineers soon began to build locomotives that appeared tilted on the flat but were horizontal on the gradient. Today many rack railways are electrified.

Railway engineering

Railway tracks, except those on mountain railways, need to be as level and as straight as possible to reduce the cost of working the railway. When building a railway, the final choice of route depends on balancing this cost against the cost of construction. Generally railways tend to go through cuttings and tunnels rather than round or over obstacles, and high bridges are often needed to cross rivers and valleys.

One form of tunnel construction is called 'cut and cover'. A trench is dug, lined, and then covered over. This method was used to build London's Metropolitan and Circle lines. However, deeper tunnels have to be bored. In the early days of railways, hard rock was tunnelled using explosives and tools. In 1861 Germain Sommeiller, while digging the MONT CENIS TUNNEL, invented the pneumatic (compressed air) drill, which made tunnelling somewhat faster. Soft ground was very difficult to tunnel until the invention of the tunnelling shield by Marc Isambard BRUNEL in 1818. As the shield moved forward, cutting into the ground, the brick lining to the tunnel was built behind it.

small, 3-hp locomotive collected current from a central third rail.

Signals, in the early days, were simple, 2-aspect semaphore signals that indicated 'stop' or 'proceed'. Later, an additional 2-aspect signal that indicated 'proceed with caution' or 'proceed at full speed' was introduced. As railways became busier, signals came closer together and these 2 kinds of signal were combined into 3-aspect signals, each with 2 semaphore arms. Today nearly all signals are coloured lights, visible in daylight. Four-aspect signals were introduced to control both fast and slow trains. On such signals a red light indicates 'stop', a green light indicates 'proceed at full speed' and a single amber light indicates 'proceed with caution' to all trains. However, a double amber light gives fast trains (over 110 km/hour) advance warning of an approaching 'caution' signal and gives them time to slow down; slower trains may proceed at full speed past a double amber light. At busy junctions there used to be large gantries of semaphore signals. These have now been replaced by single sets of lights with illuminated indicators.

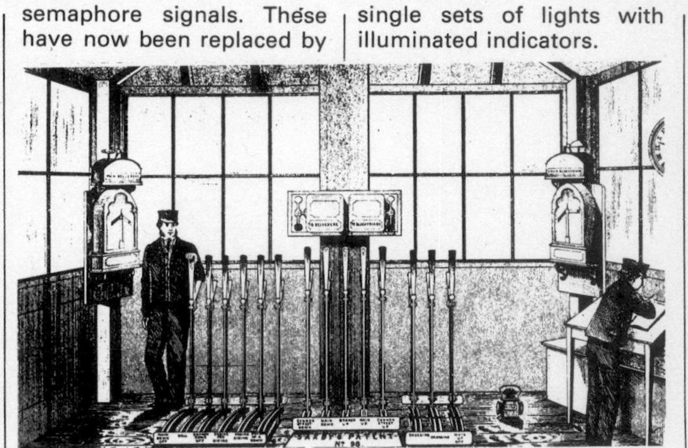

Signal box at Waterloo, England, 1865

Simplon Tunnels are the longest tunnels through the Alps. The first was bored between 1898 and 1906 and is 19.803 km long. It is a single track tunnel with a passing place in the middle. The second tunnel, completed in 1922, is 19.823 km long and, excluding London Transport's NORTHERN LINE, is the longest in the world.

Speed on the railways is increasing all the time. The current records for the various forms of propulsion are: steam – *Mallard,* 1938 – 202 km/hour; electric – French Co-Co No. 7107, 1955 – 330

Left: In mountainous country railway lines often have to follow tortuous paths to avoid steep gradients. Here, at Georgetown, USA, in order to raise the line through about 15 metres, engineers have had to construct a loop and two bridges.

Below: Sydney Harbour Bridge is one of the most famous bridges in the world. Its steel arch weighs 39,000 tonnes.

Later, a compressed air chamber was added behind the shield, and today prefabricated cast-iron or concrete sections are used instead of bricks to line tunnels.

Tunnelling is still expensive, difficult and often dangerous. Hazards that may occur, particularly in deep, hard-rock tunnels, include inrushes of water from underground springs and the occurrence of weak rock strata deep inside mountains. In long tunnels, there is the problem of ventilation, sometimes solved by driving long shafts down from the surface above.

In order to avoid long detours and to take the weight of trains, railway builders had to build bridges larger and stronger than ever before. Over long spans, ordinary beam bridges were not strong enough and between 1845 and 1850 Robert STEPHENSON built the first iron box-girder bridge, the BRITANNIA BRIDGE, across the Menai Straits in Wales. In America, where wood was cheap and easily available, many timber trestle bridges were built. Arch bridges have also been successful in all parts of the world, including solidly-built stone viaducts and long-span steel arches. However, the most massive railway bridges are the great cantilever bridges, notably the FORTH RAILWAY BRIDGE in Scotland. Suspension bridges are mostly built for road traffic as they are not suitable for withstanding the heavy loads and vibration on railways.

km/hour; diesel – British Rail HST prototype, 1973 – 230 km/hour; gas turbine (rail) – British Rail experimental APT, 1975 – 244 km/hour; gas turbine (hovertrain) – French *Aerotrain*, 1967 – 378 km/hour; linear induction motor test vehicle (USA), 1974 – 410 km/hour.

Stephenson, George (1781–1841) was an English inventor. His father was the fireman for a steam engine used to pump water out of a coal mine and he became fascinated by machines. He invented a miner's lamp (at about the same time as Sir Humphry Davy) and studied the work of WATT. He set out to devise a travelling steam engine, and although he was not the first to achieve this (*see* TREVITHICK) his machines were so successful that he is often regarded as the 'inventor' of the steam locomotive. Stephenson was also a great railway engineer and after 1825 concentrated mostly on building railways.

Stephenson, Robert (1803–59) was the son of George STEPHENSON and one of the greatest early railway engineers. He assisted his father in surveying for the Stockton and Darlington Railway and took over the management of Robert Stephenson and Company in 1823. After spending some time in Colombia, he returned to help his father build the *Rocket* and was appointed engineer of the London and Birmingham Railway in 1838. He built several great bridges, including the BRITANNIA BRIDGE, and a number of successful locomotives.

Robert Stephenson

Superheating is a method of increasing the temperature of steam after it leaves the boiler barrel via the regulator valve. It passes to a saturated steam header which leads to a number of small superheating tubes. Each tube is located within a superheater flue in the boiler and is folded so that the steam makes 4 passes down the flue. The superheated steam is then passed to the cylinders via the superheated steam header and the VALVES.

Sydney Harbour Bridge, built between 1924 and 1932, is perhaps the greatest steel arch bridge in the world because of its load-carrying capacity. Its main span is 503 metres long and carries 4 rail tracks, a roadway and 2 pedestrian walkways. The arch was erected

Steel rail
Wooden wedge
Sleeper
Bolt
Iron spike
Wooden peg
Cast iron chair

Left: A chaired bullhead rail — the type used in Britain and a few other countries until the 1950s. The rail is held in place by a wooden key. Modern rails are flat-bottomed, but a considerable amount of the older type of rail is still in use.

Right, above: Laying new track, using an automatic spikemaster to drive in the spikes that hold down the tie-plates.

Disc and crossbar
Lower quadrant semaphore
Upper quadrant semaphore

Above: Early signals were of the 'stop/go' type. The green disc and red crossbar were set at right angles and the signal turned so that the driver could see the appropriate indicator. Later came three-aspect semaphore signals. Both the upper and lower quadrant signals shown here indicate 'proceed with caution'. Most modern signals are colour lights, like the four-aspect signal here.

Right: A track-aligning machine on Alice Springs railway in Australia.

When Isambard Kingdom Brunel built the Great Western Railway, he designed all the tunnels, bridges and viaducts himself. He built them to last and the Great Western Railway remains a masterpiece of engineering. He also designed Paddington Station at the London end of the line. Built to impress the railway travellers of the time, this and the other great Victorian stations remain as monuments to the early railway builders.

Tracks and signals

When the roadbed, the ground on which the track is to be laid, has been levelled and the tunnels and bridges have been constructed, the track can be laid. On the first railways the rails were laid on stone blocks, but George Stephenson soon began to use wooden cross-ties, or sleepers. Today these are still used, although sometimes they have been replaced by steel or concrete sleepers and in places concrete slab track has been installed instead.

Sleepers are laid on a ballast of sharp stones, which in turn is laid on a bed of sand. The ballast helps to spread the load over a wide area and allows water to drain away from the track. After the sleepers have been laid more ballast is placed around them and then compacted down. Finally the rails are laid over the sleepers. In Britain and a few other countries rails that lie in supports called chairs fixed to the sleepers were used until the 1950s. Today, however, flat-bottomed rails are fixed to the sleepers using tie-plates. The length of rails also varies in different parts of the world. In Europe the standard rail is 30 metres long and in America it is 39 metres. Increasingly, however, track is being laid using continuous welded rail. Lengths up to 300 metres are laid by special trains and then welded together.

As the amount and speed of traffic on the

---

as 2 cantilevers, held back by cables until they met in the middle. Then the hangers and deck were added.

Tank locomotive is one that carries its own water and fuel supplies.

Tay Bridge, opened in 1878 across the Firth of Tay in Scotland, was a wrought iron truss bridge. It had 84 spans and during a gale on the night of December 28 1879 the central 13 spans collapsed while the Edinburgh mail train was crossing. None of the 78 people on board survived. The reason for the disaster was that the bridge had not been braced to withstand high wind pressures. The bridge was rebuilt and completed in 1887. It now has 85 spans and is the longest bridge in Europe (3.552 km).

The Tay Bridge disaster, December 28, 1879

Tees Bridge was a suspension bridge built in about 1830 to carry the Stockton and Darlington Railway across the River Tees. Within a few years it had been hammered to destruction. The weights of trains and the vibration they produce were thus shown to be too much for a flexible suspension bridge.

Tender is a special wagon, coupled more or less permanently to a steam engine, used for carrying fuel and water. Fuel was shovelled into the firebox by the fireman, but at the same time the boiler had to be replenished with water. Some early locomotives had a water pump operated by a wheel axle, with the disadvantage that the locomotive had to be in motion for it to work. Later locomotives used steam, controlled by a valve lever in the driving cab, to inject water into the boiler.

Track circuit is an electrical circuit, connected to SIGNAL lights, in which the rails form part of the circuit. The track is divided into sections, or blocks. A train on the rails of a block short circuits the

railways have increased, the problem of control has become greater. In 1889, semaphore SIGNALS, with coloured lights for night use, came into service on all British railways, and points were mechanically linked to signals to prevent accidents. Signals and points were operated by levers in signal boxes and a system of lever interlocking ensured that the signalman could not set up a route that might result in a collision.

Initially, two-aspect signals were used, but as the spacing between signals decreased three-aspect signals, indicating 'stop', 'caution' or 'proceed at full speed', were introduced. Today coloured lights, first introduced in 1920, are used in many places instead of semaphore signals.

TRACK CIRCUITS use electric currents to detect the presence of a train on the rails. They were first used in 1870 and today play an important part in controlling busy railways. The track is divided into sections and the arrival of a train in one section automatically changes the signals in the sections behind.

Track circuits have also enabled the use of illuminated track diagrams in control centres, which are gradually replacing manual signal boxes. A track diagram shows the positions of all the trains in the area and identifies them by numbers. Motorized points can be operated by buttons on the diagram and the controller can use these to set up a route. Detectors on the points check that they are locked, the route lights up on the diagram to indicate that it is clear and the signals then operate to allow a train to pass.

Electronic control is also of great importance in modern marshalling yards. In the early days, making up a goods train by shunting each wagon into a siding was a laborious and slow business. The introduction of the marshalling yard hump speeded up the process. Loaded wagons could be shunted to the top of the hump and then allowed to roll into the correct siding. Later, power operated points and wagon retarders (braking devices fixed to the tracks) allowed a few men to operate a marshalling yard from a control tower. However, judging how much braking to apply to each wagon was very difficult. Today, all the necessary information, such as the loaded weight of a particular wagon, how freely it rolls, how full the siding is already, and the wind force, is fed into a computer. The controller merely selects the desired classification siding and the wagon is

Above: Controllers at work in a modern signal control centre. The control panels are positioned to give them a clear view of the illuminated track diagram around the wall.

Below: A marshalling yard near Midway, Chicago, in America. Wagons from the hump (out of the picture to the left) are directed to the correct classification sidings by controllers in a tower.

track circuit and the 4-aspect signal immediately behind turns red. On the 3 blocks behind this the electric circuits select, in order, a single amber signal, a double amber signal and a green signal.

Trans-Siberian Express (the *Russia*) is a train that makes the longest journey in the world. Trains run every day and it takes about 8 days to travel the distance of 9,297 km. The original route, completed in 1904, runs from Moscow through Kirov, Sverdlovsk, Omsk, Krasnoyarsk, Irkutsk, Chita, Skovorodino and Khabarovsk to Vladivostok. Travellers can use 'hard' cars, in which only seats are provided, 'soft' cars, in which there are 'places for lying', or 'soft first category' cars, which are modern sleeping cars.

Trevithick, Richard (1771–1883) was the inventor of the high-pressure steam engine. He built 2 steam carriages and 4 steam locomotives. However, although his locomotives worked well he had a great deal of bad luck and had to contend with broken axles, fires and public hostility. He began the Rotherhithe Tunnel, but had to abandon the work and it was later built by Marc BRUNEL. He went to Peru, where he built 9 mine engines. But in order to return home he had to borrow money from Robert STEPHENSON, who was also in South America at the time. Trevithick died in poverty, having failed to obtain any recompense for his many ideas and achievements.

Trevithick's Common Road Passenger locomotive, 1803

U

Union Pacific Big Boys were the largest locomotives ever built. They were 4-8-8-4 MALLETS and 25 were built by the American Locomotive Company between 1941 and 1944. They weighed about 350 tonnes and the total weight, with

Left: A Japanese Hikari bullet-train crossing a steel truss bridge at high speed. Mount Fuji San is in the background.

Above: British Rail's Advanced Passenger train taking a curve at high speed. Its 'pendulum' suspension system causes it to tilt inwards to a greater degree than a conventional train, thus maintaining passenger comfort as well as safety.

slowed down by exactly the right amount to allow it to join the end of the train. Some marshalling yards even do without controllers. Electronic sensing devices 'read' colour codes on the sides of the wagons, which are then automatically directed into the correct siding.

Railways today

There are at present about 1,223,000 kilometres of track around the world and railways provide an increasingly important service. In the last few years railway planners have concentrated to a large extent on high-speed travel. Fast trains now run between major cities in many countries. British Rail's high-speed trains can reach speeds of up to 200 kilometres per hour. In Japan, the bullet trains that run on the SHINKANSEN line can achieve speeds of 210 kilometres per hour. However, above these speeds conventional trains cause discomfort to passengers when going round curves. The solution to this problem has been found by the builders of the gas-turbine-powered British Advanced Passenger Train. With its new 'pendulum' suspension system it will be able to negotiate curves at speeds of up to 240 kilometres per hour. Even this may not be the limit. In the future hovertrains may be built. They will have no wheels but will be raised off their tracks by air-cushions *(see page 126)* or MAGNETIC LEVITATION and may be propelled by LINEAR ELECTRIC MOTORS. Such trains may reach speeds of up to 480 kilometres per hour.

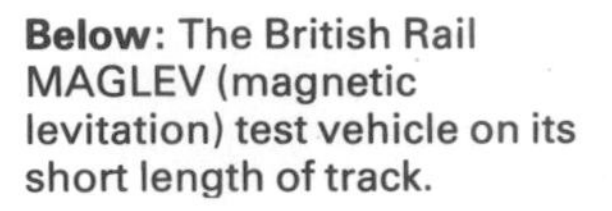

Below: The British Rail MAGLEV (magnetic levitation) test vehicle on its short length of track.

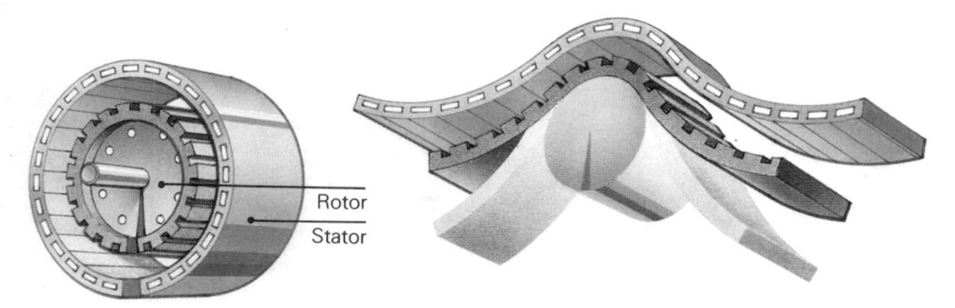

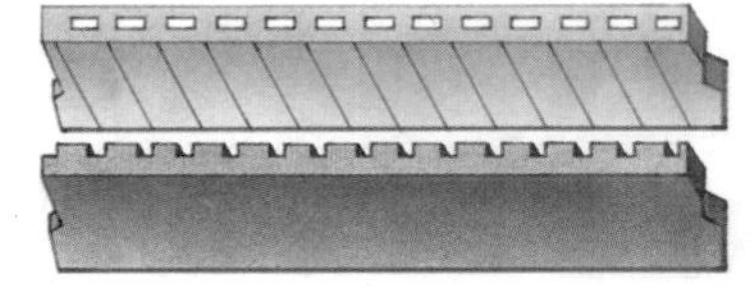

Left: A linear induction motor can be regarded as an opened-out conventional motor. In the conventional motor a rotor turns inside a fixed stator. If this is opened out and the flattened 'rotor' fixed, then the flattened 'stator' will move horizontally.

tender, was over 400 tonnes. On the level they could haul 3,000 tonnes at 112 km/hour.

V **Valves**, in the steam box of a steam locomotive, are used to pass steam into the cylinder, first to one side of the piston and then to the other. Slide valves work against a flat surface in which the ports (openings) to the cylinder are situated. In a piston valve system the ports are located in the side of a cylinder and are opened and closed by 2 circular piston valve heads. Slide valves and piston valves move back and forth at the same rate as the piston. They are driven by a complicated system of cranks, rods and links called the valve gear, the position of which can be altered to cause the locomotive to reverse. Several types of valve gear have been devised, the most popular being Egide Walschaert's, invented in 1848. Some locomotives have camshaft-driven poppet valves similar to those of an internal combustion engine.

Victoria Falls Bridge across the Zambezi River is a steel arch bridge with a main span of 152 metres. It was built in 1905 to carry Cecil Rhodes's projected Cape-to-Cairo railway at a height of 400 metres above the river, spanning the spectacular gorge.

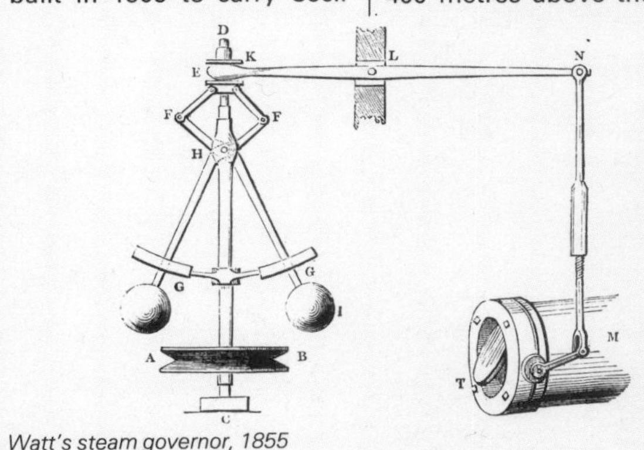

Watt's steam governor, 1855

W **Watt**, James (1736–1819) was a Scottish engineer. He not only improved NEWCOMEN'S engine but also made his own engines do more than merely act as pumps; almost any mechanical device could be powered by a Watt engine. In addition, he invented a governor – a device that controls the speed of an engine by reducing or increasing the steam supply to the cylinder as the engine goes faster or slower.

The mastery of flight, one of man's earliest dreams, is another remarkable story of technological progress. It has culminated in the supersonic airliner and rockets, which are now probing the secrets of space.

Travelling through Air

Left: According to Greek legend Icarus, not heeding his father Daedalus's warning, flew too near the Sun. His wax wings melted and the goose feathers dropped out. Icarus plummeted into the sea and Daedalus returned home alone.

There are probably few people in the world who have not dreamed of being able to copy the birds and fly. Sadly, however, man is not built for flying unaided and we have had to develop machines to get into the air. But despite our technical achievements and superb modern aircraft, we have not nor will we ever manage to match the flight control of the birds.

Early attempts and ideas

It is fortunate that man in his attempts to fly has never been discouraged by failure. Many would-be aeronauts have plunged to their deaths, as the legendary Greek Icarus is supposed to have done 4,000 years ago when his wax wings melted in the heat of the Sun. Some were luckier. In about AD 1000 a Benedictine monk is said to have launched himself from a tower. The wings attached to his arms carried him 100 metres to a crash-landing in which he broke both legs. In later years many people tried strapping on wings and launching themselves off cliffs or towers. Generally, the result was a vertical fall, and a painful, often fatal, landing.

The reason for these disasters was a failure to understand how birds fly. People believed that birds flew by flapping their wings downwards and backwards and that they got off the ground because they were inherently light. This idea, based on the theories of the Greek philosopher Aristotle (384–322 BC), persisted until the 1800s. Meanwhile ideas for flying machines had been put forward. In the 1200s Friar Roger Bacon predicted the possibility of making 'Engines for flying' with 'artificiall Wings made to beat the Aire'. During the 1400s the Italian artist and scientist Leonardo da Vinci worked on this idea and designed a number of flying machines. However, he quickly realized that muscle-power was not enough to raise such machines off the ground and began designing soaring machines. Although these were based on mistaken ideas about how birds soared, in time Leonardo might have designed a practical glider.

Into the air

While some people were failing to copy the birds others managed to get into the air in a quite different way – by making machines lighter than air. In 1709 Bartolomeu de GUSMÃO demonstrated the first hot-air balloon in Portugal. This model balloon was made of thick paper and the hot air came from a small fire burning in an earthenware bowl. The Montgolfier brothers were unaware of this experiment when they too worked out that a container of hot air would rise. They demonstrated their first successful balloon

Reference

A **Ader,** Clement (1841–1925) built the *Eole* (wingspan 14 metres) in which he made the first uncontrolled 'hop' on October 9 1890. In 1897 his second aircraft *Avion* III failed to leave the ground.

Air speed indicator is an instrument that measures the difference in pressure between air rushing past the wing and still air. The forward movement of the aircraft causes an increase in pressure in a small open-ended, forward-pointing tube on the wing-tip.

Altimeter is an instrument that measures the decrease or increase in air pressure as an aircraft climbs or descends. It works in the same way as an anaeroid barometer but measures the height above sea level.

Avro F (Britain, 1912) was the first aircraft with an enclosed cockpit. *Engine:* Viale 5-cylinder radial, 40 hp. *Wingspan:* 8.53 metres. *Speed:* 105 km/hour.

Avro Lancaster Mk. 1 (Britain, 1942) was the most famous bomber of World War II. 7,377 Lancasters were built and they carried out 156,000 missions. *Engines:* 4 Rolls-Royce Merlin X 12-cylinder V, 1,460 hp. *Wingspan:* 13.72 metres. *Max. speed:* 462 km/hour at 3,500 metres. *Crew:* 7. *Armament:* 10 machine guns, 9,980 kg of bombs.

Avro F

B **Benoist XIV,** a biplane seaplane, inaugurated the world's first scheduled passenger service between St Petersberg and Tampa, Florida, USA (34.5 km) in 1914. *Engine:* Roberts 6-cylinder in-line, 75 hp. *Wingspan:* 13.72 metres. *Max. speed:* 103 km/hour. *Crew:* 1. *Passenger:* 1.

Boeing 80A (America, 1928) was the second of the long line of successful Boeing civil transport aircraft (the first was the smaller 40A). The 80As remained in service until the arrival of the 247. *Engines:* 3 Pratt and Whitney Hornet 9-cylinder radial, 525 hp each. *Wingspan:* 24.38 metres. *Cruising speed:* 201 km/hour. *Crew:* 2-3. *Passengers:* 18.

Boeing 247 (America, 1933) made all other transport aircraft of the time obsolete. *Engines:* 2 Pratt and Whitney Wasp 9-cylinder radial, 550 hp each. *Wingspan:* 22.56 metres. *Cruising speed:* 250 km/hour. *Crew:* 2-3. *Passengers:* 10.

Boeing B-17G Flying Fortress (America, 1942) was the legendary bomber used in World War II for daylight

in France on April 25 1783 and the first manned flight in an untethered balloon was made on November 21 of the same year. Hot-air balloons were for many years replaced by gas balloons, but today hot-air ballooning has become a popular sport. Modern hot-air balloons take their fire with them, in the form of a flame fuelled by bottled gas. Thus their range is much greater than the Montgolfier balloons.

Gas balloons, filled with hydrogen or coal gas, were also invented in 1783, by Professor Jacques CHARLES. After 1852 engines and propellers were added and the first airships appeared, powered by steam or electricity. The arrival of the petrol engine in 1888 led to a rapid development of airships in the late 1800s and early 1900s, including the great German ZEPPELINS. These were dirigible (steerable) airships with the gas, usually hydrogen, contained in rigid aluminium envelopes. Many such airships were built in Germany, Britain and America. They were used by the armed forces during World War I and for carrying passengers. However, hydrogen is a dangerous, flammable gas. The inert gas helium was being used in some airships, but it was very expensive and often difficult to obtain. So, after several accidents with hydrogen-filled airships, notably the HINDENBURG and R.101 (*see* R.100) disasters, airship development came to an end.

Above: A drawing of the third Montgolfier balloon flight on September 19 1783, in the presence of Louis XVI, Marie Antoinette and the French court. A basket was attached to the balloon and in it were placed a cock, a duck and a sheep. The balloon climbed to about 550 metres and landed just over 3 km away. The cock appeared somewhat the worse for wear, but investigation showed that he had not suffered from the high altitude; his appearance was probably the result of having been trampled by the sheep.

Today, only a few non-rigid airships are used for publicity purposes.

The first aeroplanes

Sir George CAYLEY was the first to work out how to produce lift without having flapping wings. In 1804 he built and flew a model glider – the first fixed-wing aircraft. In 1849 he built a triplane glider that carried a boy for several metres, and in 1859 he launched his coachman off in another triplane. The coachman promptly resigned, unaware that as the first man to fly in an aeroplane he had made history.

Cayley and others did considerable work on aerofoils and Otto Lilienthal used the knowledge so far gained to build a number of gliders between 1891 and 1896. Lilienthal had an understanding of bird flight and his aim was to build a workable ornithopter (a flapping-wing aircraft). To learn more about stability and control he built 18 different types of glider, all with bird-like, arched wings. Although he worked a great deal on control ideas, he never quite achieved full control of his gliders. In 1896, after failing to recover from a stall, he crashed, injuring himself fatally.

By this time there were a number of people searching for a method of powered flight. Many ideas were put forward and although some of these were, to us, outrageously funny, others were more practical. Clement ADER'S bat-like *Eole* was propelled by a steam engine. In 1890 this aircraft managed to hop 50 metres and was the first powered aircraft to take off. But it could not achieve sustained flight. The main problem with this and other experimental aircraft was lack of control. Aircraft builders were concentrating on stability and did not appear to see the need for control surfaces.

The final breakthrough came from the work of glider designers. Octave CHANUTE built a biplane hang-glider in which the wings were strongly braced. Although he too gave priority to the idea of stability, he gave great encouragement and assistance to two other glider enthusiasts, Wilbur and Orville Wright. Influenced by Chanute's design, these two brothers set about systematically designing aircraft in America.

In 1899 they built a glider that had a front elevator to control pitching and a fixed tailplane. The most important feature was the system of

---

raids over Europe. *Engines:* 4 Wright R-1820-97 Cyclone 9-cylinder radial, 1,200 hp. *Wingspan:* 31.62 metres. *Crew:* 10. *Armament:* 13 machine guns, 7.985 kg of bombs.

Boeing B-29 Superfortress (America, 1944) was the best strategic bomber of World War II and B-29s were operational during the Korean war. They are best remembered for two missions in 1949, when the first atomic bombs were dropped over Hiroshima and Nagasaki. *Engines:* 4 Wright R-3350-57 Cyclone 18-cylinder radial, 2,200 hp. *Wingspan:* 43.05 metres. *Max. speed:* 576 km/hour at 7,620 metres. *Crew:* 10. *Armament:* 20-mm cannon, 10 machine guns, 9,090 kg of bombs.

Boeing 707 (America, 1954) was the first jet airliner built by Boeing. *Engines:* 4 Pratt and Whitney JT3C turbojets, 5,625 kg thrust each. *Wingspan:* 39.9 metres. *Cruising speed:* 917 km/hour. *Passengers:* 179.

Boeing 747 (America, 1969) was the world's first wide-bodied airliner. *Engines:* 4 Pratt and Whitney JT9D turbojets, 22,680 kg thrust each. *Wingspan:* 59.64 metres. *Cruising speed:* 937 km/hour. *Passengers:* 385.

Boeing 707

Bristol Beaufighter Mk. X (Britain, 1943) was used for anti-submarine warfare and ground attack for the last years of World War II. *Engines:* 2 Bristol H Hercules XVII, 14-cylinder radial, 1,770 hp each. *Wingspan:* 17.63 metres. *Max. speed:* 531 km/hour at 400 metres. *Crew:* 2. *Armament:* 4 20-mm cannon, 964 kg torpedo, 226 kg of bombs.

Bubble sextant, unlike the marine SEXTANT *(see page 79)* does not need to be lined up with the horizon. Instead, a small bubble acts as an 'artificial horizon' and the index mirror is adjusted so that the Sun or star appears to sit in the centre of the bubble.

C

Cayley, Sir George (1773–1857) was a British inventor. He built a successful glider and is now regarded as the 'father of aviation'.

Chanute, Octave (1832–

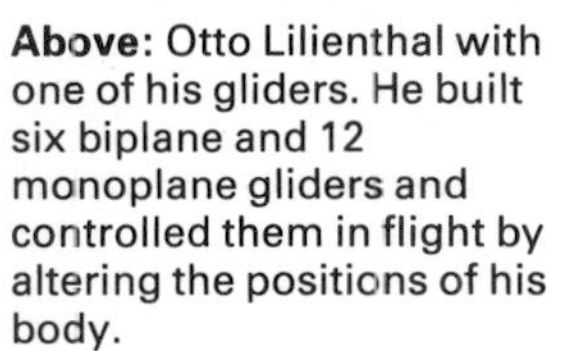

Above: Otto Lilienthal with one of his gliders. He built six biplane and 12 monoplane gliders and controlled them in flight by altering the positions of his body.

Right: A modern hang glider banking to the right. The pilot uses the same method of control as Lilienthal.

Below: On December 14 1903 Wilbur Wright (1867–1912) first tried the Wright Flyer I. He crashed soon after take-off, slightly damaging the front elevator On December 17 Orville Wright (1871–1948) made the first controlled powered flight. *Engine:* Wright 4-cylinder in-line, 12 hp. *Wingspan:* 12.29 metres. *Speed:* 48 km/hour.

wing warping (twisting) that enabled the glider to bank. Moving the control lever to the left caused cables to pull the trailing edge of the right wing-tip down and the trailing edge of the left wing-tip up. The result of this was that the aircraft banked to the left. However, they discovered that twisting the wing-tip downwards caused extra drag, resulting in a tendency for the glider to turn away from the bank. To counteract this effect they eventually added a rudder to the tail-plane.

By now the Wright brothers had more experience in flight than anyone before, and they built the Wright Flyer I, with a four-cylinder engine designed by themselves. At about 10.30 a.m. on December 17 1903, after an unsuccessful attempt three days before, Orville Wright flew a distance of 36.5 metres. Four more flights were made that day and on the last Wilbur flew a distance of 260 metres. The Wright brothers built three more aircraft and in 1905 made a flight of 39 kilometres. In 1908 Wilbur took a Wright Flyer IV to Europe.

Powered flight in Europe

During the years 1901–7 only a few Europeans were making much progress. Ferdinand Ferber was flying copies of Wright gliders and Gabriel VOISIN was experimenting with box-kite gliders.

1910) was an American railway engineer who became a glider designer and began building gliders in 1896.

Charles, Professor Jacques (1746–1823) designed and built the first hydrogen-filled balloon. It made its first ascent on August 27 1783 and on December 1, he and Marie-Noel Robert made the first ascent in a gas-filled balloon.

Consolidated B-24 Liberator (America, 1942) was the most widely used aircraft of World War II. Its main role was as a bomber and Liberators dropped about 635,000 tonnes of bombs and downed 4,189 enemy aircraft. *Engines:* 4 Pratt and Whitney R-1830-43 Twin Wasp 14-cylinder radial, 1,200 hp each. *Wingspan:* 33.52 metres. *Max. speed:* 488 km/hour at 7,620 metres. *Crew:* 8-10. *Armament:* 10 machine guns, 4,000 kg of bombs.

Curtiss, Glen (1878–1930) was an American pioneer aircraft builder. His aeroplanes won a number of competitions and he was noted particularly for his seaplanes. A Curtiss biplane made the first take-off from the deck of a ship. The Curtiss Aeroplane and Motor Company became famous for its naval aircraft.

Curtiss P-40B (America, 1941) was the most important American fighter in 1941-2 and was also used by the RAF, who designated it the Tomahawk Mk. II. *Engine:* Allison V-1710-33 12-cylinder V, 1,040 hp. *Wingspan:* 11.38 metres. *Max. speed:* 566 km/hour at 4,500 metres. *Crew:* 1. *Armament:* 4 machine guns (later ones carried bombs).

Curtiss P-40B, 1941

D de Havilland, Sir Geoffrey (1882–1965) was a British aviation designer pioneer whose name is now linked with many aircraft.

de Havilland D.H.4A (Britain, 1919) was a modified version of the biplane D.H.4 bomber and was used to inaugurate the passenger service from London to Amsterdam. *Engine:* Rolls Royce Eagle VIII 12-cylinder V, 350 hp. *Wingspan:* 12.93 metres. *Max. speed:* 195 km/hour. *Crew:* 1. *Passengers:* 2.

Above: A replica of the Bleriot XI designed by Louis Bleriot (1872–1936). It used wing-warping for control and the first flight was made at Issy in France on January 23 1909. Raymond Saulnier later added several modifications that considerably improved the aircraft's performance. In the early morning of July 25 1909 Bleriot made the first cross-Channel flight from Calais to Dover in 36 minutes. *Engine:* Anzani 3-cylinder semi-radial, 22-25 hp. *Wingspan:* 7.8 metres. *Speed:* 58 km/hour.

However, the first powered flight in Europe was made by the Brazilian Alberto SANTOS-DUMONT, who had previously pioneered the development of the airship. On November 12 1906 he made six flights in his 14-*bis*, the last of which covered 220 metres. His aircraft was a curious affair with the tailplane at the front and the propeller at the rear, and it was the first aeroplane to use ailerons. However, this idea did not catch on and Santos-Dumont went on to design the Demoiselles, small monoplanes with wing-warping for control.

By 1907 Gabriel Voisin and Henri FARMAN were achieving success with box-kite aeroplanes. Louis Blériot, who had been Voisin's assistant, was building monoplanes that resembled modern aeroplanes, with tails at the rear and puller propellers. However, they were all still a long way behind the Wright brothers. In 1908 Wilbur amazed everyone with two-hour flights in his Wright Flyer IV. This, at last, convinced the Europeans that control was necessary and in 1908 Farman added ailerons to a Voisin-Farman aircraft and flew a distance of 27 kilometres. On July 25 1909 Blériot flew his XI across the English Channel to accomplish the first-ever crossing.

How aeroplanes fly

The wings of an aeroplane are described as aerofoils because they produce lift and it is these aerofoils that cause the craft to fly. In cross-section an aerofoil can be seen to have a particular shape. Air travelling over the upper, curved surface has farther to go than air travelling over the flatter, lower surface. As a result, the air travelling over the upper surface moves faster, its pressure is reduced and the consequent higher pressure on the lower surface of the wing produces lift.

The angle at which the wing meets the air (angle of attack) is important. Increasing the angle increases the lift up to a certain point. If the angle is increased beyond this point, the air

de Havilland D.H.82A Tiger Moth (Britain, 1931) was developed from the 1925 D.H.60 Moth. A total of 7,300 Tiger Moths were built and were used by the RAF for 15 years. Many ended up in civil aviation and there are still a number of airworthy planes. *Engine:* de Havilland Gypsy Major 4-cylinder in-line, 130 hp. *Wingspan:* 8.94 metres. *Max. speed:* 167 km/hour. *Crew:* 2.

Tiger Moths

de Havilland Mosquito Mk. I (Britain, 1941) was built as a fast reconnaissance aircraft. Because of its success, later versions were built as fighter bombers, the best of which was the Mk. VI. *Engines:* 2 Rolls Royce Merlin XXI 12-cylinder V, 1,460 hp each. *Wingspan:* 16.51 metres. *Max. speed:* 611 km/hour. *Crew:* 2. *Armament:* 4 20-mm cannon, 4 machine guns, 907 kg of bombs.

de Havilland Comet I (Britain, 1952) was the world's first turbojet airliner. *Engines:* 4 de Havilland Ghost turbojets, 2,018 kg thrust each. *Wingspan:* 35.05 metres. *Cruising speed:* 789 km/hour. *Passengers:* 36. A later version, the Comet 4, made the first transatlantic jet crossing in 1958.

Dornier Do 217 (Germany, 1940) was the successor to the 1939 Do 17 bomber and various versions were used as reconnaissance aircraft, torpedo-carriers and night fighters. The Do 217E-1 was a bomber. *Engines:* 2 BMW 801 MA 14-cylinder radial, 1,580 hp each. *Wingspan:* 19.00 metres. *Max. speed:* 515 km/hour at 5,200 metres. *Crew:* 4. *Armament:* 5 machine guns, 15-mm cannon, 2,000 kg of bombs.

Douglas DC-2 (America, 1934) was the first of a long line of successful low-wing monoplanes, leading to the modern jets, the DC-8 and the wide-bodied DC-10. *Engines:* 2 Wright Cyclone F.3 9-cylinder radial, 710 hp each. *Wingspan:* 25.91 metres. *Cruising speed:* 273 km/hour. *Crew:* 2-3. *Passengers:* 4.

Drag is caused when some of the energy of a forward-moving aircraft is absorbed, and it is produced in 3 ways. First, energy is absorbed by the action of pushing the air aside as the aircraft moves forward; streamlining helps to reduce this. Second, the inner layer of air tends to stick to the surface of the aircraft, forming a boundary layer of slow-moving air, and energy is absorbed by friction. Third, energy is ab-

moving over the upper surface becomes turbulent, causing a loss of lift, and the aircraft stalls.

Three other forces act on a wing. In constant, level flight the lift is balanced by the weight of the aircraft, as gravity tries to pull it towards the ground. As well as producing lift the forward movement of the wing through the air produces DRAG, which resists the forward motion. This is balanced by the thrust produced by the aircraft engines.

The control surfaces are operated by cables attached to the control column in the cockpit. The ailerons on each wing are used in keeping the aircraft level and in turning. For example, lowering the port (left) aileron raises the port wing by increasing its curvature and causing it to generate more lift. The aircraft thus banks (tilts) to the right. If, at the same time, the rudder is pushed to starboard (right) the aircraft turns in that direction. On the tail-plane the elevators act in the same way as the ailerons, but they are both used together in order to make the aircraft dive or climb.

Since 1935 several more control surfaces have been added, particularly to high-speed aircraft. Flaps on the trailing edges of wings are used to generate more lift during take-off and to increase drag during landing. Leading edge flaps reduce

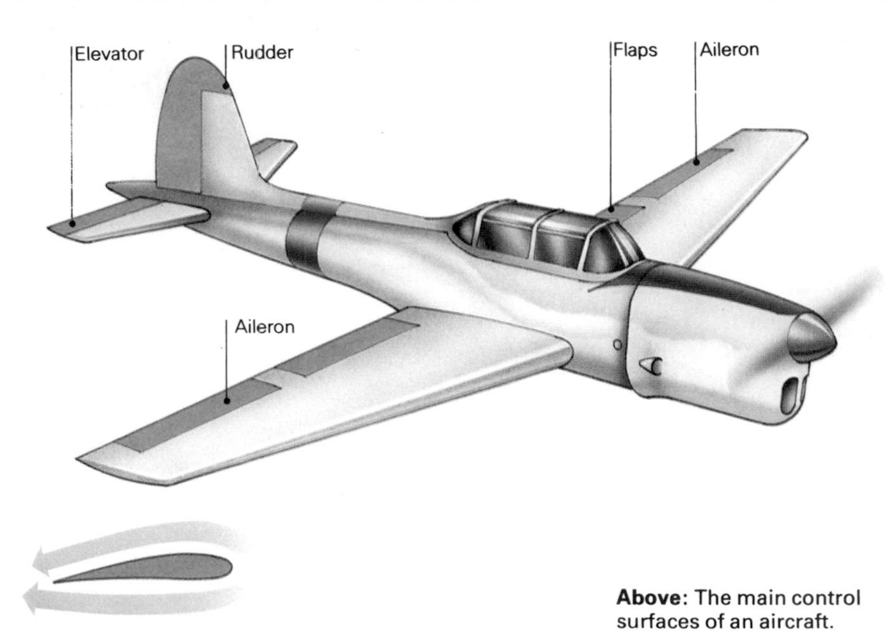

Above: The main control surfaces of an aircraft.

Above: A cross-section of an aerofoil.

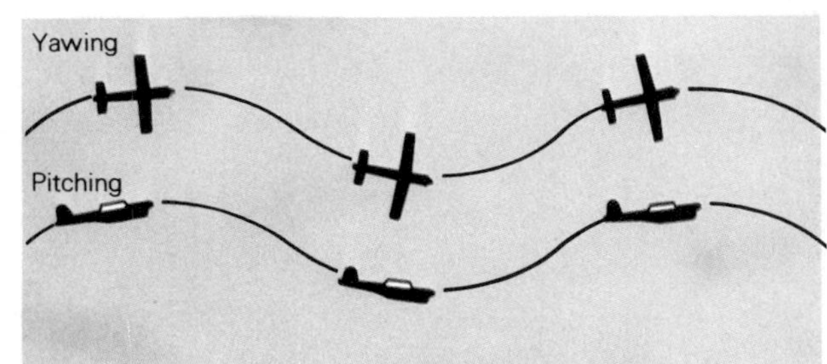

Right: Yawing, a side-to-side movement of the aircraft, is prevented by the sides of the fuselage and the tail fin. Pitching, an up-and-down movement of the aircraft, is prevented by the horizontal surface of the tailplane.

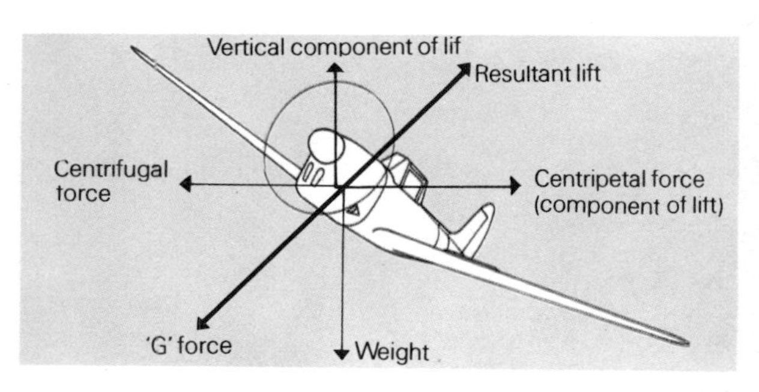

Right: In a balanced turn the wings have to provide extra lift to produce the centripetal force that holds the aircraft in the turn as well as balancing its weight. The combination of the centrifugal force (the reaction to the centripetal force) and the weight of the aircraft produces a force sometimes known as the 'G force' which balances the overall lift.

Above: To turn the aircraft the pilot uses the rudder and the ailerons. In turning to starboard (right) the rudder is deflected to the right. At the same time the aircraft is banked to the right by lowering the port (left) aileron (causing an increase in lift on the port wing) and raising the starboard aileron.

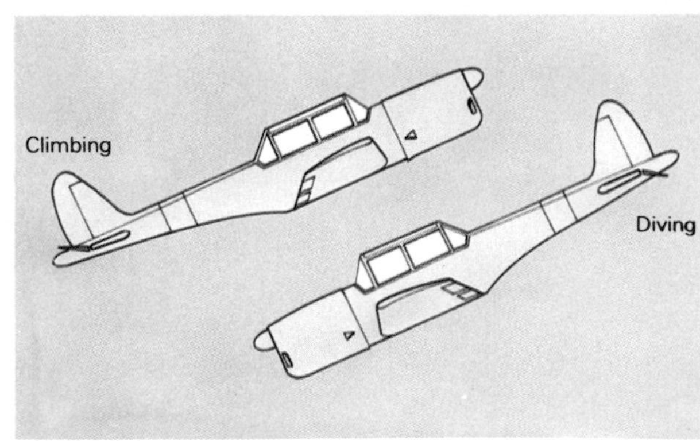

Right: The elevators on the tailplane are used to make the aircraft climb or dive. If the control column is pulled back, the elevators are deflected upwards. The lift on the tailplane therefore decreases and the nose rises. When the control column is pushed forward, the elevators are deflected downwards. The lift on the tailplane therefore increases and the nose drops.

sorbed by the vortices (spirals) of air created at the wing-tips.

F **Farman,** Henri (1874–1958) was a British-born aircraft builder. He built his first aircraft in France in 1909 and became a naturalized French citizen in 1937.

Farman F.60 Goliath (France, 1919) was a highly successful civil transport aircraft that broke several world records. *Engines:* 2 Salmson CM 9-cylinder radial, 260 hp each. *Wingspan:* 26.46 metres. *Cruising speed:* 120 km/hour at 2,000 metres. *Crew:* 2. *Passengers:* 12.

Focke-Wulf Fw2 190A-1 (Germany, 1941) was an even better fighter than the Messerschmitt Bf 109. A later version was a fighter bomber. *Engine:* BMW 801C-1 14-cylinder radial, 1,600 hp. *Wingspan:* 10.50 metres. *Max. speed:* 626 km/hour at 5,500 metres. *Crew:* 1. *Armament:* 4 machine guns, 2 20-mm cannon.

Fokker, Anthony (1890–1939) was the great Dutch aircraft designer who founded companies in Germany and Holland and became famous for his World War I aircraft.

Fokker E III (Germany, 1915), a monoplane fighter, was named the 'Scourge' because at the time it was the only aircraft with a synchronized, forward-firing machine gun. *Engine:* Oberusel U1 9-cylinder rotary, 100 hp. *Wingspan:* 9.41 metres. *Max. speed:* 140 km/hour. *Crew:* 2. *Armament:* 1-2 machine guns.

Fokker Dr I (Germany, 1917) was a triplane fighter with outstanding manoeuvrability. *Engine:* Thulin-built Le Rhône 9J 9-cylinder rotary, 110 hp. *Wingspan:* 7.19 metres. *Max. speed:* 165 km/hour at 4,000 metres. *Crew:* 1. *Armament:* 2 machine guns.

Fokker D VII (Germany, 1918), a biplane, was the finest fighter of World War I. *Engine:* Mercedes D ILI 6-cylinder in-line, 160 hp.

Fokker D VII

Wingspan: 8.9 metres. *Max. speed:* 200 km/hour. *Crew:* 1. *Armament:* 2 machine guns.

G **Gloster Meteor** (Britain, 1943) was the first British jet fighter. Meteors never saw action against the Messerschmitt 262, but the Mk. I version was used to shoot down V-1 flying bombs. The Mk. III took part in a few missions over Germany in 1945. *Engines:* 2 Rolls Royce Derwent 1,907 kg thrust. *Wingspan:* 13.11 metres. *Max. speed:* 667 km/hour at 3,000 metres.

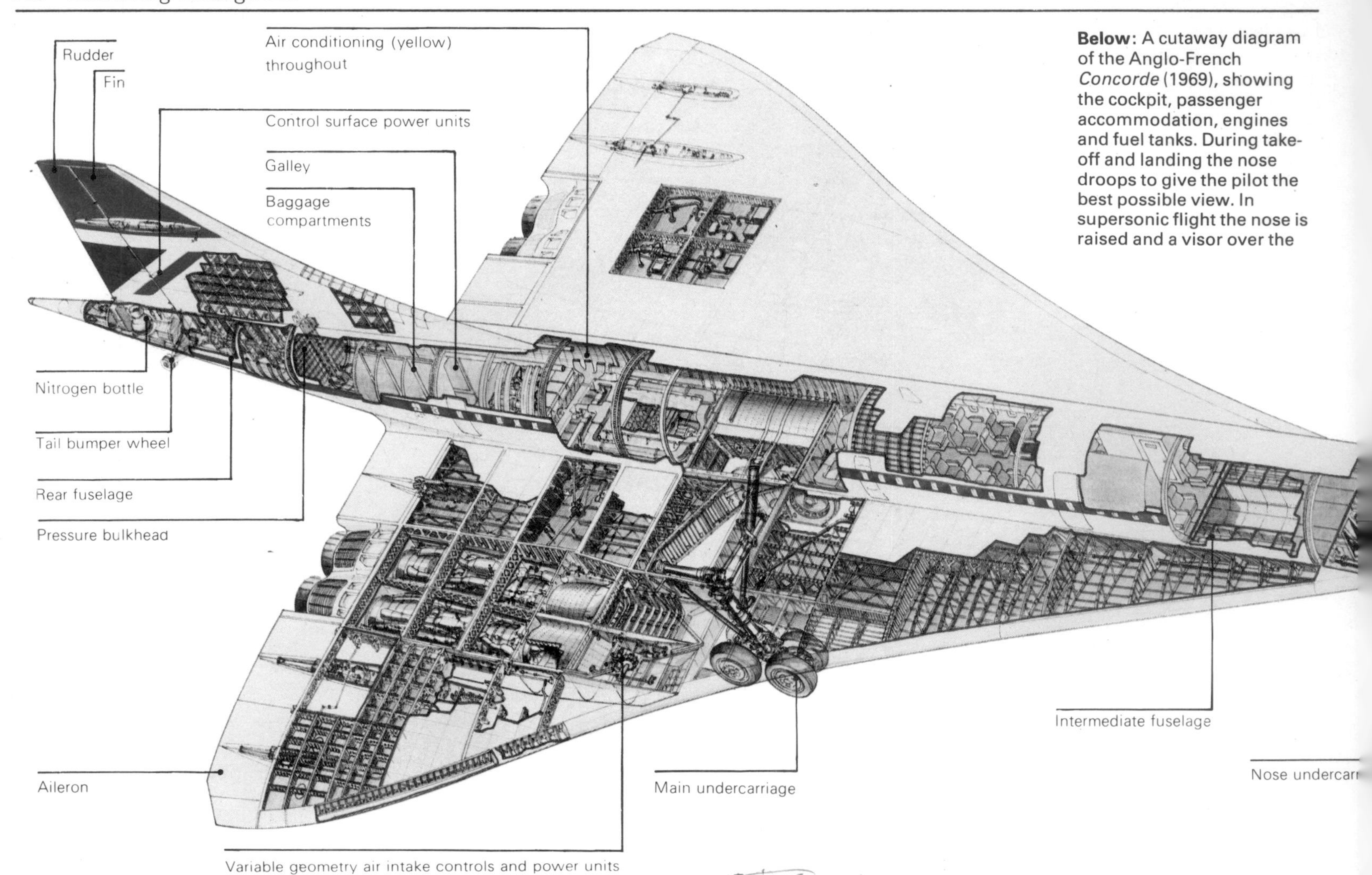

Below: A cutaway diagram of the Anglo-French *Concorde* (1969), showing the cockpit, passenger accommodation, engines and fuel tanks. During take-off and landing the nose droops to give the pilot the best possible view. In supersonic flight the nose is raised and a visor over the

air turbulence allowing aircraft to achieve higher angles of attack without stalling. Spoilers are vertical flaps near the trailing edge of a wing used to assist in banking and to increase drag during landing.

Aircraft development

From 1909 to 1914 aircraft were used mostly for sport and trials. Racing competitions, such as the Gordon Bennett Aviation Cup (1909) and the Schneider Trophy (1913), were organized and speed records were constantly being broken. The first manufacturing company had been set up in France by Gabriel and Charles Voisin. In America Glen CURTISS was building seaplanes and in Europe Geoffrey DE HAVILLAND and Anthony FOKKER, two men who were to become

Left: 1. The Royal Aircraft Factory S.E.5a (Britain, 1917) was one of the best fighters in the last year of World War I. *Engine:* Wolseley W.4 Viper 8-cylinder in-line V, 200 hp. *Wingspan:* 8.12 metres. *Max. speed:* 222 km/hour. *Crew:* 1. *Armament:* 2 machine guns.

Crew: 1. *Armament:* 4 20-mm cannon.

Graf Zeppelin (Germany, 1928) was a luxurious inter-continental passenger airship. Designated LZ 227, the Graf Zeppelin completed 590 flights, including 140 across the Atlantic. By 1940, when the airship was scrapped, it had flown 1,695,252 km. *Engines:* 5 Machbach, 550 hp each. *Length:* 236.60 metres. *Diameter:* 30.5 metres. *Cruising speed:* 109 km/hour. *Crew:* 45. *Passengers:* 20.

Gusmão, Bartolomeu de (1686–1727) was a Brazilian priest. On August 8 1709, in Portugal, he demonstrated the first hot-air balloon, which rose 3.6 metres off the ground.

H

Handley Page 0/400 (Britain, 1916) was a heavy bomber used for both day and night raids in World War I. *Engines:* 2 Rolls Royce Eagle VIII 12-cylinder in-line V, 360 hp each. *Wingspan:* 30.48 metres. *Max. speed:* 157 km/hour. *Crew:* 4. *Armament:* 4-5 machine guns, 907 kg of bombs.

Handley Page HP42

Handley Page HP42E (Britain, 1931) was a large biplane passenger transport used by Imperial Airways until 1939, even after it had been superseded by newer aircraft. *Engines:* 4 Bristol Jupiter XIF, 9-cylinder radial, 550 hp each. *Wingspan:* 39.62 metres. *Cruising speed:* 160 km/hour. *Crew:* 2. *Passengers:* 24.

Handley Page Halifax (Britain, 1944) was one of the finest bombers of World War I. It was also used for dropping paratroops and towing gliders. One of the best versions was the Mk. II. *Engines:* 4 Bristol Hercules XVI 14-cylinder radial, 1,615 hp. *Wingspan:* 31.75 metres. *Max. speed:* 454 km/hour at 4,100 metres. *Crew:* 7. *Armament:* 9 machine guns, 5,890 kg of bombs.

Hawker Hurricane Mk. I (Britain, 1937) was the RAF's first monoplane fighter. During 1940 Hurricanes accounted for more than half the German aircraft shot down. Later versions were built as night fighters and fighter bombers. *Engine:* Rolls Royce Merlin II 12-cylinder V, 1,030 hp. *Wingspan:* 12.19 metres. *Max.*

cockpit window helps in streamlining. *Engines:* 4 Rolls Royce/SNECMA Olympus 593 Mk 602 turbojets, 17,259 kg thrust each. *Wingspan:* 25.55 metres. *Max. cruising speed:* 2,179 km/hour. *Crew:* 3. *Passengers:* 128–144.

Right: A Douglas DC-3 (America, 1936) still in use. The DC-3 was the successor to the DC-2 and during World War II it became the most versatile of all transport aircraft. *Engines:* 2 Pratt and Whitney 9-cylinder radial, 1,200 hp each. *Wingspan:* 28.96 metres. *Cruising speed:* 290 km/hour. *Crew:* 2. *Passengers:* 14–32.

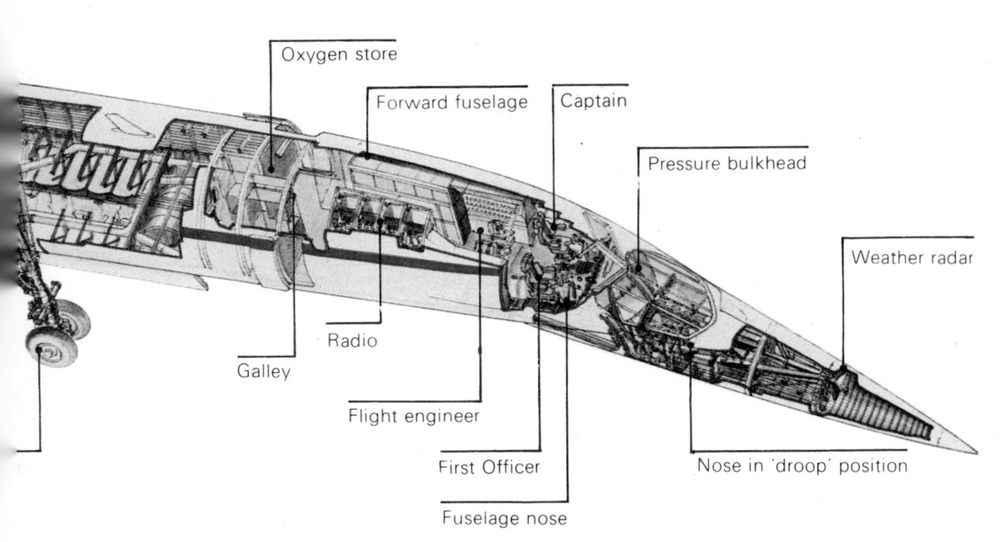

2. The Heinkel He IIIH-2 (Germany, 1939) was the Luftwaffe's main bomber at the beginning of World War II. *Engines:* 2 Junkers Jumo 211A-3 12-cylinder V, 1.100 hp each. *Wingspan:* 22.60 metres. *Max. speed:* 405 km/hour. *Crew:* 5. *Armament:* 6 machine guns, 2,495 kg of bombs.
3. A Navy/Curtiss NC-4 (America, 1919) was the first aeroplane to complete the Atlantic crossing in May 1919, from Trepassy Bay, Newfoundland, to Lisbon, Portugal, stopping at the Azores on the way. *Engines:* 4 Liberty 12 12-cylinder V, 400 hp each. *Wingspan:* 38.40 metres. *Max. speed:* 137 km/hour. *Crew:* 6.

famous, were also building aeroplanes. In 1912 the British designer Alliot Roe built the AVRO F, which was the first aeroplane to have an enclosed cockpit. In Russia Igor SIKORSKY built the first four-engined aeroplane, the *Russkii Vitiaz*, which flew in 1913.

At the beginning of World War I aircraft were seen as only being useful for reconnaissance and a number of aircraft were built for this purpose. However, war in the air soon escalated. The first aircraft equipped with machine guns were the French MORANE-SAULNIER Ls in 1915. In the same year the Germans introduced the 'Scourge' — the monoplane Fokker E III (Eindekker) — which had a forward-firing machine-gun synchronized to fire in between the blades of the propeller. By the end of World War I both sides had a number of fighters, including the Royal Aircraft Factory S.E.5a, the SOPWITH Camel, the FOKKER D VII and the Fokker Dr I (the aircraft flown by Manfred von Richthofen, the 'Red Baron'). For bombing the Germans used Zeppelin airships until 1917, when they were replaced with strategic bombers, such as the Zeppelin (Staaken) R VI. British bombers in use by 1918 included the HANDLEY PAGE 0/400 and the VICKERS Vimy.

The valuable experience gained during World War I was put to more constructive use during the years that followed. In 1919 John Alcock and Arthur Whitten Brown made their historic non-stop crossing of the Atlantic Ocean in a Vickers Vimy. By now engines were more powerful, and it had become possible to build aircraft to carry passengers. After the war, airline services opened in Germany, France and Britain and the first international service was established by a British company, Aircraft Transport and Travel, using war-surplus bombers, DE HAVILLAND D.H.4As.

The first all-metal aircraft, the Junkers J 1, had been built in 1915. After the war the Germans were forbidden by the terms of the Versailles treaty to build large aeroplanes and so they turned their attention to high-performance aircraft. They built the JUNKERS F 13, successor to the J 1, and this excellent four-passenger seaplane saw service with about 30 airlines around the world. Another important civil aeroplane of the time was the FARMAN F.60 Goliath.

By the early 1930s airlines were carrying passengers to many parts of the world, and aircraft with large passenger-carrying capacities had been built. Many experimental designs

speed: 515 km/hour at 6,100 metres. *Crew:* 1. *Armament:* 8 machine guns.
Heinkel He 178 (Germany, 1939) was the world's first jet aircraft. *Engine:* HeS 3B, 498 kg thrust. *Wingspan:* 7.09 metres. *Max. speed:* 708 km/hour.
Heinkel He 219A-2/R1 (Germany, 1943) was the Luftwaffe's finest night fighter. The first prototype shot down at least 25 British bombers in 10 days. The He 219 was also the first aircraft equipped with ejector seats. *Engines:* 2 Daimler Benz DB 603A 12-cylinder V, 1,750 hp each. *Wingspan:* 18.50 metres. *Max. speed:* 670 km/hour at 7,000 metres. *Crew:* 2. *Armament:* 6 20-mm cannon.
Hindenburg (Germany, 1936), designated LZ 129, was the world's largest airship. For about a year it operated a passenger service between Frankfurt and New York. On May 6 1937 it caught fire over Lakehurst, New Jersey, and 35 of the 97 people on board were killed. *Engines:* 4 Daimler Benz, 1,320 hp each. *Length:* 245 metres. *Diameter:* 41.20 metres. *Max. speed:* 135 km/hour. *Crew:* 40. *Passengers:* 50.

Hindenburg

J **Junkers F 13** (Germany, 1919) was a highly-successful all-metal small transport aircraft. *Engine:* BMW 111A 6-cylinder in-line, 185 hp. *Wingspan:* 17.75 metres. *Cruising speed:* 140 km/hour. *Crew:* 2. *Passengers:* 4.
Junkers Ju 87D1 (Germany, 1938) was the well-known and much-feared 'Stuka' dive-bomber of World War II. *Engine:* Junkers Jumo 211 J-1 12-cylinder V, 1,400 hp. *Wingspan:* 13.79 metres. *Max. speed:* 410 km/hour at 4,100 metres. *Crew:* 2. *Armament:* 4 machine guns, 1,800 kg of bombs.
Junkers Ju 88 (Germany,1939) was the Luftwaffe's most versatile aircraft. More than 16,000 were built and were used for reconnaissance, ground-attack, bombing, dive-bombing and torpedo bombing. The Ju 88G-7 was a formidable night fighter. *Engines:* 2 Junkers Jumo 213E 12-cylinder V, 1,725 hp each. *Wingspan:* 20.80

Turbo-jet

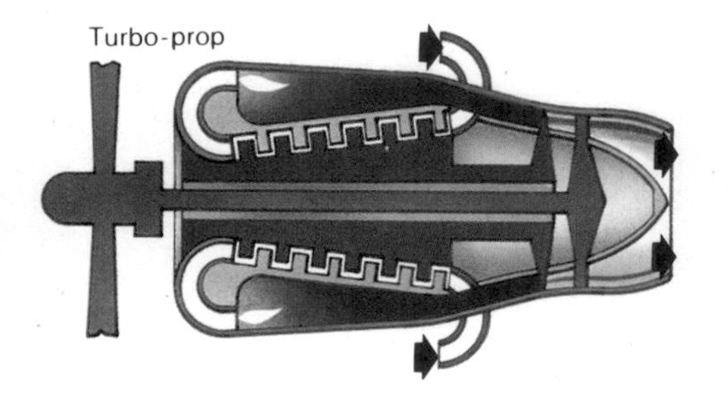

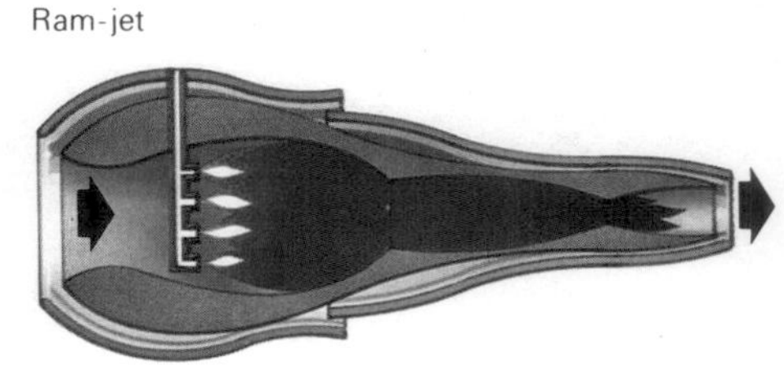

Above: Four types of jet engine. In a turbojet the compressor draws air into the engine and compresses it. Fuel is injected into the combustion chamber, where it ignites in the hot, compressed air. The expanding gases drive the turbine as they leave the rear of the engine. In a turboprop engine the compressor again compresses the air before fuel is injected into the combustion chambers. However, the turbine not only drives the compressor but also a propeller, which helps to power the aircraft. The pulse jet and ramjet have no compressors. Instead, the air is compressed by the forward movement of the engine itself. As a result they can only be used when the craft is already moving. The pulse jet, which was the type used on the German V-1 flying bombs in World War II, has valves that open and close by pressure in the combustion chamber, and combustion takes the form of a series of explosions. The ramjet produces a continuous stream of hot gases.

appeared and some of these had short lives; others even failed to get off the ground. However, there were also many successes. By 1935 famous passenger transport included the enormous flying boats, such as Imperial Airways' SHORT S.8 Calcutta and Pan American's MARTIN M-130 China Clippers and Sikorsky S-42s. Land aircraft included the Handley Page H.P.42s, which were virtually synonymous with Imperial Airways for many years, and the Boeing 80A. These aircraft were superseded by the Boeing 247 and Douglas DC-2. These all-metal, low-wing monoplanes, with their new technology and advanced design and performance, marked the beginning of a new era.

Alongside the success of large commercial aircraft there had also been considerable development of smaller types, including fighters, trainers, civil transporters, racers and private aircraft. Among the most famous of these were the Ryan NYP SPIRIT OF ST LOUIS, in which Charles Lindbergh made the first solo, non-stop Atlantic crossing in 1927, the Fokker F VIIb-3m SOUTHERN CROSS, in which Charles Kingsford Smith made the first crossing of the Pacific in 1928, and the de Havilland D.H.82A Tiger Moth, which is probably the best-known training aircraft in the world.

By the beginning of World War II the biplane fighters of the early 1930s had been outdated by the new sleek monoplanes. In 1939 the German Luftwaffe had 1,200 fighters, including the formidable MESSERSCHMITT Bf 109 (popularly known as the Me 109), which could outclass all other fighters. In Britain the RAF had fewer aeroplanes, but despite this disadvantage the squadrons of the now-famous HAWKER Hurricanes and SUPERMARINE SPITFIRES won the Battle of Britain. Later well-known British World War II fighters included the multirole de Havilland Mosquito and the BRISTOL BEAUFIGHTER, which was equipped with an early version of radar for night interception. In Germany the Bf 109 was joined by the FOCKE-WULF Fw 190, possibly Germany's best fighter, and the HEINKEL He 219. New fighters appeared throughout the world and the best of these included the American Lockheed P-38 Lightnings, which destroyed more enemy aircraft in the Pacific than any other fighter, and the outstanding Japanese MITSUBISHI A6M2 Reisen, or Zero.

Strategic bombing played an important part in World War II. Among the bombers that carried out missions over Britain were the Heinkel He 111, the DORNIER Do 217, many of which were later converted into night fighters, and the multirole Junkers Ju 88, which was the German equivalent of the Mosquito. In Britain a number of bombers were developed, including the well-known VICKERS WELLINGTON and the later Avro Lancaster, which took part in some famous raids, such as the 'Dam Busters' mission in May 1943. American bombers included the Boeing B-17 Flying Fortress, which was also used for long-range reconnaissance, and the CONSOLIDATED B-24 Liberator, of which 18,000 were built, more than any other American aircraft.

The jet age

As in all forms of transport the development of aircraft has progressed with the development of engines. The invention of the internal combustion engine enabled the Wright brothers to fly the first successful powered aircraft. Progressively more advanced engines were developed over the years and the size, endurance and speed of aircraft increased accordingly.

However, the greatest leap forward came with the invention of the gas turbine jet engine, developed by Sir Frank WHITTLE between 1928 and 1939. The jet engine has a much greater

metres. *Max speed:* 626 km/hour at 9,000 metres. *Crew:* 4. *Armament:* 6 20-mm cannon, 1 machine gun.

Lockheed P-38 Lightning (America, 1942) was a powerful high-altitude interceptor fighter. The best version was the P-38J. *Engines:* 2 Allison V-1710-91 12-cylinder V, 1,425 hp each. *Wingspan:* 15.85 metres. *Max speed:* 666 km/hour at 7,600 metres. *Crew:* 1. *Armament:* 20-mm cannon, 14 machine guns, 1,450 kg of bombs.

Lockheed L-1011 Tristar (America, 1970) is a wide-bodied jet airliner. *Engines:* 3 Rolls Royce RB. 211 turbofan, 19,050 kg thrust. *Wingspan:* 47.34 metres. *Cruising speed:* 901 km/hour at 10,700 metres. *Passengers:* 400.

Mach number is the speed of an aircraft expressed as a multiple of the speed of sound; i.e. Mach 2 is twice the speed of sound. It is expressed in this way because the speed of sound varies with the temperature and pressure of the air and is lower at high altitudes than at sea level. The speed of sound at 0°C at sea level is 1,194 km/hour.

Lockheed L-1011 Tristar

Martin M-130 China Clippers (America, 1930) were 3 flying boats flown overseas. *Engines:* 4 Pratt and Whitney Twin Wasp 14-cylinder radial, 830 hp each. *Wingspan:* 39.70 metres. *Cruising speed:* 266 km/hour. *Crew:* 5. *Passengers:* 48.

McDonnell-Douglas Phantom F-4 (America, 1958) is a modern jet fighter of which many versions have been built. *Engines:* 2 General Electric J-19-17 turbojets. *Wingspan:* 11.7 metres. *Max. speed:* 1,464 km/hour. *Crew:* 2. *Armament:* 4 AIM Sparrow missiles, 20-mm multibarrel cannon, 7,620 kg of bombs.

Messerschmitt Bf 109 (Germany, 1935) was for 5 years the best fighter in the world. More than 35,000 were built in several versions. The 1939 version, which fought in the Battle of Britain, was the Bf 109E-1. *Engine:* Daimler Benz DB 601A 12-cylinder V 1,050 hp. *Wingspan:* 9.87 metres. *Max. speed:* 550 km/hour at

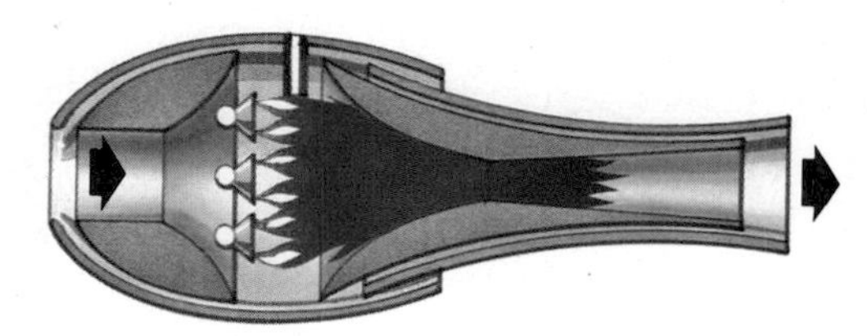

power than a piston engine and, when jet engines began to be introduced into aircraft during the last years of World War II, the scene was set for reaching hitherto undreamed of speeds.

The first jet aircraft to fly was the HEINKEL He 178 in 1939 and the first jet fighter used in action was the Messerschmitt 262. The GLOSTER METEOR, the direct result of Sir Frank Whittle's work, took to the air in 1943.

Today, most aircraft are powered by jet engines. New designs, such as arrow-shaped swept wings and delta wings, have been introduced to cope with increasing speeds. Jet fighters built since World War II include the American MCDONNELL-DOUGLAS Phantoms and the Russian Mikoyan MiGs. The first jet airliner was the de Havilland D.H.106 Comet I and this aircraft showed up a major problem of high-altitude flying. In 1953 and 1954 three Comet Is broke up in the air. Eventually metal fatigue of the pressure cabin was diagnosed as being the cause. In recent years the most successful jet airliners have been the Boeings, beginning with the Boeing 707s in 1958 and leading to the gigantic, wide-bodied Boeing 747 Jumbos in 1973, the largest of which can hold 500 passengers.

Above left: The Hawker Siddeley Harrier GR 3 (Britain, 1966) is a ground attack aircraft, capable of taking off vertically and of slow, low-level flight as well as high-speed flight. *Engine:* Bristol Siddeley Pegasus 103 vectored thrust turbofan, 1,752 kg thrust. *Wingspan:* 7.69 metres. *Max. speed:* 1,158 km/hour. *Crew:* 1. *Armament:* 2 30-mm guns and up to 2,270 kg of bombs, including 68-mm SNEB rockets or sidewinder missiles.

Above: A helicopter can be used almost anywhere. In this case one is being employed to transport skiers to the upper slopes of a snow-covered mountain.

The first jet aircraft that could fly faster than the speed of sound (*see* MACH NUMBER) in level flight was the NORTH AMERICAN F-100 Super Sabre in 1953. Today there are many supersonic aircraft, including the latest jet airliners, the Russian TUPOLEV Tu 144 and the Anglo-French Concorde. Supersonic flight has also involved changes in design. The shock wave generated by an aircraft travelling near, at or faster than the speed of sound causes additional drag. To overcome this designers have used thin, highly swept back wings and symmetrical aerofoil sections with pointed leading and trailing edges. Some supersonic aircraft, such as the American ROCKWELL B-1 bomber, have variable geometry, or swing wings. At high speeds the wings are swung back to reduce the amount of lift and the effects of the shock waves.

Vertical take-off and landing

The idea of taking off vertically is nearly as old as the idea of flight. Leonardo da Vinci designed helicopters, George Cayley built models, and a number of other inventors designed machines, most of which were totally impractical. The first helicopter capable of carrying a man was

4,000 metres. *Crew:* 1. *Armament:* 2 machine guns, 2 20-mm cannon.

Messerschmitt 262A-1a (Germany, 1944) was the first jet fighter to see action. Faster than any other fighter of its time it might have changed the course of World War II if it had been used as a daytime fighter. Instead it was used for reconnaissance and as a bomber and night fighter. *Engines:* 2 Junkers Jumo 004B-1, 898 kg thrust. *Wingspan:* 12.48 metres. *Max. speed:* 869 km/hour at 6,000 metres. *Crew:* 1. *Armament:* 4 30-mm cannon.

Messerschmitt Bf 109

Mitsubishi A6M2 Reisen (Japan, 1940) was used in the attack on Pearl Harbor (1941) and later versions were used in suicide missions in 1945. *Engine:* Nakajima NK1C Sakee 12 14-cylinder radial, 950 hp. *Wingspan:* 12.10 metres. *Max. speed:* 534 km/hour at 4,500 metres. *Crew:* 1. *Armament:* 2 20-mm cannon, 2 machine guns, 120 kg of bombs.

Morane-Saulnier L (France, 1913) was a monoplane fighter. In 1915 it was fitted with a machine gun that fired through the propeller area. The blades had steel plates to deflect bullets. *Engine:* Gnome rotary, 80 hp. *Wingspan:* 11.2 metres. *Max. speed:* 115 km/hour at 2,000 metres. *Crew:* 1-2. *Armament:* 1-2 machine guns, a few bombs.

N

Nakajima Ki-84-1a Hayate (Japan, 1943) was one of the finest Japanese fighters of World War II. *Engine:* Nakajima Ha-45 18-cylinder radial, 1,900 hp. *Wingspan:* 11.23 metres. *Max. speed:* 631 km/hour at 6,100 metres. *Crew:* 1. *Armament:* 2 20-mm cannon, 500 kg of bombs.

North American P-51D Mustang (America, 1944) was one of America's best World War II fighters. A total of 7,966 were built. *Engine:* Packard V-1650-7 12-cylinder V, 1,510 hp. *Wingspan:* 11.28 metres. *Max. speed:* 703 km/hour at 7,600 metres. *Crew:* 1. *Armament:* 6 machine guns, 907 kg of bombs.

designed by the Frenchman Paul Cornu in 1907. However, it had engineering and stability problems that could not be overcome at the time. In 1942 in America Igor Sikorsky built the first practical helicopter, the VS-316A (which was designated R-4 by the military).

The blades of a helicopter have an aerofoil cross-section to provide lift. The pitch of the blades (the angle at which they meet the air) can be altered to generate more lift in climbing or less lift when descending. The reaction of a helicopter to the torque (twisting force) produced by a horizontally-spinning rotor is a tendency to spin uncontrollably in the opposite direction. The various configurations of helicopters are designed to overcome this by balancing the torque using a vertical rotor at the rear, a second horizontal rotor spinning in the opposite direction or a single horizontal rotor propelled by jets at the tips of the blades.

Today there are many uses for helicopters, generally in situations where it is impossible or inconvenient to use conventional aircraft. They are used in air-sea rescue, aerial-crane work, offshore drilling operations, passenger transport across cities, and warfare.

Research into Vertical Take-Off and Landing (VTOL) aircraft began seriously in the 1950s, when Rolls Royce tried out two jet engines on a test machine made of tubular steel, called the Flying Bedstead. The success of this experimental craft led to the first VTOL aircraft, the Short Brothers and Harland SC-1, which had separate lift and propulsion jets. Over the years various systems have been tried, including jets that can be tilted and propellers mounted on tilt-wings. By far the most successful is the system of vectored (directed) thrust first devised in 1960. In this system the jet engines are fixed and the hot gases are directed either downwards or backwards through swivelling nozzles. The first fully-operational VTOL aircraft was the Hawker Siddeley Harrier, which entered service in 1968 and is the world's most versatile aircraft.

Aircraft navigation

In the early days of flying there were no sophisticated navigational aids. A pilot would often find his position by flying low and reading the name of a railway station (later, some stations had their names written in large white letters on the roof). One way of navigating was to fly from one landmark to the next and to follow railway lines.

It was soon discovered that the wind affected the flight path of an aircraft, causing it to drift off course. Thus, a navigational system known as dead reckoning had to be used. In this system a pilot, knowing the wind speed and direction, works out the necessary compass heading of the aircraft for it to travel along the correct course. Modern aircraft use electronic methods of determining their rate of drift.

As in ships the compass was and still is an essential navigational instrument. However, the marine sextant proved too difficult to use and a new, slightly less accurate instrument called a BUBBLE SEXTANT was devised. Other new instruments included the ALTIMETER, AIR SPEED INDICATOR and the artificial horizon, which informs the pilot of the 'attitude' (angle compared with the horizontal) of his aircraft. Today, the pilot has the aid of an array of instruments including a rate of climb/descent indicator and a turn/slip indicator.

The artificial horizon uses a gyroscope and this device is also used in the GYROCOMPASS *(see page 74)*, which works in conjunction with the magnetic compass. Gyroscopes are also used in automatic pilots, or INERTIAL GUIDANCE SYSTEMS *(see page 74)*, which can fly an aircraft along a preset course, automatically correcting any changes in height or attitude.

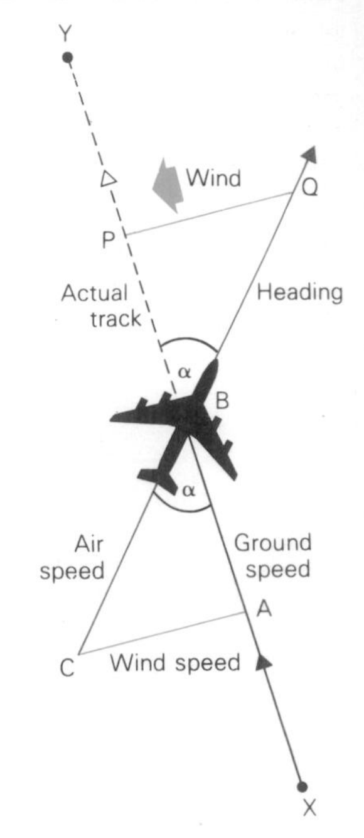

Above: Dead reckoning is used to determine the necessary compass heading of an aircraft in order to compensate for drift caused by the wind. In this instance the pilot wishes to fly from X to Y. If he knows his ground speed (AB) and the speed and direction of the wind (AC) he can plot a triangle on the map to find the airspeed and direction he must adopt (CB). The angle α is known as the angle of drift.

North American F-100 Super Sabre (America, 1954) was the first aircraft to fly faster than the speed of sound. *Engine:* Pratt and Whitney J57 turbojet, 7,700 kg thrust. *Wingspan:* 11.81 metres. *Max. speed:* 1,350 km/hour. *Crew:* 1. *Armament:* 4 20-mm cannon.

R

R 34 (Britain, 1919) made the first airship crossing of the Atlantic in July 1919 from East Fortune, Scotland, to Mineola, Long Island, in 110 hours. *Engines:* 5 Sunbeam Maori, 250 hp each. *Length:* 196 metres. *Diameter:* 24 metres. *Max. speed:* 100 km/hour. *Crew:* 23.

R 100 (Britain, 1929) was a highly successful airship. However, only 1 trip was made, to Canada and back in July/August 1930. The R 100 was scrapped after the disastrous crash of the R 101 at Beauvais in France on October 4 1930, during a test flight (47 out of the 54 people on board were killed). *Engines:* 6 Rolls Royce Condor IIIA. *Length:* 219.4 metres. *Diameter:* 41.55 metres. *Max. speed:* 131 km/hour. *Crew:* 44.

Rockwell B-1 (America, 1968), a jet bomber, is the largest aircraft with variable geometry (swing wings). It can take off from small airbases and yet fly supersonically at altitude. *Engines:* 4 General Electric F101-100 turbofan, 13,610 kg thrust each. *Wingspan:* max. 41.7 metres, min. 23.8 metres. *Max. speed:* 2,135 km/hour. *Crew:* 5. *Armament:* 24-32 AGM-69B missiles, 34,020-52,614 kg of bombs.

Rockwell B1 jet bomber

S

Santos-Dumont, Alberto (1873–1932) was a Brazilian aviation pioneer who lived in Paris. He built several airships, the most famous of which was the *No. 6*. In 1898 it flew 11.3 km from St Cloud round the Eiffel Tower in Paris and back again.

Santos-Dumont 14-*bis* (France, 1906) was a box-kite aeroplane. It was the first European aircraft to make a controlled powered flight.

Right: Behind the cockpit of a USAF B-52 bomber the navigator and ECM operator sit at their controls. ECM stands for Electronic Counter Measures and the operator's task is to reduce the effect of enemy radar and other electromagnetic apparatus.

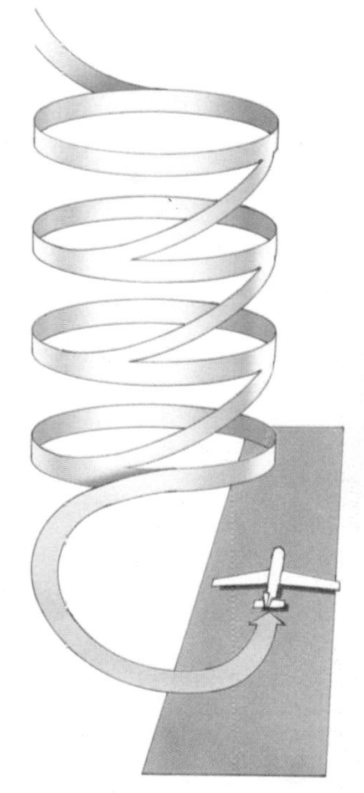

Above: At a busy airport aircraft may need to wait in a holding pattern or 'stack' before landing. Aircraft enter above those already circling and descend as others leave the holding pattern to land. The lowest holding altitude is 2,133 metres. Above this aircraft are held at 305-metre intervals, up to 8,840 metres. However, to avoid enormous delays, aircraft are usually diverted to other airports when the holding pattern reaches 6,096 metres.

Left: A Boeing 747 being loaded with passengers and luggage at an airport terminal. The 747 was the first wide-bodied jet airliner to enter service.

The first radio navigational aid used in aircraft was the D/F loop *(see page 79)*. One of the first ADF (Automatic Direction Finding) systems was called VOR (Very high-frequency Omnidirectional Radio-range). This uses two transmitting stations to fix the position of an aircraft. However, it can only be used in narrow airways and additional DME (Distance Measuring Equipment) is used to show the distances of radio stations. Modern ADF systems include DECCA and LORAN *(see pages 72 and 75)*. Beams of radio waves are also used in the various kinds of blind landing systems now available, which are designed for use in conditions of low or nil visibility.

RADAR *(see page 78)* is used in aircraft navigation in a number of ways. DOPPLER *(see page 73)* navigational radar uses four narrow beams aimed at the ground and compares the differences in frequency between the emitted and reflected signals to determine the speed and track of the aircraft over the ground. Radar on board an aircraft is also used to detect other aircraft and storm clouds. In ground control stations radar is used to keep track of the ever-increasing air traffic, particularly near busy airports.

Guiding missiles

Guided missiles are today taking over many of the former roles of strike aircraft, as it is cheaper and safer to launch a missile from a distance and guide it to its target. The first missile used in warfare was the German V-1 flying bomb. It was guided by a magnetic compass and an automatic pilot. At a preset range the fuel supply to the jet engine was cut off and the V-1 dived to the ground. The V-2 was the first rocket-powered ballistic missile. It was aimed by calculating the correct launching angle for its trajectory to end on the target.

Since World War II there has been worldwide development of various kinds of missiles, including ballistic, anti-ballistic, air-to-air, air-to-ground, ground-to-air, anti-tank and anti-ship missiles, many of which are armed with nuclear warheads. Propulsion methods include liquid and solid-fuelled, single and multistage rockets and ram-jets *(see page 120)*.

There are various guidance systems. Many early missiles were radio-controlled. Ballistic missiles rely on their trajectory and inertial guidance. Some anti-tank missiles are guided by a wire that carries electrical signals from an

Engine: Antoinette 8-cylinder in-line V, 50 hp. *Wingspan:* 11.2 metres. *Speed:* 40 km/hour. *Crew:* 1.

Short S.8 Calcutta (Britain, 1928) was a passenger seaplane used by Imperial Airways in the Mediterranean. *Engines:* 3 Bristol Jupiter XIF 9-cylinder radial, 540 hp each. *Wingspan:* 28.34 metres. *Cruising speed:* 148 km/hour. *Crew:* 3. *Passengers:* 15.

Sikorsky, Igor (1881–1972) was a Russian pioneer aircraft builder. After the Russian revolution of 1905 he emigrated to America and there designed aircraft and helicopters.

Sikorsky Ilya Muromets (Russia, 1915), the successor to the 4-engined *Russkii Vitiaz* (Russian Knight), was the first heavy bomber. Three were built. *Engines:* 4 Sunbeam 8-cylinder in-line V, 150 hp each. *Wingspan:* 29.8 metres. *Max. speed:* 121 km/hour. *Crew:* 4-7. *Armament:* 3-7 machine guns, 522 kg of bombs.

Sikorsky S-42 (America, 1935) was the first of the great flying boats. *Engines:* 4 Pratt and Whitney Hornet 9-cylinder radial, 750 hp each. *Wingspan:* 34.79 metres. *Cruising speed:* 274 km/hour. *Crew:* 5. *Passengers:* 32.

Short S8 Calcutta flying boat

Sopwith F.1 Camel (Britain, 1917) was one of the most famous biplane fighters of World War I. Camels shot down 1,294 enemy aircraft. *Engine:* Clerget 9B 9-cylinder rotary, 130 hp. *Wingspan:* 8.53 metres. *Max. speed:* 185 km/hour at 2,000 metres. *Crew:* 1. *Armament:* 2 forward-firing, synchronized machine guns.

Southern Cross (Holland, 1928) was the Fokker F VIIb-3m that on May 31 set out on the first crossing of the Pacific. Charles Kingsford Smith, with a 3-man crew, completed the journey from San Francisco to Brisbane in 83 hours 28 minutes. *Engines:* 3 Wright Whirlwind 9-cylinder radial, 300 hp each. *Wingspan:* 21.71 metres. *Cruising speed:* 178 km/hour. *Crew:* 2. *Passengers:* 8-10.

Spirit of St Louis, Ryan NYP (America, 1927) was the aircraft flown by Charles Lindbergh on his solo non-stop crossing of the Atlantic in 1927. The journey from Mineola, New York, to Le Bourget, France, took 33

Left: The American ALCM (Air Launched Cruise Missile). This missile requires no aiming; it is merely released from the aircraft and then flies to its target under the control of a computer program. It is powered by a turbofan engine and can fly at tree-top level, avoiding obstacles and enemy radar and defences.

Below left: An ALCM under trial, just after being launched from a Boeing B-52 bomber low over the desert.

optical sight. Some missiles home in on the infra-red (heat) rays of their targets. Others home in on radar pulses emitted either from a ground station or their own radar systems and reflected from the target. Laser and television guidance systems are also used.

The latest type of missile, and the most difficult to combat, is the cruise missile, such as the Boeing ALCM (Air-Launched Cruise Missile) and the US Navy Tomahawk. Powered by turbofan engines, both these missiles can fly a precise course to their targets at tree-top level, avoiding radar. They are guided by a computer program that enables them to avoid defence positions and follow the terrain, continuously comparing it to the information in the computer.

Navigation in space

When a satellite is launched into space it goes into orbit around the Earth. Its position is monitored at all times by ground stations, which use radio beams to follow its track. The satellite remains in its orbit owing to the combined

hours 39 minutes. *Engine:* Wright Whirlwind J-5-C 9-cylinder radial, 200 hp. *Wingspan:* 14.02 metres. *Cruising speed:* 180 km/hour. *Crew:* 1.

Spirit of St Louis

Supermarine Spitfire (Britain, 1936) was one of the most famous aircraft of World War II and over 20,300 were built in about 30 versions. One of the fastest was the 1944 Mk. XIV. *Engine:* Rolls Royce Griffon 65 12-cylinder V, 2,050 hp. *Wingspan:* 11.22 metres. *Max. speed:* 721 km/hour at 7,900 metres. *Crew:* 1. *Armament:* 2 20-mm cannon, 4 machine guns, 454 kg of bombs.

Supersonic flight begins when an aircraft exceeds the speed of sound. As an aircraft moves through the air it creates disturbances. Below the speed of sound these form a pattern of pressure waves continually moving away from the aircraft at the speed of sound. Above the speed of sound the aircraft leaves the pressure waves behind and a cone-shaped shock wave forms. As the shock wave passes a listener on the ground he hears a 'sonic boom'. The shock wave is an additional source of DRAG and supersonic aircraft are specially designed to reduce this.

T **Tupolev ANT-9/M-17** (Russia, 1932) was a successful passenger transport. *Engines:* 2 M-17 12-cylinder V, 680 hp each. *Wingspan:* 23.73 metres. *Cruising speed:* 175 km/hour. *Crew:* 2. *Passengers:* 9.

Tupolev Tu 144 (Russia, 1968) was the first supersonic airliner to fly (Dec. 31 1968). *Engines:* 4 Kuznetsov NK-144 turbofan, 20,000 kg thrust each. *Wingspan:* 28.80 metres. *Max. cruising speed:* 2,500 km/hour at 17,000 metres. *Crew:* 3-4. *Passengers:* 140.

V **Vickers Vimy** (Britain, 1918) was designed as a bomber, but later versions were used as commercial transports. On June 14-15 1919 John Alcock and Arthur Whitten Brown made the first non-stop Atlantic cros-

Left: The Astronaut Manoeuvering Unit tested on Gemini 9 in 1966. On the astronaut's back is a life-support, communications, propulsion and control system. An emergency system is at the front. Controls for the 12 small jets are on the arm rests.

Right: The interior of an Apollo command module showing the astronauts at the controls used for navigating the spacecraft.

effects of the Earth's gravity and its own speed.

The pull of gravity is insufficient to control satellites that orbit thousands of kilometres away from Earth. One method of controlling such satellites is by using a system of photoelectric cells and a suitably positioned shade. The cells are locked on the Sun or a star and if the satellite wanders off course, the shade moves in between the cells and the light source. Small rockets on the satellite then correct its position. A similar system is used on a satellite nearer the Earth to keep its solar power cells facing the Sun.

Before sending a spacecraft to the Moon or a planet the course has to be worked out very carefully. This is because such destinations are moving through space. Therefore the spacecraft has to be aimed at a point some distance ahead of the planet's position at take-off.

An unmanned spacecraft is navigated by the ground station, using information sent back by the spacecraft's instruments. In a manned spacecraft the astronauts keep a constant check on their position. When orbiting the Earth, tracking stations send information by radio. Deeper in space navigation is accomplished by using inertial guidance systems and sextants and special telescopes to observe stars and planets. The information they provide is fed into computers, which inform the pilot of any course changes necessary.

sing. They flew their Vickers Vimy from St John's, Newfoundland, to Clifden, Ireland, in 16 hours 27 minutes. *Engines:* 2 Rolls Royce Eagle VIII 12-cylinder in-line V, 360 hp each. *Wingspan:* 20.47 metres. *Max. speed:* 166 km/hour. *Crew:* 3. *Armament* (removed for the Atlantic crossing): 2-4 machine guns, 1,123 kg of bombs.

Vickers Wellington (Britain, 1938) was the RAF's main bomber at the beginning of World War II. *Engines:* 2 Bristol Pegasus XVIII 9-cylinder radial, 1,000 hp each. *Wingspan:* 26.26 metres. *Max. speed:* 378 km/hour at 4,700 metres. *Crew:* 6. *Armament:* 6 machine guns, 2,000 kg of bombs.

Vickers Wellington

Vickers-Armstrong Viscount (Britain, 1948) was the first successful turboprop airliner. *Engines:* 4 Pratt and Whitney Rolls Royce Dart propeller-turbines, 1,740 hp each. *Wingspan:* 28.56 metres. *Cruising speed:* 521 km/hour. *Passengers:* 40-47.

Voisin, Charles (1882–1912) and Gabriel (1886–) were French pioneer aircraft builders.

W

Whittle, Sir Frank (1907 –) is the British designer of the jet engine. In 1937 he ran the first turbojet and a later version of this jet powered the GLOSTER METEOR.

Z

Zeppelin, Ferdinand von (1838–1917) was a German airship builder. His first rigid airship, the *Luftschiff Zeppelin* I (LZ 1) flew in 1900.

Zeppelin (Staaken) R VI (Germany, 1917) was a large heavy bomber used to bomb London towards the end of World War I. *Engines:* 4 Mercedes D IVa 6-cylinder in-line, 260 hp each. *Wingspan:* 42.2 metres. *Max. speed:* 129 km/hour. *Crew:* 7. *Armament:* 4-7 machine guns, 2,000 kg of bombs.

The hovercraft is one of the most adaptable of all forms of transport, because it can travel over land and water. The air-cushion principle has many applications. It has even been used for lawn mowers.

Hovercraft and Hydrofoils

Right: The SR-N4 Mk. III, or Super 4, is the world's largest hovercraft. It is 56 metres long, 28 metres wide and fully loaded weighs over 300 tonnes. It is powered by 4 turboprop engines which drive the lift fans and propellers via gears. The craft is steered by movements of the propeller pylons and the 2 tail fins.

Man has continually tried to achieve higher and higher speeds on land, in the air and on water. Streamlining and powerful engines are the main methods of increasing speed and these work well for cars, trains and aeroplanes. Boats, however, have to contend with the 'stickiness' or viscosity (resistance to flow) of water. Bow and stern waves and the friction between water and the hull all tend to slow a boat down. Speed-boats, which at high speeds remove all but their sterns from the water, solve this problem to some extent, but are extremely uncomfortable. Hovercraft and hydrofoils solve the problem by having hulls that rise completely clear of the water, and so provide comfortable high-speed travel.

Hovercraft

A hovercraft is the most versatile of all forms of transport. It rides on a cushion of air and so can go almost anywhere, over land and water.

Reference

A **ACT-100,** built by Arctic Engineers and Construction, is a 250-tonne air-cushion transporter. When it was first towed slowly across ice it was found to act as an icebreaker. The air-cushion depressed the water beneath the ice which thus lost its support and cracked. This principle is now used to allow conventional ships to pass through ice. An air cushion is pushed ahead of the ship and a plough-like deflector pushes the broken ice to the sides.

Air-cushion landing systems (ACLS) are used on some aircraft. The first of these was the Lake LA-4 single-engined light amphibious aircraft built by Bell Aerospace. It has a doughnut-shaped bag skirt, from which the air escapes through thousands of tiny holes. These can be sealed to allow the craft to float.

B **Bell,** Alexander Graham (1847–1922), the American inventor, under a licence acquired from FORLANINI, built a number of 'hydrodromes' in America. One of these set a new world water-speed record of 114.03 km/hour. It was a torpedo-shaped craft with a pair of wings. It had 4 hydrofoils, 1 under each wing, 1 at the bow and 1 at the stern, and 2 air-propellers were mounted on the wings.

Alexander Bell

Boeing PGH-2 Tucumcari was the US Navy's first hydrofoil gunboat. It has sideways retracting foils and, propelled by water-jets, its top speed is 72 km/hour. Improved versions of the Tucumcari have been built, including the PHM (patrol hydrofoil missile). The Boeing Jetfoil is a 106-tonne commercial hydrofoil propelled by water-jets. Its top speed is 83 km/hour and it has a carrying capacity of 250 passengers. Both the PHM and Jetfoil have stern main foils that can be retracted by swinging them backwards.

F **Forlanini,** Enrico (1848—1930), an Italian helicopter and airship pioneer, built the first successful hydrofoil in 1905. His canoe-shaped craft, which he called a hydro-aeroplane, was equipped with 3 sets of ladder-foils, 2 in front and 1

The idea of a craft riding on an air-cushion was thought of many years ago. For example, in 1879 John Thorneycroft, a British naval architect, patented the idea of a vessel riding on 'a layer of air between the bottom of the vessel and the surface of the water'. However, the technology of the time was not sufficiently advanced for him to be able to achieve this. Many inventors took up the idea, but it was not until the 1950s that the problem of keeping the air-cushion in place was solved. In 1954 Sir Christopher Cockerell, an English engineer, tried an experiment with two coffee tins, one inside the other, and an industrial drier. He discovered that forcing the air through the gap in between the tins provided an air curtain that created and maintained a central air-cushion. Using this principle, now called the annular jet, he designed the first workable hovercraft, the SR-N1, which was completed in 1959.

Small hovercraft can use Cockerell's system, but air-cushions produced in this way do not give large hovercraft sufficient clearance over obstacles. This problem was solved by adding a flexible skirt as an extension of the annular jet. It was soon discovered that this arrangement increased the clearance by up to 10 times, and the skirt automatically adjusted itself over waves or rough ground. In addition, the ring of spray around the hovercraft was found to be considerably reduced.

The air-cushion on a hovercraft is produced by one or more large fans. The main forms of forward propulsion include air-propellers, water-propellers and water-jets. Steering is achieved by rudders, either in the airstream generated by air-propellers or in the water as in a conventional boat.

Hovercraft are not only used for high speed travel over water, but also for carrying loads in factories, along roads and across swamps and ice (*see* ACT-100, VOYAGEUR). Some aircraft are fitted with AIR-CUSHION LANDING SYSTEMS so that they can land and take off almost anywhere. HOVERTRAINS are also being designed.

Hydrofoils

A stationary hydrofoil looks, at first glance, just like any other boat, with a hull that floats in the water. However, below the waterline there are foils. These have the same cross-sectional shape

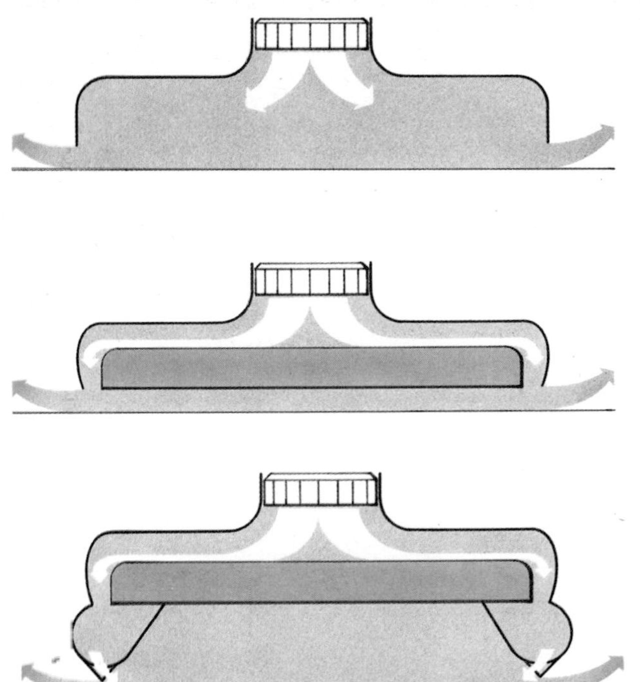

Left: The open plenum arrangement uses a single large chamber which is kept filled with air under pressure. However, this arrangement requires a considerable amount of power to maintain sufficient pressure.

Left: Sir Christopher Cockerell's design used a peripheral, or annular, jet in which air is pumped round the edge of the hovercraft. The air cushion under the whole craft is maintained using much less power.

Left: The addition of a flexible skirt to the peripheral jet arrangement increased the clearance of the hovercraft by about ten times. As a result hovercraft could negotiate rougher water or ground and became generally more useful.

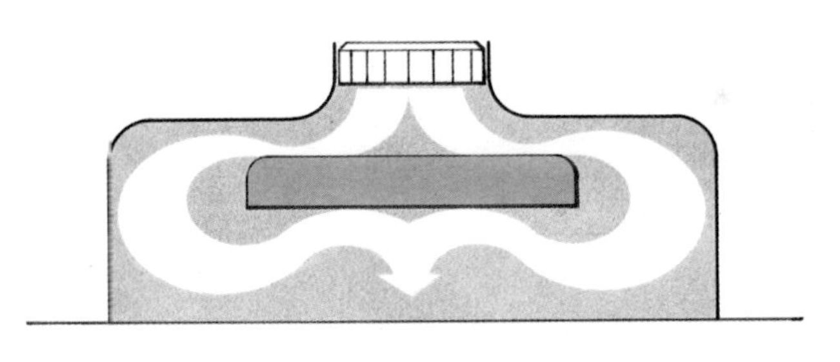

Left: A fixed sidewall type of hovercraft is also known as a CAB (Captured Air Bubble) ship. It has a flexible skirt at the front and the fixed sidewalls effectively seal in the air cushion. They are more economical than hovercraft with flexible sidewalls, but they can only travel over water.

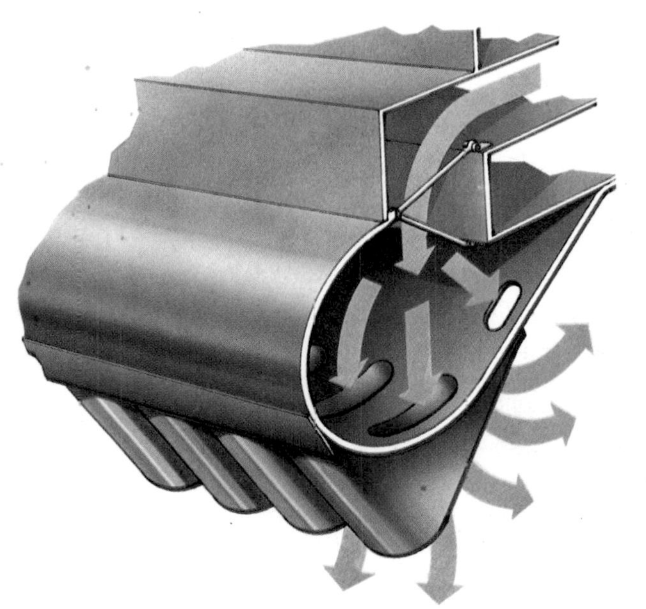

Left: A common type of skirt is one with a flexible bag to which flexible projections, known as fingers, are attached. The fingers make contact with the surface and form a seal round the air cushion. The advantage of using fingers is that if one gets pushed upwards by an obstacle or becomes damaged, the efficiency of the other fingers remains unaffected. Hovercraft with this type of skirt are comfortable to ride in and can travel at relatively high speeds in rough conditions.

behind, and was powered by fore and aft air propellers. It could reach speeds of up to 70 km/hour.

Grumman AGEH-1 Plainview hydrofoil

G **Grumman AGEH-1 Plainview** is the world's largest hydrofoil. An experimental ASW (anti-submarine warfare) craft, it weighs 320 tonnes and has fully-submerged foils that can be retracted sideways.

H **Hovertrains,** or tracked air-cushion vehicles (TACV), such as the Bertin Aerotrain being developed in France, will run on T-shaped guideways. They may be propelled by turbines or LINEAR ELECTRIC MOTORS *(see page 103)*. Speeds of over 480 km/hour have been forecast for hovertrains.

Hydrofin was invented by Sir Christopher Hook, a British organ builder, in the 1940s. It consists of a feeler arm projecting forward with a subfoil at the front end. Attached to the subfoil, which rides clear of the water, is a heel or sensor flap, which rides on the waves. Movement of this sensor is passed to the feeler arm and thence to the main foil via linkages. Thus the pitch of the main foil is continually altered to meet each wave at the angle that keeps it at the required depth.

L **Ladder-foils** are arranged in tiers, like the rungs of a ladder. A ladder-foil is a form of SURFACE-PIERCING FOIL, and rolling and pitching are automatically controlled. If the ladder becomes more submerged, more foils generate lift and the ladder is pushed upwards again. However, flat foils have the disadvantage that a sudden drop in the amount of lift occurs as each foil breaks the surface. This problem is overcome by using foils which have a V-shape.

S **SR-Ns** were the first series of hovercraft to be

1

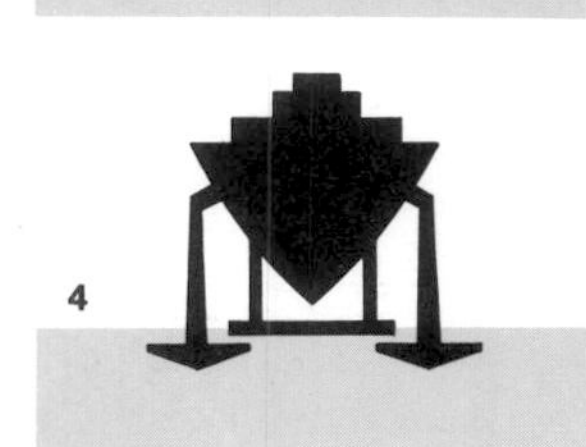

Above: Four types of hydrofoil.
1. Ladder V-foils in a conventional, or aeroplane, arrangement, with two foils at the front and a single foil at the rear. In some hydrofoils this arrangement is reversed and is called a canard arrangement.
2. V-foils arranged in tandem, a common form of surface-piercing hydrofoil.
3. Fully-submerged foils in tandem. The angle at which the foils meet each wave is controlled electronically to prevent the craft pitching backwards and forwards.
4. Fully-submerged foils in a conventional arrangement.

Below: The Grumman PGH-1 Flagstaff is a hydrofoil specially designed for the US Navy as a gunboat. This one is in operation with the coastguard and lacks guns. The design of this hydrofoil provides an unusual stability in rough seas, even at speeds of over 90 km/hour. The craft is 9.27 metres long and weighs about 71 tonnes. It has 3 fully-submerged, retractable foils and a Z-drive from the engine which is linked to a propeller on the rear foil.

Right: A Boeing Jetfoil, propelled by water jets. Two pumps drive water out through backward pointing nozzles. The pumps are operated by gas turbine engines and each one delivers over 101,300 litres of water a minute and has a thrust of over 8,100 kg. The Jetfoil weighs 108 tonnes and cruises at about 83 km/hour. It has an automatic stabilizing system and the control surfaces are computer-operated. The foils are retractable.

as a wing and behave in water in exactly the same way as a wing does in air *(see page 116)*. At low speeds the hull moves through the water in the ordinary way, but as the hydrofoil increases speed, the foils generate lift and the hull rises out of the water.

The first full-sized hydrofoil craft was built by Enrico FORLANINI and tried out on Lake Maggiore in 1905. In 1918, Alexander BELL, the inventor of the telephone, built a craft in America that established a new water-speed record of over 114 kilometres per hour. Despite this achievement, interest in hydrofoils began to wane, due partly to lack of experience with lightweight engines and hulls and partly to lack of support from governments. However, in 1927 Baron Hanns von Schertel, a German engineer, designed the first hydrofoil stable enough to be used in rough weather.

Schertel used the V-foil, or dihedral SURFACE-PIERCING configuration. The advantage of this type of foil is that rolling and pitching are automatically controlled. Various types of surface-piercing foil are used today, including W-shapes and trapeze arrangements. However, the very stability of a surface-piercing foil makes it unsuitable for use in heavy seas, because the craft tends to follow the contours of high waves. Thus, surface-piercing foils are used mostly on inland waters, while sea-going craft use fully submerged foils.

Fully submerged foils do not have a natural stability and so have to be 'instructed' how to behave. The first method of doing this was a mechanical system called the HYDROFIN, devised by Christopher Hook in the 1940s. Today, most submerged foils are controlled electronically using radar, sonar and gyroscopes.

Most hydrofoils are powered by diesel engines. Power is transmitted to the propellers by inclined shafts, V-drives or Z-drives, which are all different methods of linking the engine to a propeller far beneath the hull. Some hydrofoils use water-jet propulsion systems.

---

built. The initials stand for Saunders Roe-Nautical, who built the SR-N1 according to Sir Christopher Cockerell's design. Launched in 1959, it was 9.14 metres long and 7.31 metres wide, with a top speed of 46 km/hour. Two long ducts provided the propulsion jets and it was steered by two tall rudders. In July 1959, it crossed the English Channel. Westland Aircraft, who took over Saunders Roe in 1962, built the SR-N2. This improved hovercraft was equipped with a skirt and two air-propellers set in line with a single rudder behind them. It could carry 70 passengers at a speed of 129 km/hour. The largest SR-N is the SR-N4. Five of these 200-tonne hovercraft, with cruising speeds of 111 km/hour are used to operate car/passenger ferries across the English Channel.

Surface-piercing foils have the advantage that rolling and pitching are corrected automatically. If the craft rolls to one side, more foil area becomes submerged, more lift is therefore generated and the craft is pushed upright again. In the same way if the bow pitches forward, the bow foil generates more lift and the bow is raised again. *See also* LADDER-FOIL.

V

Voyageur, built by Bell Aerospace (Canada), is a workhorse air-cushion vehicle in use in northern Canada. Basically a flat platform on an air-cushion propelled by two air propellers, it can haul loads of 25 tonnes at speeds of up to 87 km/hour. The ice-breaking capabilities of fast-moving air-cushion vehicles were discovered during its trials. The craft produces standing waves in the ice half a length astern and the ice cracks at the wavecrests. The 12-metre wide platform can leave a trail of broken ice 36-metres wide. There are plans to use an air-cushion vehicle as a pipe-laying vessel in ice-bound waters.

Voyageur *air cushion vehicle*

Index

Page numbers in **bold** type refer to the reference sections. Page numbers in *italics* refer to the illustrations.

Acknowledgements

Contributing artists
Marian Appleton, Charles Bannerman, Raymond Brown, Chris Forsey, Ivan Lapper, Jim Marks, Nigel Osborne, Alan Suttie, Ken Warner.

The Publishers also wish to thank the following:
A.E.I. 60B
Allsport/Don Morley 87T
W. David Askham 105TR, 116T
Australian Information Service 102B
Barnaby's 6B, 7B, 8B, 9B, 21B, 40B, 53TL, 94B, 108T, 110C, 125T,
BBC 9T
Bell Aerospace 128B
Biblioteca Ambrosiana, Milan, 29TL
Birlec Limited 49B
Bodleian Library 34T, 35TL
Boeing Aerospace Limited 124T, 124C, 128TR
British Airways 118B, 123B
British Library 38T
British Museum (Natural History) 3B
British Transport Films 112TR, 112C
Camera Press 92T
Camera Press/J. H. Pickerell 123T
Camera Press/L. Willinger 93C
J. Allan Cash 86B, 89B
Central Office of Information 121L
Bruce Coleman Ltd/Jeff Foott 3T
Cooper Bridgeman Library 99T, 101T
R. J. Davis Contents BR, 119T
Daily Telegraph Colour Library/A. Howarth 122T
Robert Estall 43TL, 58T, 96T
Mary Evans Picture Library 47C, 48TR, 71B, 73B, 74B, 76B, 79B, 119B, 126B
Fiat 91B
Fisons 57TR
Ford Motor Co. Ltd 90T, 90B
Fotomas Index 16T, 77T, 114T
James Gilbert 117B, 120B, 121B, 124B
Glaxo Ltd 57TL
R. A. Gould 5T
Gould Instruments 63C
Grand Western Houseboat Co. 95T
Sally and Richard Greenhill 12T, 87C, 105TL, 107C
Grumman Aerospace Corporation 127B, 128TL
Sonia Halliday 24T, 26T, 26C
Hamlyn Group Picture Library 27T, 31T
Robert Harding/National Museum of Archaeology, Lima 19C
Hawker Siddeley Aviation Ltd 113B
Hoverlloyd 126T
Alan Hutchinson Library Contents, 41TR, 73C
Michael Holford 18T, 22T, 23T, 40T, 45C, 48TL, 53TR, 69C, 73T, 75T, 80BR
I.B.M. 28T, 29TR, 29CL, 29BR
I.C.I. 56TL
Japan Information Centre 112TL
Macdonald Educational 42T, 44T
Macquitty Collection 30T
Mansell Collection 26B, 29CR, 29BL, 94T, 103C, 115TL
Ministry of Defence/Crown Copyright 70B, 72B, 116B
Margaret Murray 8C
National Maritime Museum 77B
National Motor Museum, Beaulieu 85T (Courtesy Museu do Automoval Do Ceramulo, Portugal)
Novosti 108C
Photographic Library of Australia 109C
Photri 62T, 108BL, 110T, 111C
Picturepoint 107TR
Pilkington Glass 56TR
Popperfoto 4B, 75B, 81BL, 81BR, 87BL, 93B, 114B, 115B
Raleigh Ltd 87BR
Rockwell International 122B
Rolls Royce Motors Ltd 89B
Ann Ronan Picture Library 5B, 10B, 11B, 12B, 13B, 14T, 14B, 15B, 16B, 17B, 18B, 19T, 19B, 20B, 21T, 22B, 23B, 24B, 25B, 27B, 28B, 29B, 31B, 32B, 33B, 34CR, 34B, 35TR, 35B, 36TR, 36B, 37B, 38B, 39B, 41TL, 41B, 42B, 43B, 44B, 45B, 46B, 47B, 48B, 50B, 51B, 52B, 53B, 54B, 55B, 56B, 58B, 59B, 61B, 62B, 63B, 64B, 67B, 68B, 69B, 88B, 95B, 96B, 97B, 98B, 99C, 99B, 100BL, 100BR, 101B, 103B, 104B, 105B, 106B, 107B, 108BR, 109B, 110B, 111B, 112B, 113T
Ann Ronan/E. P. Goldschmidt & Co. Ltd 30B
Scala 43TR
Science Museum 115C
Scottish Tourist Office 85B
Seajet 128TR
Shell Photo Services Contents, 61T, 61C
600 Group Ltd 50T
Snark International 36TL, 37T
Societe Nationale des Chemins de Fer (SNCF) 107TL, 111T
Spectrum 49T, 80T
Swiss National Tourist Office 106T
T.B.A. Industrial Products Ltd 57B
P.N. Trotter 103T
Vickers Ltd 125B
Volkswagen GB Ltd 88T, 89T
Walker Art Gallery, Liverpool 98T
John Watney 34CL
Josiah Wedgwood & Sons Ltd 47T
Yachts and Yachting 80BL
Jerry Young 7T, 115TR
Zefa 62C, 78B, 121B